Space-Time Codes and MIMO Systems

For a listing of recent titles in the
Artech House Universal Personal Communications Series,
turn to the back of this book.

Space-Time Codes and MIMO Systems

Mohinder Jankiraman

Artech House
Boston • London
www.artechhouse.com

Library of Congress Cataloging-in-Publication Data
Jankiraman, Mohinder.
    Space-time codes and MIMO systems / Mohinder Jankiraman.
      p. cm. — (Artech House universal personal communications series)
    Includes bibliographical references and index.
    ISBN 1-58053-865-7 (alk. paper)
    1. Space time codes. 2. MIMO systems. 3. Wireless communication
    systems. 4. Antenna arrays. I. Title. II. Series.

    TK5103.4877.J36 2004
    621.382—dc22   2004050668

British Library Cataloguing in Publication Data
Jankiraman, Mohinder
    Space-time codes and MIMO systems. — (Artech House universal personal
    communications library)
    1. Coding theory   2. Wireless communication systems
    I. Title
    621.3'822

ISBN 1-58053-865-7

Cover design by Yekaterina Ratner

International Standard Book Number: 1-58053-865-7
Library of Congress Catalog Card Number: 2004050668

10 9 8 7 6 5 4 3 2 1

Dedicated to my late father
Kuppuswamy Janakiraman
and to my mother
Gnanambal Janakiraman
Their extraordinary sacrifices made it possible for me to come this far

Contents

Preface

This book is intended to introduce space-time coding and multiantenna systems. The endeavor is to impart a working knowledge of the subject not just for students and researchers but for the entire wireless community.

The birth of multiantenna systems is the direct result of the long-standing struggle to achieve data rates without compromising the quality of the reception. Indeed this has been the case since the inception of wireless communications. A binding constraint in the evolution of high data rate systems is the stringent limitation imposed on the available spectrum. This, in turn, has given rise to more efficient signaling techniques. Recent study has shown that multiple antennas yield substantial increases in channel capacity. Toward this end, multiple-input multiple-output (MIMO) systems have been constructed comprising multiple antenna arrays at both ends of the wireless link. Space-time coding, as the name suggests, involves coding across space and time and is aimed at approaching the capacity limits of MIMO channels. Today space-time coding and MIMO systems are widely regarded as the most likely candidates for futuristic high data rate systems and are already being designed by many companies for the high data rate market.

This book is intended for postgraduate students, practicing engineers, and researchers. It is assumed that the reader is familiar with basic digital communications, linear algebra, and probability theory.

In view of the fact that space-time coding theory has become such a widely discussed and researched topic in recent times, it is not possible to cover all aspects in any detail. Therefore, an effort has been made to impart to the reader a "flavor" of the subject just enough to whet his/her appetite, prompting further detailed study of areas of particular interest. Toward this end, the style of writing has been kept as simple as possible and technical clichés and jargon have been kept to a minimum. All effort has been made to explain the basics in a cogent and conversational manner.

The reader is also introduced to a new technique of interfacing CDMA to OFDM systems called "Hybrid OFDM/CDMA." This approach is different from the popular OFDM/CDMA systems proposed by Fazel et al [1], wherein an MC-CDMA system transmits each bit using OFDM modulation. In the approach suggested by the author a CDMA system is directly interfaced to an OFDM system. This approach yields a CDMA system capable of handling high throughputs, bandwidth permitting, along with the added bonus of user separation based on CDMA codes. It is, to the best of the author's knowledge and belief, the first time such a concept has been discussed anywhere and had formed part of the author's Ph.D. thesis [2]. This concept has been recently proved in the field by NTT DoCoMo of Japan, wherein the company tested such a system for outdoor use using a bandwidth of 100 MHz to achieve a throughput of 100 Mbps [3] in the downlink. Further details are not known at the time of going to press. The company plans to introduce this, as part of its 4G effort, by 2010.

The teaching effort in this book is aided by a set of accompanying software. This software has been divided into a set of two broad classes—narrowband and wideband. The narrowband software is distributed on the basis of chapters and directly pertains to topics discussed in those chapters. The wideband software is orthogonal frequency division multiplexing (OFDM) based and is included as part of Chapters 7, 8, and 9. The entire coding has been kept simple and sometimes very unprofessional to enable readers to clearly understand the various steps involved in the implementation of the program. If the reader finds this annoying, the error is mine and is deeply regretted! The entire coding has been implemented at baseband and the radio frequency (RF) aspects of coding have been avoided for similar reasons. Because the software is basically intended as a "skeleton," the user is encouraged to modify it in any manner or means by adding to its RF capability and so on to suit one's convenience. This is an excellent method to learn the subject. The software presupposes a sound understanding of MATLAB® and SIMULINK® and has been tested on MATLAB® Version 6.0 (with Signal Processing and Communication Toolboxes) and above with the SIMULINK® option with DSP and Communication Blocksets. It is important to note that in a technology of this nature the best way to assimilate the subject is by programming. Coding an operation forces the user to look at all aspects of the subject. This is similar to learning mathematics through solving problems.

A consistent set of notations has been used throughout the book and excessive mathematics has been avoided. Emphasis is placed on imparting to the reader a physical understanding of the subject so that the reader has a clear grasp of the processes involved.

There will be errors even though every effort has been made to detect and eliminate them. Any inconvenience to the readers as a result is deeply regretted.

References

[1] Fazel, K., and L. Papke, "On the Performance of Convolutionally-Coded CDMA/OFDM for Mobile Radio Communication Systems," *PIMRC*, 1993.

[2] Jankiraman, M., "Wideband Multimedia Solution Using Hybrid CDMA/OFDM/SFH Techniques," *Ph.D. Thesis*, University of Aalborg, Denmark, September 2000.

[3] "NTT DoCoMo Successsfully Completes 4G Mobile-Communications Experiment Including 100 Mbps Transmission," *nG Japan*, Vol. 1, No. 11, November 25, 2002, InfoCom Research Inc., Tokyo, Japan.

Acknowledgments

It is not possible to generate a work of this nature without invaluable help from many other participants, since no man is an island. It is not possible to list all of them but I gratefully acknowledge those who helped in the preparation of this book. This invaluable help came in two principal areas—the preparation of the text and the coding of the software. I gratefully thank Dr. John Terry of Nokia Research Center, Dallas, Texas, and Michelle Johnson of Pearson Education for permission to reproduce certain figures in this book. I also thank Professor Ted Rappaport of the University of Texas at Austin for his help with obtaining permission for figures from his book, as well as Emily McGee of Pearson Education. My thanks are also due to Dr. Branka Vucetic of the University of Sydney, Australia, for her patience with my many questions. I also thank Duncan James of John Wiley & Sons for permission to use certain figures in this book. Special thanks are also due to Steve Thanos of Pentek for permission to reproduce Figure 1.3. I thank Dr. Donald Shaver of Texas Instruments for permission to reproduce Figure 1.2, and Dr. Peter Rysavy of Rysavy Research for permission to reproduce Figure 1.4(b). Thanks are also due to Dr. Richard van Nee of Airgonetworks, Netherlands, for permission to use certain figures from his book. My thanks are due to Professor Arogyaswami Paulraj of Stanford University and Ted Gerney of Cambridge University Press for permission to use certain figures in this book. Finally, I thank Jacqueline Hansson and Claudio Stanziola of IEEE for their kind permission to reproduce certain papers and figures in this book. You all have been very patient with my requests!

In addition, I thank Devendra Prasad and Albena Mihovska of the University of Aalborg, Denmark, for their invaluable help with this book. Without their help, this effort would not have been possible.

The software developed for this book has many participants. Thanks are due to Beza Negash Getu of the University of Aalborg, Denmark, for his help with the software in Chapters 2 and 4. We had many hours of fruitful discussion. Thanks are also due to Dr. Persifoni Kyritsi of Stanford University for her help with the software in Chapter 6. Persa and I thoroughly analyzed the MIMO channel. I also thank Kamil Anis of the Department of Radio Engineering, Faculty of Electrical Engineering, at Czech Technical University in Prague for his invaluable help with the space-time trellis coding in Chapter 5. In fact many of his programming ideas are incorporated in the software of Chapter 5. I also thank Dr. John Terry and Juha Heiskala of Nokia Research Center in Dallas for helping me understand the coding aspects of OFDM-based MIMO. They were very understanding and patient. I also thank Martin Clark of Mathworks, the makers of MATLAB®, for his help in obtaining permission from Mathworks to use his software in Chapter 8. Finally, I thank Efrayim Metin of Versatel Nederland B.V., Amsterdam, Netherlands, for his invaluable help with my many software problems. I am truly grateful.

A work of this nature also requires salubrious surroundings. I thank my sister Uma and her husband, Dr. M. J. Rangaraj of Monroe, Louisiana, for inviting me to their wonderful home on the Louisiana bayou. There is no better spot than sitting on the banks of the bayou on a sunny day and mulling over technical problems surrounded by cormorants, butterflies, and noisy frogs. I solved many of my knotty problems in Louisiana. I also thank my children, Pavan and Pallavi, for their patience and encouragement.

Finally, I wish to acknowledge the advice and comments from my anonymous reviewers. I thank them all. Thanks are also due to Christine Barnaby Daniele, Tiina Ruonamaa, Julie Lancashire, Judi Stone, and Jill Stoodley of Artech House for their patience with my many corrections and repeated "final versions." I acknowledge their superb support and efficient handling of this publication project.

CHAPTER 1

Introduction

"If I have seen farther, it is by standing on the shoulders of giants."
—Sir Isaac Newton

1.1 The Crowded Spectrum

It is all about spectrum. Marconi pioneered the wireless industry 100 years ago. Today life does not seem possible without wireless in some form or the other. In fact wireless permeates every aspect of our lives. The demands on bandwidth and spectral availability are endless. Currently wireless finds its widest expression in fixed and mobile roles. In the fixed role, wireless is used extensively for data transfer, especially from desktop computers and laptops. In the mobile role, wireless networks provide mobility for use from fast vehicles for both voice and data. Consequently wireless designers face an uphill task of limited availability of radio frequency spectrum and complex time varying problems in the wireless channel, such as fading and multipath, as well as meeting the demand for high data rates. Simultaneously, there is an urgent need for better quality of service (QoS), compared with that obtainable from DSL and cable.

1.2 Need for High Data Rates

The gradual evolution of mobile communication systems follows the quest for high data rates, measured in bits/sec (bps) and with a high spectral efficiency, measured in bps/Hz. The first mobile communications systems were analog and are today referred to as systems of the first generation. In the beginning of the 1990s, the first digital systems emerged, denoted as *second generation* (2G) systems. In Europe, the most popular 2G system introduced was the *global system for mobile communications* (GSM), which operated in the 900 MHz or the 1,800-MHz band and supported data rates up to 22.8 kbit/s. In many parts of the world today, GSM is still in vogue. Basically GSM is a cellular system [i.e., it typically uses a single *base transceiver station* (BTS), which marks the center of a cell and which serves several *mobile stations* (MS), meaning the users]. In the United States, the most popular 2G system is the *TDMA/136*, which is also a digital cellular system. TDMA stands for time-division multiple access.

To accomplish higher data rates, two add-ons were developed for GSM, namely *high-speed circuit switched data* (HSCSD) and the *general packet radio service* (GPRS), providing data rates up to 38.4 kbit/s and 172.2 kbit/s, respectively.

The demand for yet higher data rates forced the development of a new generation of wireless systems, the so-called *third generation* (3G). 3G systems are characterized by a maximum data rate of at least 384 kbit/s for mobile and 2 mbit/s for indoors.

One of the leading technologies for 3G systems is the now well-known *universal mobile telephone system* (UMTS) [also referred to as *wideband code-division multiplex* (WCDMA) or UTRA FDD/TDD]. UMTS represents an evolution in terms of services and data speeds from today's "second generation" mobile networks. As a key member of the global family of 3G mobile technologies identified by the International Telecommunication Union (ITU), UMTS is the natural evolutionary choice for operators of GSM networks, currently representing a customer base of more than 850 million end users in 195 countries and representing over 70% of today's digital wireless market. UMTS is already a reality. Japan launched the world's first commercial WCDMA network in 2001, and WCDMA networks are now operating commercially in Austria, Italy, Sweden and the United Kingdom, with more launches anticipated during 2004. Several other pilot and precommercial trials are operational in the Isle of Man, Monaco, and other European territories. UMTS is also a cellular system and operates in the 2-GHz band. Compared with the 2G systems, UMTS is based on a novel technology. To yield the 3G data rates, an alternative approach was made with the *enhanced data rates for GSM evolution* (*EDGE*) *concept.* The EDGE system is based on GSM and operates in the same frequency bands. The significantly enhanced data rates are obtained by means of a new modulation scheme, which is more efficient than the GSM modulation scheme. As for GSM, two add-ons were developed for EDGE, namely *enhanced circuit switched data* (ECSD) and the *enhanced general packet radio service* (EGPRS). The maximum data rate of the EDGE system is 473.6 kbit/s, which is accomplished by means of EGPRS. EDGE was introduced in the United States as a generic air interface to enhance the TDMA/136 system. Some 200 operators worldwide are also giving their customers a taste of faster data services with so-called 2.5G systems based on GPRS technology, a natural evolutionary stepping-stone toward UMTS.

The new IEEE and High Performance Radio Local Area Network (HIPERLAN) standards specify bit rates of up to 54 mbit/s, although 24 mbit/s will be the typical rate used in most applications. Such high data rates impose large bandwidths, thus pushing carrier frequencies for values higher than the UHF band. HIPERLAN has frequencies allocated in the 5- and 17-GHz bands; multimedia broadcasting systems (MBS) will occupy the 40- and 60-GHZ bands; and even the infrared band is being considered for broadband wired local area networks (WLANs).

A comparison of several systems, based on two of the key features (mobility and data rate) is shown in Figure 1.1 [1], where it is clear that no competition exists between the different approaches.

The applications and services of the various systems are also different. IEEE 802.11 is mainly intended for communications between computers (thus being an extension of WLANs). Future wireless *broadband applications* are likely to require

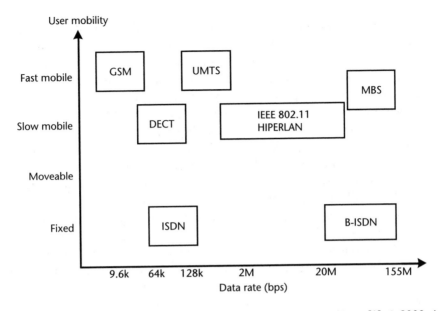

Figure 1.1   Comparison of mobility and data rates for several systems. (*From:* [1]. © 2000, Artech House. Reprinted with permission.)

data rates that are hundreds of megabits per second—up to 250 times the maximum data rate promised for UMTS. Such a broadband service could, for example, be wireless high-quality video conferencing (up to 100 mbit/s) or wireless virtual reality (up to 500 mbit/s, when allowing free body movements).

Therefore, the goal of the next generation of wireless systems—the *fourth generation* (4G)—is to provide data rates yet higher than the ones of 3G while granting the same degree of user mobility. 4G is the short term for fourth generation wireless, the stage of broadband mobile communications that will follow the still-burgeoning 3G that is expected to reach maturity between 2003–2005. 4G services are expected to be introduced first in Japan, as early as 2006—four years ahead of the previous target date. The major distinction of 4G over 3G communications is increased data transmission rates, just as it is for 3G over 2G and 2.5G (the current state of wireless services, hovering somewhere between 2G and 3G). According to *NTT-DoCoMo*, the leading Japanese wireless company, the current download speed for i-Mode (mobile Internet service) data is, theoretically, 9.6 kbit/s, although in practice the rates tend to be slower. 3G rates are expected to reach speeds 200 times higher, and 4G to yield further increases, reaching 20–40 mbit/s (about 10–20 times the current rates of ADSL service). 4G is expected to deliver more advanced versions of the same improvements promised by 3G, such as enhanced multimedia, smooth streaming video, universal access, and portability across all types of devices. Industry insiders are reluctant to predict the direction that less-than-immediate future technology might take, but 4G enhancements are expected to include world-wide "roaming" capability. As was projected for the ultimate 3G system, 4G might actually connect the entire globe and be operable from any location on—or above—the surface of the earth. This aspect makes it distinctly different from the techno-logies developed until now. These technologies were built for or overlaid onto proprietary networking equipment. In fact the outlook for 3G is uncertain. Network providers in Europe and North America maintain separate standards bodies (3GPP for Europe and Asia; 3GPP2 for North America). These different standards bodies reflect the difference in air interface technologies. In addition to 3G's technical challenges, there are problems from a financial aspect, such as justifying the large expense of building systems based on less-than-compatible 2G technologies. In contrast, 4G wireless networks that are Internet Protocol (IP)-based have an intrinsic advantage over their predecessors. IP tolerates a variety of radio protocols. It allows you to design a core network that gives you complete flexibility as to what shape the access network will take. One can support diverse technologies like 802.11, WCDMA, Bluetooth, HIPERLAN and so on, as well as some new CDMA protocols. A 4G IP network also has certain financial advantages. Equipment costs are much lower than what they used to be for 2G and 3G systems. Wireless service providers would no longer be bound by single-system vendors of proprietary equipment. An all-IP wireless core network would enable services that are sufficiently varied for consumers. This implies improved data access for mobile Internet devices. Currently wireless communications are heavily biased toward voice. However, studies indicate that growth in wireless data traffic is rising exponentially relative to the demand for voice traffic. This is especially true of 802.11 data transfer protocol, a wireless LAN standard developed by IEEE, as a distinct data access technology that can work in a variety of radio spectrums, including infrared. Chapters 8 and 9 are exclusively devoted to this technology. Because an all-IP core layer is easily accessible, it is ideally suited to meet this challenge. It will not be surprising if we leapfrog directly into 4G from the current 2.5G.

WLANs are already characterized by data rates significantly higher than 2 mbit/s. WLANs cover comparably small areas, often hot spots in shopping malls, airports, hotels or office buildings. Typical properties of WLANs are comparably small cells and mobile terminals moving, at most, with pedestrian speed.

The *Bluetooth* system, already commercially available, operates in the 2-GHz band and provides data rates of <1 mbit/s. Bluetooth is normally employed to interconnect certain electronic devices rather than to establish a complete wireless LAN. A wireless Bluetooth interconnection for example between a laptop computer and a *personal digital assistant* (PDA), is typically characterized by a short range of about 5–10m. The U.S. standard *IEEE 802.11b,* as implemented in current WLAN products, operates in the same band and provides data rates up to 11 mbit/s over a distance of 50–100m. The succeeding standard *IEEE 802.11a* delivers data rates up to 54 mbit/s using the 5-GHz band. The European standard corresponding to IEEE 802.11a is *HIPERLAN Type 2*. Data rates in WLANs could be significantly enhanced by exploiting yet higher frequency bands. In this context, for example, the 60-GHz band is the subject of current research activities. In particular, a new WLAN standard for 60 GHz could be developed such that it is compatible with IEEE 802.11a or HIPERLAN/2. This would enable 60-GHz systems to use the 5-GHz band as a fallback option, in case the channel quality at 60 GHz becomes insufficient.

The first 4G systems are likely to be an *integration* of 3G systems and WLAN systems. By this means, considerable data rates can be granted at hot spots. On the other hand, the interworking of WLAN and 3G systems will provide a good degree of mobility, given that *seamless* handover is accomplished between several heterogeneous systems.

Figure 1.2 reveals the possible candidates for 4G systems. The figure is self-explanatory. There are, however, a few interesting points. The "hottest" candidates for 4G appear to be:

- *BWIF:* The Broadband Wireless Internet Forum (BWIF) [3] is the principal organization chartered with creating and developing next generation fixed wireless standards. The broadband wireless specifications are based on *vector orthogonal frequency division multiplexing* (VOFDM) technology and *data over cable service interface specification* (DOCSIS). BWIF was formed to address the needs of the quickly emerging wireless broadband market. Further, through BWIF, members establish product road maps that lower product costs, simplify deployment of advanced services, and ensure the availability of interoperable solutions. BWIF extends the partnership model to all companies offering expanded broadband wireless technology to multiple markets. At the core of the partnerships, membership includes:
  - *ASIC semiconductor* companies, which develop new ASICs based on VOFDM technology;
  - *Customer premise equipment* (CPE) companies, which use the chips to build new subscriber equipment;
  - *Systems integrators,* which design and deploy the networks based on these products;
  - *Service providers,* which incorporate VOFDM products and technology into their network infrastructure to offer to new services customers;
  - *RF/ODU manufacturers,* which supply subsystems to the total wireless solution offering.

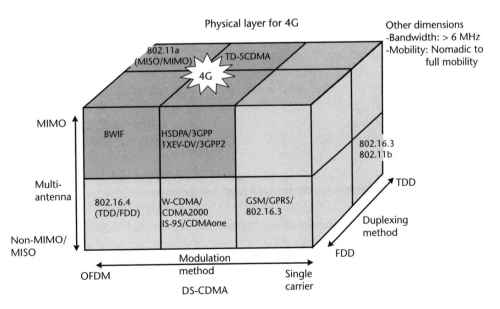

Figure 1.2   Possible candidates for 4G systems. (*From:* [2]. Reprinted with the permission of Donald Shaver, Texas Instruments.)

- Many companies associated with wireless technology see the market potential for new applications, products, and services. BWIF members have committed to the VOFDM specification for optimized and open broadband fixed wireless access. Founding companies and promoting members do not collect royalties for the intellectual property they contribute to support VOFDM technology. In addition, BWIF is organized as a program of the *IEEE Industry Standards and Technology Organization* (IEEE-ISTO), which acts as the managing body for this forum.

- *TD-SCDMA:* Time-division synchronous CDMA (TD-SCDMA) is the Chinese contribution to the ITU's IMT-2000 specification for 3G wireless mobile services. It endeavors to integrate with the existing GSM system. It is designed to manage both symmetric circuit-switched services, such as speech or video, as well as asymmetric packet-switched services, such as mobile Internet data flows. TD-SCDMA combines two leading technologies—an advanced TDMA system with an adaptive CDMA component—to overcome this challenge. Further details are beyond the scope of this book. The interested reader is advised to refer to [4].

- *HSDPA: High-speed downlink packet access* (HSDPA) is a packet-based data service in WCDMA downlink with data transmission up to 8–10 mbit/s [and 20 mbit/s for multiple-input multiple-output (MIMO) systems] over a 5-MHz bandwidth in WCDMA downlink. HSDPA implementations include adaptive modulation and coding (AMC), MIMO, hybrid automatic request (HARQ), fast cell search, and advanced receiver design. In *third generation partnership project* (3GPP) standards, Release 4 specifications provide efficient IP support, enabling provision of services through an all-IP core network, and Release 5 specifications focus on HSDPA to provide data rates up to approximately 10 mbit/s to support packet-based multimedia services. MIMO systems are the work item in Release 6 specifications, which will support even higher data transmission rates up to 20 mbit/s. HSDPA is evolved from and backward compatible with Release 99 WCDMA systems. The interested reader is referred to [5].

It can be seen that there are many approaches toward 4G. The final standard has not been defined as yet, but it is expected to be finalized by 2007. The common theme in all these proposals, as can be seen from Figure 1.2, is increasing spectrum efficiency using MIMO techniques. These techniques are the subject of this book.

1.3 Multiple-Input Multiple-Output Systems

This technique is mainly based on the theoretical work developed by Teletar [6] and Foschini [7]. The core of this idea is to use multiple antennas both for transmission and reception. This increases the capacity of the wireless channel. Capacity is expressed as the maximum achievable data rate for an arbitrarily low probability of error. Hence, the thrust has been toward the development of codes and schemes that would enable systems to approach their Shannon capacity limit [8]. This technology received a fillip when Tarokh et al. [9] introduced their space-time trellis coding techniques and Alamouti introduced his space-time block coding techniques to improve link-level performance based on diversity [10]. It received another boost when Bell Laboratories introduced its Bell Laboratories Layered Space-Time (BLAST) coding technique [11], demonstrating spectral efficiencies as high as 42 bit/s/Hz. This represents a tremendous boost in spectral efficiency compared with the current 2–3 bit/s/Hz achieved in cellular mobile and wireless LAN systems. There is, therefore, a need for communication engineers to understand this remarkable technology. This book has been expressly written to fulfill such a need.

We will now discuss a MIMO system pioneered by AT&T Labs-Research in Middletown, New Jersey, [12]. It conducted field tests to characterize the mobile MIMO radio channel. The company measured the capacity of a system with four antennas on a laptop computer and four antennas on a rooftop base station. The field tests showed that close to the theoretical fourfold increase in capacity over a single antenna system can be supported in a 30-KHz channel with dual polarized spatially separated base station and mobile terminal antennas. Figure 1.3 shows the arrangement. Note the mounting of the four antennas on the laptop computer and the rooftop antennas.

Figure 1.3   Upper: Transmitter with four antennas on a laptop and the 1,900-MHz coherent transmitters. Lower: The four receivers with real-time baseband processing and rooftop
antennas. (*From:* [12]. Reprinted with the permission of AT&T Labs-Research and Pentek
Inc.)

The base station rooftop antenna array used dual-polarized antennas separated by 11.3 feet, which is approximately 20 wavelengths apart and a multibeam antenna. The laptop-mounted antennas included a vertically polarized array and a dual-polarized array with elements spaced half a wavelength apart. Different signals were transmitted out of each antenna simultaneously in the same bandwidth and then separated at the receiver. With four antennas at the transmitter and receiver, this has the potential to provide four times the data rate of a single antenna system without an increase in transmit power or bandwidth, provided the multipath environment is rich enough. This means that high capacities are theoretically possible unless there is a direct line of sight between transmitter and receiver. These and other concepts will be examined in this book.

We discussed earlier the aspect of 4G being IP-based. We now examine briefly as to what this entails, as wireless IP will be a principal beneficiary of MIMO.

1.4   Internet Protocol

The Internet is a collection of thousands of networks interconnected by gateways. It was developed under the U.S. Department of Defense Advanced Research Projects Agency (DARPA) support as an effort to connect the myriad of local area networks at U.S. universities to the ARPANET. The protocol developed to interconnect the various networks was called the IP.

The individual networks comprising the Internet are joined together by gateways. A gateway appears as an external site to an adjacent network, but appears as a node to the Internet as a whole. In terms of layering, the Internet sublayer is viewed as sitting on top of the various network layers. The Internet sublayer packets are called datagrams and each datagram can pass through a path of gateways and individual networks on the way to its final destination.

1.4.1   Routing Operations

The IP gateway makes routing decisions based on the routing list. If the destination host resides in another network, the IP gateway must decide how to route to the other network. If multiple hops are involved, then each gateway must be traversed and the gateway must make decisions about the routing.

Each gateway maintains a routing table that contains the next gateway on the way to the final destination network. In effect, the table contains an entry for each reachable network. These tables could be static or dynamic. The IP module makes a routing decision on all the datagrams it receives.

1.4.2   The Transmission Control Protocol

IP is not designed to recover from certain problems nor does it guarantee the delivery of traffic. IP is designed to discard datagrams that are outdated or have exceeded the number of permissible transit hops in an Internet. Certain user applications require assurance that all datagrams have been delivered safely to the destination. Furthermore, the transmitting user may need to know that the traffic has been delivered at the receiving host. The mechanisms to achieve these important services reside in the Transmission Control Protocol (TCP).

TCP resides in the transport layer of the layered model. It is situated above IP and below the upper layers. Figure 1.4 illustrates that TCP is not loaded into a gateway. It is designed to reside in the host computer or in a machine that is tasked with end-to-end integrity of the transfer of user data.

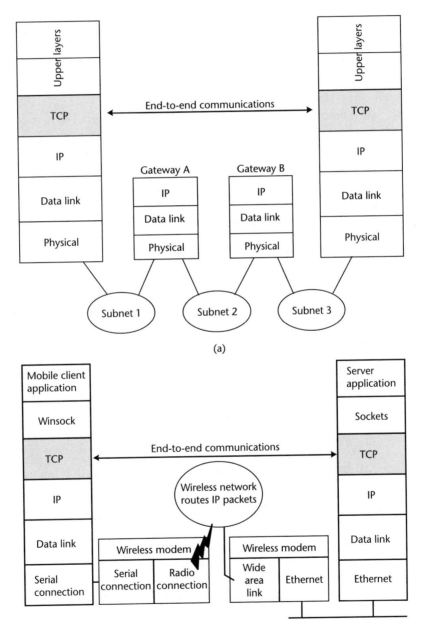

Figure 1.4   (a) TCP structure. (b) Wireless network IP that is packet-based. (*From:* [13]. Reprinted with the permission of Peter Rysavy, Rysavy Research.)

The figure also shows that TCP is designed to run over the IP. Since IP is a connectionless network, the tasks of reliability, flow control, sequencing, opens, and closes are given to TCP. TCP and IP are tied together so closely that they are used in the same context. Many of the TCP functions (such as flow control, reliability, sequencing and so on) could be handled within an application program. But it makes little sense to code these functions into each application. The preferred approach is to develop generalized software that provides community functions applicable to a wide range of applications and then invoke these programs from the application software. This allows the application programmer to concentrate on solving the application problem and it relieves the programmer from the details and problems of networks.

TCP provides the following principal services to the upper layers:

- *Connection-oriented data management:* This term refers to the fact that TCP maintains status and state information about each user data stream flowing into and out of the TCP module. Used in this context, the term means TCP is responsible for the end-to-end transfer of data across multiple networks (i.e., the three networks in Figure 1.2) to a receiving user application.
- *Reliable data transfer:* To ensure reliability, TCP uses checksum routines and timers to ensure that the lapse of time is not excessive before remedial measures are taken for either transmission of acknowledgements from the receiving site and/or retransmission of data at the transmitting site.
- *Stream-oriented data transfer:* Stream-oriented transmission is designed to send individual characters and not blocks, frames, datagrams and so on. The bytes are sent on a stream basis, byte by byte. When they arrive at the TCP layer, the bytes are grouped into TCP segments. These segments are then passed to the IP.
- *Flow control:* The receiver's TCP module is able to flow control the sender's data to prevent buffer overruns. This is based on a transmission "window" that allows a specific number of bytes.
- *Full-duplex transmission:* TCP provides full-duplex transmission between two TCP entities. This permits simultaneous two-way transmission without having to wait for a turnaround signal.
- *Multiplexing:* TCP also has a very useful facility for multiplexing multiple user sessions within a single host computer. This is accomplished through a simple naming convention for ports and sockets between TCP and IP modules.
- *Precedence and security:* TCP also provides the user with the capability to specify levels of security and precedence (priority level) for the connection.
- *Graceful close:* TCP provides a graceful close to the logical connection between two users. A graceful close ensures that all traffic has been acknowledged before the virtual circuit is removed.

The preceding paragraphs were a brief overview of IP. Until now, IP has been implemented as the foundation of the Internet and virtually all multivendor private Internet works. The currently deployed version of IP is actually IP version 4.

Previous versions of IP (1 through 3) were successively defined and replaced to reach IPv4. This protocol is reaching the end of its useful life and a new protocol known as IPv6 (IP version 6) has been defined to ultimately replace IP. The driving motivation for the adoption of a new version of IP was the limitation imposed by the 32-bit address field in IPv4. In addition, IP is a very old protocol and new requirements in the areas of security, routing flexibility, and traffic support have developed. To meet these needs, IPv6 has been defined [14, 15] and includes functional and formatting enhancements over IPv4.

1.5   Wireless Internet Protocol

Having briefly examined IP, we now extend this concept to wireless IP, which is illustrated in Figure 1.4(b).

In an ideal world, a computer connected over a wireless network would work just like a computer on a LAN. But wireless networks currently operate at lower speeds with higher latency and connections can be lost at any moment, especially when mobile. Experiments conducted in the field showed that the basic bottleneck was low data rates, which made downloading of pages inordinately long, as much as 15 seconds for screen updates. Therefore, a need exists to develop a modulation scheme well-suited to this requirement. The chosen scheme should not only support high bit rates, but should have a very fast synchronization scheme suited to packet transmission in fading channels. Toward this end, multiple antenna systems were developed with a view to increasing spectral efficiencies.

In Chapter 2, we examine the capacities of MIMO systems both in deterministic as well as random channels. The material has been drawn substantially from the outstanding pioneering research work conducted over the past six years by Dr. Paulraj and his team at Stanford University and discussed in their books and publications [16]. I have retained in this book the symbols and expressions they used in their papers and publications. I wish to add that I did this after a lot of deliberation, as over the years, these symbols and expressions have *de facto* passed into the *lingua franca* of MIMO technology. Hence, since this purports to be a book for the beginner as well, it is best to familiarize the reader with these symbols and expressions early on in the learning stage itself, so as to enable the reader to follow MIMO technical papers in the future for further understanding of the subject.

This chapter also examines the channel capacities under conditions when the channel is narrowband and is known to the transmitter and when it is unknown to the transmitter. It then examines frequency selective channels for the same two cases. We also discuss the ergodic and outage capacities of random channels.

Chapter 2 examines the MIMO wireless channel and determines its capacity under different conditions.

Chapter 3 provides an overview of the fading channel models and channel propagation.

Chapter 4 introduces the reader to space-time block coding techniques. We also examine the performance of space-time block codes in the presence of imperfect channel estimation and antenna correlation. We conclude with dominant eigenmode transmission.

Chapter 5 introduces the reader to space-time trellis codes and its design and performance evaluation in fast and slow fading channels.

Chapter 6 discusses layered space-time codes and BLAST algorithms.

Chapter 7 introduces the reader to orthogonal frequency division multiplexing (OFDM) techniques, which have become an integral part of MIMO technology. We also examine synchronization and channel estimation of OFDM signals.

Chapter 8 introduces the reader to the IEEE 802.11a standard, which, as we discussed earlier, has a lot of potential as a packet transmission system for MIMO systems.

Chapter 9 discusses the behavior of the space-time algorithms in a broadband wireless channel. We also discuss the capacity of spatially multiplexed signals and their performance. In addition, the reader is also introduced to a new topic called "hybrid OFDM/CDMA." This is a new hardware-oriented technique intended primarily for field engineers and is totally unlike existing OFDM/CDMA techniques. The idea here is to extend the capabilities of the current CDMA systems (used in cell phones) using the high throughput capability imparted by OFDM. The CDMA-OFDM system is then coupled with MIMO to yield a powerful solution to the ever-increasing requirement of high throughput cell phones.

Chapter 10 discusses topics for further reading in the MIMO field.

References

[1]   van Nee, R., and R. Prasad, *OFDM for Wireless Multimedia Communications*, Norwood, MA: Artech House, 2000.

[2]   Donald Shaver, Texas Instruments.

[3]   Broadband Wireless Internet Forum, www.bwif.org.

[4]   TD-SCDMA Forum, www.td-scdma-forum.org.

[5]   UMTS Web site, www.umtsworld.com/technology/hsdpa.htm.

[6]   Telatar, E., "Capacity of Multiantenna Gaussian Channels," *European Transactions on Telecommunications,* Vol. 10, No. 6, November/December 1999, pp. 585–595.

[7]   Foschini, G. J., and M. J. Gans, "On Limits of Wireless Communications in a Fading Environment When Using Multiple Antennas," *Wireless Personal Communications,* Vol. 6, 1998, pp. 311–335.

[8]   Shannon, C. E., "A Mathematical Theory of Communication," *Bell Syst. Tech. J.,* Vol. 27, October 1948, pp. 379–423 (Part One), pp. 623–656 (Part Two), Reprinted in Book Form, Urbana, IL: University of Illinois Press, 1949.

[9]   Tarokh, V., N. Seshadri, and A. R. Calderbank, "Space-Time Codes for High Data Rate Wireless Communication: Performance Criterion and Code Construction," *IEEE Trans. Inform. Theory,* Vol. 44, No. 2, March 1998, pp. 744–765.

[10] Alamouti, S. M., "A Simple Transmit Diversity Technique for Wireless Communications," *IEEE Journal Select. Areas Commun.,* Vol. 16, No. 8, October 1998, pp. 1451–1458.

[11] Foschini, G. J., "Layered Space-Time Architecture for Wireless Communications in a Fading Environment When Using Multiple Antennas," *Bell Labs. Tech. J.,* 1996, Vol. 6, No. 2, pp. 41–59.

[12] "Smart Antenna Experiments for 3G and 4G Cellular Systems," *The Pentex Pipeline,* Summer 2001, Vol. 10, No. 2, AT&T Labs-Research and Pentek Inc., Upper Saddle River, NJ.

[13]   Rysavy, Peter, "Wireless IP: A Case Study," www.pcsdata.com.

[14]   Stallings, W., *Data and Computer Communications,* 5th edition, Upper Saddle River, NJ: Prentice Hall, 1996.

[15]   Huitema, C., *IPv6: The New Internet Protocol,* Upper Saddle River, NJ: Prentice Hall, 1996.

[16]   Paulraj, Arogyaswami, Rohit Nabar, and Dhananjay Gore, *Introduction to Space-Time Wireless Communications,* Cambridge, UK: Cambridge University Press, 2003.

CHAPTER 2

The MIMO Wireless Channel

2.1 Introduction

It is best to begin at the beginning. We will start by examining a few terms that are part of the MIMO antenna systems. These terms will be used throughout this book and, in fact, will become part of the MIMO vocabulary. We shall first discuss the MIMO system model. This will be followed by a study of frequency-flat MIMO channel capacities for cases when the channel is known to the transmitter and when it is unknown to the transmitter. This study will be carried out both for deterministic as well as for random channels. We shall then investigate the effects of the physical parameters of the channel, such as fading correlation, line-of-sight problems and cross-polarization discrimination (XPD) problems of MIMO antennas and their impact on channel capacity. This will be followed by a study of the phenomenon of keyhole effect or degenerate channels and the effect of such channels on MIMO capacity. We conclude with a similar study of MIMO capacity on frequency-selective channels.

2.2 Preliminaries

2.2.1 Multiantenna Systems

Figure 2.1 illustrates different antenna configurations used in defining space-time systems. Single-input single-output (SISO) is the well-known wireless configuration, single-input multiple-output (SIMO) uses a single transmitting antenna and multiple (M_R) receive antennas, multiple-input single-output (MISO) has multiple (M_T) transmitting antennas and one receive antenna, MIMO has multiple (M_T) transmitting antennas and multiple (M_R) receive antennas and, finally, MIMO-multiuser (MIMO-MU), which refers to a configuration that comprises a base station with multiple transmit/receive antennas interacting with multiple users, each with one or more antennas. We now examine the meaning of certain terms.

2.2.2 Array Gain

Array gain is the average increase in the signal-to-noise ratio (SNR) at the receiver that arises from the coherent combining effect of multiple antennas at the receiver or transmitter or both. If the channel is known to the multiple antenna transmitter, the transmitter will weight the transmission with weights, depending on the channel

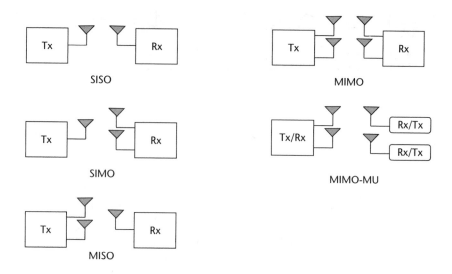

Figure 2.1 Different antenna configurations in space-time systems.

coefficients, so that there is coherent combining at the single antenna receiver (MISO case). The array gain in this case is called transmitter array gain. Alternately, if we have only one antenna at the transmitter and no knowledge of the channel and a multiple antenna receiver, which has perfect knowledge of the channel, then the receiver can suitably weight the incoming signals so that they coherently add up at the output (combining), thereby enhancing the signal. This is the SIMO case. This is called receiver array gain. Basically, multiple antenna systems require perfect channel knowledge either at the transmitter or receiver or both to achieve this array gain.

2.2.3 Diversity Gain

Multipath fading is a significant problem in communications. In a fading channel, signals experience fades (i.e., they fluctuate in their strength). When the signal power drops significantly, the channel is said to be in a fade. This gives rise to high bit error rates (BER). We resort to diversity to combat fading. This involves providing replicas of the transmitted signal over time, frequency, or space. There are three types of diversity schemes in wireless communications.

- *Temporal diversity:* In this case replicas of the transmitted signal are provided across time by a combination of channel coding and time interleaving strategies. The key requirement here for this form of diversity to be effective is that the channel must provide sufficient variations in time. It is applicable in cases where the coherence time of the channel is small compared with the desired interleaving symbol duration. In such an event, we are assured that the interleaved symbol is independent of the previous symbol. This makes it a completely new replica of the original symbol.
- *Frequency diversity:* This type of diversity provides replicas of the original signal in the frequency domain. This is applicable in cases where the

coherence bandwidth of the channel is small compared with the bandwidth of the signal. This assures us that different parts of the relevant spectrum will suffer independent fades.

- *Spatial diversity:* This is also called antenna diversity and is an effective method for combating multipath fading. In this case, replicas of the same transmitted signal are provided across different antennas of the receiver. This is applicable in cases where the antenna spacing is larger than the coherent distance to ensure independent fades across different antennas. The traditional types of diversity schemes are [1] *selection diversity, maximal ratio diversity,* and *equal gain diversity.* These schemes will be investigated in Chapter 3. Space-time codes exploit diversity across space and time. These will be examined in Chapters 4, 5, and 6.

Basically the effectiveness of any diversity scheme lies in the fact that at the receiver we must provide *independent* samples of the basic signal that was transmitted. In such an event we are assured that the probability of two or more relevant parts of the signal undergoing deep fades will be very small. The constraints on coherence time, coherence bandwidth, and coherence distance ensure this. The diversity scheme must then optimally combine the received diversified waveforms so as to maximize the resulting signal quality. We can also categorize diversity under the subheading of spatial diversity, based on whether diversity is applied to the transmitter or to the receiver.

- *Receive diversity:* Maximum ratio combining is a frequently applied diversity scheme in receivers to improve signal quality. In cell phones it becomes costly and cumbersome to deploy. This is one of the main reasons transmit diversity became popular, since transmit diversity is easier to implement at the base station.
- *Transmit diversity:* In this case we introduce controlled redundancies at the transmitter, which can be then exploited by appropriate signal processing techniques at the receiver. Generally this technique requires complete channel information at the transmitter to make this possible. But with the advent of space-time coding schemes like Alamouti's scheme [2], discussed in Chapter 4, it became possible to implement transmit diversity *without* knowledge of the channel. This was one of the fundamental reasons why the MIMO industry began to rise. Space-time codes for MIMO exploit both transmit as well as receive diversity schemes, yielding a high quality of reception.

Therefore, in MIMO we talk a lot about receive antenna diversity or transmit antenna diversity. In receive antenna diversity, the receiver that has multiple antennas receives multiple replicas of the same transmitted signal, assuming that the transmission came from the same source. This holds true for SIMO channels. If the signal path between each antenna pair fades independently, then when one path is in a fade, it is extremely unlikely that all the other paths are also in deep fade. Therefore, the loss of signal power due to fade in one path is countered by the same signal but received through a different path (route). This is like a line of soldiers. When one soldier falls in battle, another is ready to take his place. Hence, extending this analogy further, the more the soldiers, the stronger the line. The same is the argument in diversity. The more the diversity, the easier we can combat fades in a channel. Diversity is characterized by the number of independent fading branches, or paths (routes). These paths are also known as diversity order and are equal to the number of receive antennas in SIMO channels. Logically, the higher the diversity order (independent fading paths, or receive antennas), the better we combat fading. If the number of receive antennas tends to infinity, the diversity order tends to infinity and the channel tends to additive white Gaussian noise (AWGN). This is illustrated in Figure 2.2 [3]. In the figure, the sharp drops in power are called "fading margins." Note that with rising diversity order, the fading margins come down in intensity. This has been measured over a time period of 900 samples.

In the category of spatial diversity there are two more types of diversity that we need to consider. These are:

- *Polarization diversity:* In this type of diversity horizontal and vertical polarization signals are transmitted by two different polarized antennas and received correspondingly by two different polarized antennas at the receiver. Different polarizations ensure that there is no correlation between the data streams, without having to worry about coherent distance of separation between the antennas.
- *Angle diversity:* This applies at carrier frequencies in excess of 10 GHz. At such frequencies, the transmitted signals are highly scattered in space. In such an event the receiver can have two highly directional antennas facing in totally different directions. This enables the receiver to collect two samples of the same signal, which are totally independent of each other.

2.2.4 Data Pipes

The term *data pipe* is derived from fluid mechanics. Pipes are used to transfer water to a tank/reservoir. The more the number of pipes, the greater the quantum

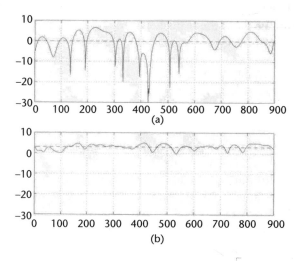

Figure 2.2 Rayleigh fading margins for (a) 1-input 1-output and (b) 2-input 2-output system. (*From:* [3]. Reprinted with permission. © CRC Press, Boca Raton, Florida.)

of flow of water into a tank/reservoir. This is similar to data pipes, but the analogy of communications with fluid mechanics ends there. We consider a case of two data pipes between the transmitter and receiver. In this situation there are two cases; either the data in the data pipes are identical to each other or they are independent samples, completely different from each other. In the former case, effectively the data going through is as if it is going through one data pipe, with the other pipe merely being a replica of the first one. This is a case of full correlation and because of this correlation, we do not get any throughput (bits per second) advantage. However, we do get a diversity advantage of two. The latter case deals with a situation where there is absolutely no correlation between the data carried by the two pipes. The data streams are independent. Hence, there is no diversity, but the throughput (output in bit/s) is definitely higher than in the first case. Therefore, the more the data pipes, the higher the throughput, provided the signals in the data pipes are not replicas of each other or correlated. In such an event the same signal is going through both pipes, so no new information is getting transferred. Therefore, correlation is not a good thing and it does reduce capacity, as we shall see. Remember that transmit diversity comes at the cost of throughput and vice versa. If we wish to eat the cake and still have it, then one way out is to sacrifice transmit diversity at the cost of throughput and incorporate diversity in the receiver (receive diversity). This way we at least have receive diversity, rather than no diversity at all in the system. This is what is done in spatial multiplexing.

2.2.5 Spatial Multiplexing

Spatial multiplexing offers a linear (in the number of transmit-receive antenna pairs or $\min(M_R, M_T)$ increase in the transmission rate (or capacity) for the same bandwidth and with no additional power expenditure. It is only possible in MIMO channels [4, 5]. Consider the case of two transmit and two receive antennas. This can be extended to more general MIMO channels.

The bit stream is split into two half-rate bit streams, modulated and transmitted simultaneously from both the antennas. The receiver, having complete knowledge of the channel, recovers these individual bit streams and combines them so as to recover the original bit stream. Since the receiver has knowledge of the channel it provides receive diversity, but the system has no transmit diversity since the bit streams are completely different from each other in that they carry totally different data. Thus spatial multiplexing increases the transmission rates proportionally with the number of transmit-receive antenna pairs.

This concept can be extended to MIMO-MU. In such a case, two users transmit their respective information simultaneously to the base station equipped with two antennas. The base station can separate the two signals and can likewise transmit two signals with spatial filtering so that each user can decode his or her own signal correctly. This allows capacity to increase proportionally to the number of antennas at the base station and the number of users.

2.2.6 Additional Terms

Automatic request for repeat (ARQ).   This is an error control mechanism in which received packets that cannot be corrected are retransmitted. This is a type of temporal diversity.

Forward error correction (FEC).   This is a technique that inserts redundant bits during transmission to help detect and correct bit errors during reception.

Coding gain.   The improvement in SNR at the receiver because of FEC is called coding gain.

Interleaving.   A form of data scrambling that spreads bursts of bit errors evenly over the received data allowing efficient forward error correction.

Multiplexing gain.   Capacity gain at no additional power or bandwidth consumption obtained through the use of multiple antennas at both sides of a wireless link.

2.3   MIMO System Model

We consider a MIMO system with a transmit array of M_T antennas and a receive array of M_R antennas. The block diagram of such a system is shown in Figure 2.3.

The transmitted matrix is a $M_T \times 1$ column matrix s where s_i is the ith component, transmitted from antenna i. We consider the channel to be a Gaussian channel such that the elements of s are considered to be independent identically distributed (i.i.d.) Gaussian variables. If the channel is unknown at the transmitter, we assume that the signals transmitted from each antenna have equal powers of E_s/M_T. The covariance matrix for this transmitted signal is given by

$$R_{ss} = \frac{E_s}{M_T} I_{M_T} \tag{2.1}$$

where E_s is the power across the transmitter irrespective of the number of antennas M_T and I_{M_T} is an $M_T \times M_T$ identity matrix. The transmitted signal bandwidth is so narrow that its frequency response can be considered flat (i.e., the channel is

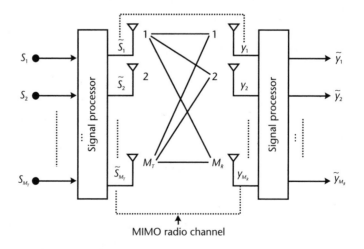

Figure 2.3   Block diagram of a MIMO system.

memoryless). The channel matrix $\mathbf{H}$ is a $M_R \times M_T$ complex matrix. The component $h_{i,j}$ of the matrix is the fading coefficient from the jth transmit antenna to the ith receive antenna. We assume that the received power for each of the receive antennas is equal to the total transmitted power E_s. This implies we ignore signal attenuation, antenna gains, and so on. Thus we obtain the normalization constraint for the elements of $\mathbf{H}$, for a deterministic channel as

$$\sum_{j=1}^{M_T} |h_{i,j}|^2 = M_T, \, i = 1, 2, \ldots, M_R \tag{2.2}$$

If the channel elements are not deterministic but random, the normalization will apply to the expected value of (2.2).

We assume that the channel matrix is known at the receiver but unknown at the transmitter. The channel matrix can be estimated at the receiver by transmitting a training sequence. If we require the transmitter to know this channel, then we need to communicate this information to the transmitter via a feedback channel. The elements of $\mathbf{H}$ can be deterministic or random.

The noise at the receiver is another column matrix of size $M_R \times 1$, denoted by $\mathbf{n}$. The components of $\mathbf{n}$ are zero mean circularly symmetrical complex Gaussian (ZMCSCG) variables. The covariance matrix of the receiver noise is

$$\mathbf{R_{nn}} = E\{\mathbf{nn}^H\} \tag{2.3}$$

If there is no correlation between components of $\mathbf{n}$, the covariance matrix is obtained as

$$\mathbf{R_{nn}} = N_0 \mathbf{I}_{M_R} \tag{2.4}$$

Each of the M_R receive branches has identical noise power of N_0.

The receiver operates on the maximum likelihood detection principle over M_R receive antennas. The received signals constitute a $M_R \times 1$ column matrix denoted by $\mathbf{r}$, where each complex component refers to a receive antenna. Since we assumed that the total received power per antenna is equal to the total transmitted power, the SNR can be written as

$$\gamma = \frac{E_s}{N_0} \tag{2.5}$$

Therefore, the received vector can be expressed as

$$\mathbf{r} = \mathbf{Hs} + \mathbf{n} \tag{2.6}$$

The received signal covariance matrix defined as $E\{\mathbf{rr}^H\}$, is given by [using (2.6)]

$$\mathbf{R}_{\mathbf{rr}} = \mathbf{H}\mathbf{R}_{\mathbf{ss}}\mathbf{H}^H \tag{2.7}$$

while the total signal power can be expressed as $\mathrm{tr}\,(\mathbf{R}_{\mathbf{rr}})$.

2.4 MIMO System Capacity

The system capacity is defined as the maximum possible transmission rate such that the probability of error is arbitrarily small [6].

We assume that the channel knowledge is unavailable at the transmitter and known only at the receiver.

The capacity of MIMO channel is defined as [4, 5]

$$C = \max_{f(\mathbf{s})} I(\mathbf{s}; \mathbf{y}) \tag{2.8}$$

where $f(\mathbf{s})$ is the probability distribution of the vector $\mathbf{s}$ and $I(\mathbf{s}; \mathbf{y})$ is the mutual information between vectors $\mathbf{s}$ and $\mathbf{y}$. We note that

$$I(\mathbf{s}; \mathbf{y}) = H(\mathbf{y}) - H(\mathbf{y}\,|\,\mathbf{s}) \tag{2.9}$$

where $H(\mathbf{y})$ is the differential entropy of the vector $\mathbf{y}$, while $H(\mathbf{y}\,|\,\mathbf{s})$ is the conditional differential entropy of the vector $\mathbf{y}$, given knowledge of the vector $\mathbf{s}$. Since the vectors $\mathbf{s}$ and $\mathbf{n}$ are independent, $H(\mathbf{y}\,|\,\mathbf{s}) = H(\mathbf{n})$. From (2.9),

$$I(\mathbf{s}; \mathbf{y}) = H(\mathbf{y}) - H(\mathbf{n}) \tag{2.10}$$

If we maximize the mutual information $I(\mathbf{s}; \mathbf{y})$ reduces to maximizing $H(\mathbf{y})$. The covariance matrix of $\mathbf{y}$, $R_{\mathbf{yy}} = \epsilon\{\mathbf{y}\mathbf{y}^H\}$, satisfies

$$\mathbf{R}_{\mathbf{yy}} = \frac{E_s}{M_T}\mathbf{H}\mathbf{R}_{\mathbf{ss}}\mathbf{H}^H + N_0\mathbf{I}_{M_R} \tag{2.11}$$

where $\mathbf{R}_{\mathbf{ss}} = \epsilon\{\mathbf{s}\mathbf{s}^H\}$ is the covariance matrix of $\mathbf{s}$. Among all vectors $\mathbf{y}$ with a given covariance matrix $\mathbf{R}_{\mathbf{yy}}$, the differential entropy $H(\mathbf{y})$ is maximized when $\mathbf{y}$ is ZMCSCG. This implies that $\mathbf{s}$ must also be ZMCSCG vector, the distribution of which is completely characterized by $\mathbf{R}_{\mathbf{ss}}$. The differential entropies of the vectors $\mathbf{y}$ and $\mathbf{n}$ are given by

$$H(\mathbf{y}) = \log_2\left(\det\left(\pi e \mathbf{R}_{\mathbf{yy}}\right)\right) \text{ bps/Hz} \tag{2.12}$$

$$H(\mathbf{n}) = \log_2\left(\det\left(\pi e \sigma^2 \mathbf{I}_{M_R}\right)\right) \text{ bps/Hz} \tag{2.13}$$

Therefore, $I(\mathbf{s}; \mathbf{y})$ in (2.10) reduces to [4]

$$I(\mathbf{s}; \mathbf{y}) = \log_2 \det\left(\mathbf{I}_{M_R} + \frac{E_s}{M_T N_0}\mathbf{H}\mathbf{R}_{\mathbf{ss}}\mathbf{H}^H\right) \text{ bps/Hz} \tag{2.14}$$

and from (2.8), the capacity of the MIMO channel is given by

$$C = \max_{\text{Tr}(\mathbf{R}_{ss}) = M_T} \log_2 \det\left(\mathbf{I}_{M_R} + \frac{E_s}{M_T N_0} \mathbf{H} \mathbf{R}_{ss} \mathbf{H}^H\right) \text{ bps/Hz} \qquad (2.15)$$

The capacity C in (2.15) is also called error-free spectral efficiency or data rate per unit bandwidth that can be sustained reliably over the MIMO link. Thus if our bandwidth is W Hz, the maximum achievable data rate over this bandwidth using MIMO techniques is WC bit/s [7].

2.5 Channel Unknown to the Transmitter

If the channel is unknown to the transmitter, then the vector $\mathbf{s}$ is statistically independent (i.e., $\mathbf{R}_{ss} = \mathbf{I}_{M_T}$). This implies that the signals are independent and the power is equally divided among the transmit antennas. The capacity in such a case is, (from 2.15)

$$C = \log_2 \det\left(\mathbf{I}_{M_R} + \frac{E_s}{M_T N_0} \mathbf{H} \mathbf{H}^H\right) \qquad (2.16)$$

The reader is cautioned that this is not Shannon capacity since it is possible to outperform $\mathbf{R}_{ss} = \mathbf{I}_{M_T}$, if one has the channel knowledge. Nevertheless we shall refer to (2.16) as capacity. Now $\mathbf{H}\mathbf{H}^H$ is an $M_R \times M_R$ positive semidefinite Hermitian matrix. The eigendecomposition of such a matrix is given by $\mathbf{Q}\Lambda\mathbf{Q}^H$ [8], where $\mathbf{Q}$ is an $M_R \times M_R$ matrix satisfying $\mathbf{Q}^H\mathbf{Q} = \mathbf{Q}\mathbf{Q}^H = \mathbf{I}_{M_R}$ and $\Lambda = diag\{\lambda_1 \lambda_2 \ldots \lambda_{M_R}\}$ with $\lambda_i \geq 0$. We assume that the eigenvalues are ordered so that $\lambda_i \geq \lambda_{i+1}$. Then

$$\lambda_i = \begin{cases} \sigma_i^2, & \text{if } i = 1, 2, \ldots r \\ 0 & \text{if } i = r, r+1, \ldots M_R \end{cases} \qquad (2.17)$$

where σ_i are the singular values obtained as $\Sigma = diag\{\sigma_1 \quad \sigma_2 \quad \ldots \quad \sigma_r\}$ from the singular value decomposition of $\mathbf{H} = \mathbf{U}\Sigma\mathbf{V}^H$. Then the capacity of the MIMO channel is given by

$$C = \log_2 \det\left(\mathbf{I}_{M_R} + \frac{E_s}{M_T N_0} \mathbf{Q}\Lambda\mathbf{Q}^H\right) \qquad (2.18)$$

Using $\det(\mathbf{I}_m + \mathbf{A}\mathbf{B}) = \det(\mathbf{I}_n + \mathbf{B}\mathbf{A})$ for matrices $\mathbf{A}(m \times n)$ and $\mathbf{B}(n \times m)$ and $\mathbf{Q}^H\mathbf{Q} = \mathbf{I}_{M_R}$, (2.18) simplifies to

$$C = \log_2 \det\left(\mathbf{I}_{M_R} + \frac{E_s}{M_T N_0} \Lambda\right) \qquad (2.19)$$

or

$$C = \sum_{i=1}^{r} \log_2 \left(1 + \frac{E_s}{M_T N_0} \lambda_i \right) \qquad (2.20)$$

where r is the rank of the channel and λ_i $(i = 1, 2, \ldots, r)$ are the positive eigenvalues of $\mathbf{H}\mathbf{H}^H$. Equation (2.20) expresses the capacity of the MIMO channel as a sum of the capacities of r SISO channels, each having a power gain of λ_i $(i = 1, 2, \ldots, r)$ and transmit power E_s / M_T.

This means that the technique of multiple antennas at the transmitter and receiver opens up multiple scalar spatial data pipes between the transmitter and receiver. Furthermore, equal transmit energy is allocated to each spatial data pipe. This is for the case when the channel is unknown at the transmitter.

We define the squared Frobenius norm of $\mathbf{H}$, as $\|\mathbf{H}\|_F^2 = \text{Tr}(\mathbf{H}\mathbf{H}^H) = \sum_{i=1}^{M_R} \sum_{j=1}^{M_T} |h_{i,j}|^2$. Frobenius norm is interpreted as the total power gain of the channel.

Also $\|\mathbf{H}\|_F^2 = \sum_{i=1}^{M_R} \lambda_i$ where λ_i $(i = 1, 2, \ldots, M_R)$ are the eigenvalues of $\mathbf{H}\mathbf{H}^H$. We fix this total power so that $\|\mathbf{H}\|_F^2 = \beta$. Then if the channel matrix is of full rank such that $M_T = M_R = M$, the capacity C in (2.20) is maximized when $\lambda_i = \lambda_j = \beta/M$ $(i, j = 1, 2, \ldots, M)$ (remember, the channel is unknown, so equal power distribution). To achieve this, $\mathbf{H}\mathbf{H}^H = \mathbf{H}^H\mathbf{H} = (\beta/M)\mathbf{I}_M$, (i.e., the channel matrix $\mathbf{H}$ should be orthogonal). This gives

$$C = M \log_2 \left(1 + \frac{\beta E_s}{N_0 M^2} \right) \qquad (2.21)$$

If the elements of $\mathbf{H}$ have ones along the diagonal, then $\|\mathbf{H}\|_F^2 = M^2$ and

$$C = M \log_2 \left(1 + \frac{E_s}{N_0} \right) \qquad (2.22)$$

The capacity of an orthogonal MIMO channel is therefore M times the scalar channel capacity [7]. This conclusion is once again verified using a different approach for OFDM-based spatial multiplexing, as discussed in Chapter 9.

2.6   Channel Known to the Transmitter

It is possible by various means, which will be discussed in Chapter 4, to learn the channel state information (CSI) at the transmitter. In such an event the capacity can be increased by resorting to the so-called "water filling principle" [4], by assigning various levels of transmitted power to various transmitting antennas. This power is assigned on the basis that the better the channel gets, the more power it gets and vice versa. This is an optimal energy allocation algorithm.

2.6.1   Water-Pouring Principle

Consider a MIMO channel where the channel parameters are known at the transmitter. The "water-pouring principle" or "water-filling principle" can be derived by maximizing the MIMO channel capacity under the rule that more power is allocated to the channel that is in good condition and less or none at all to the bad channels.

This analysis is taken from [7]. Consider a ZMCSCG signal vector $\tilde{s}$ of dimension $r \times 1$ where r is the rank of the channel $\mathbf{H}$ to be transmitted. We note from Figure 2.4 that the vector is multiplied by a matrix $\mathbf{V}$ prior to transmission (based on the fact that $\mathbf{H} = \mathbf{U}\Sigma\mathbf{V}^H$ through singular value decomposition). At the receiver, the received signal vector $\mathbf{y}$ is multiplied by the matrix $\mathbf{U}^H$.

The input-output relationship for this operation is given by

$$\tilde{\mathbf{y}} = \sqrt{\frac{E_s}{M_T}} \, \mathbf{U}^H \mathbf{H} \mathbf{V} \tilde{s} + \mathbf{U}^H \mathbf{n} \tag{2.23}$$

$$= \sqrt{\frac{E_s}{M_T}} \, \Sigma \tilde{s} + \tilde{n}$$

where $\tilde{\mathbf{y}}$ is the transformed received signal vector of size $r \times 1$ and $\tilde{n}$ is the ZMCSCG transformed noise vector of size $r \times 1$ with the covariance matrix $\epsilon\{\tilde{n}\tilde{n}^H\} = N_0\mathbf{I}_r$. The vector $\tilde{s}$ satisfies $\epsilon\{\tilde{s}\tilde{s}^H\} = M_T$ to constrain the total transmit energy. Equation (2.23) shows us that with channel knowledge at the transmitter, $\mathbf{H}$ can be explicitly decomposed into r parallel SISO channels satisfying

$$\tilde{y}_i = \sqrt{\frac{E_s}{M_T}} \, \sqrt{\lambda_i} \tilde{s}_i + \tilde{n}_i, \; i = 1, 2, \ldots, r \tag{2.24}$$

The capacity of the MIMO channel is the sum of the individual parallel SISO channel capacities and is given by

$$C = \sum_{i=1}^{r} \log_2\left(1 + \frac{E_s \gamma_i}{M_T N_0} \lambda_i\right) \tag{2.25}$$

where $\gamma_i = \epsilon\{|s_i|^2\}$ $(i = 1, 2, \ldots, r)$ is the transmit energy in the ith subchannel such that $\sum_{i=1}^{r} \gamma_i = M_T$.

To maximize mutual information, the transmitter can access the individual subchannels and allocate variable power levels to them. Hence, the mutual information maximization problem becomes,

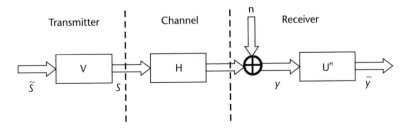

Figure 2.4   Decomposition of $\mathbf{H}$ when the channel is known to the transmitter and receiver.

$$C = \max_{\Sigma_{i=1}^r \gamma_i} \sum_{i=1}^r \log_2\left(1 + \frac{E_s \gamma_i}{M_T N_0}\lambda_i\right)$$ (2.26)

Using Lagrangian methods, the optimal energy allocation procedure is

$$\gamma_i^{opt} = \left(\mu - \frac{M_T N_0}{E_s \lambda_i}\right), i = 1, 2, \ldots, r \text{ and}$$ (2.27)

$$\sum_{i=1}^r \gamma_i^{opt} = M_T$$

where μ is a constant and $(x)_+$ implies

$$(x)_+ = \begin{cases} x & \text{if } x \geq 0 \\ 0 & \text{if } x < 0 \end{cases}$$ (2.28)

We determine this optimal energy allocation iteratively through the "water-pouring algorithm" [9]. We now describe the algorithm.

We set the iteration count p to 1 and calculate the constant μ in (2.27):

$$\mu = \frac{M_T}{(r - p + 1)}\left[1 + \frac{N_0}{E_s}\sum_{i=1}^{r-p+1}\frac{1}{\lambda_i}\right]$$ (2.29)

Using this value of μ, the power allocated to the ith subchannel is calculated as

$$\gamma_i = \left(\mu - \frac{M_T N_0}{E_s \lambda_i}\sum_{i=1}^{r-p+1}\frac{1}{\lambda_i}\right), i = 1, 2, \ldots, r - p + 1$$ (2.30)

If the power allotted to the channel with the lowest gain is negative (i.e., $\lambda_{r-p+1} < 0$), we discard this channel by setting $\gamma_{r-p+1}^{opt} = 0$ and rerun the algorithm with the iteration count p incremented by 2. The optimal power allocation strategy, therefore, allocates power to those spatial subchannels that are non-negative. Figure 2.5 illustrates the water-pouring algorithm. Obviously, since this algorithm only concentrates on good-quality channels and rejects the bad ones during each channel realization, it is to be expected that this method yields a capacity that is equal or better than the situation when the channel is unknown to the transmitter.

Channel capacity for the case when the channel is unknown to the transmitter and receiver is an area of ongoing research [10, 11].

2.6.2 Capacity When Channel Is Known to the Transmitter

This has been already discussed and is given by (2.26)

$$C = \max_{\Sigma_{i=1}^r \gamma_i} \sum_{i=1}^r \log_2\left(1 + \frac{E_s \gamma_i}{M_T N_0}\lambda_i\right)$$ (2.31)

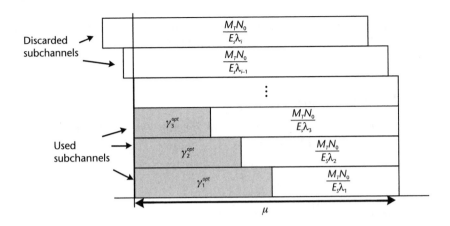

Figure 2.5   Schematic of the water-pouring algorithm.

2.7   Deterministic Channels

2.7.1   SIMO Channel Capacity

In a SIMO channel, $M_T = 1$ and there are M_R receive antennas. In such a case the channel matrix is a column matrix

$$H = \left(h_1 \, h_2 \, \ldots \, h_{M_R} \right)^T \tag{2.32}$$

where $(\cdot)^T$ denotes matrix transpose. Since $M_R > M_T$, (2.16) is modified as

$$C = \log_2 \det \left(\mathbf{I}_{M_T} + \frac{E_s}{M_T N_0} \mathbf{H}^H \mathbf{H} \right) \tag{2.33}$$

Now $\mathbf{H}^H \mathbf{H} = \sum_{i=1}^{M_R} |h_i|^2$ and $M_T = 1$. Hence,

$$C = \log_2 \det \left(1 + \sum_{i=1}^{M_R} |h_i|^2 \frac{E_s}{N_0} \right) \tag{2.34}$$

If the channel matrix elements are equal and normalized as

$$|h_1|^2 = |h_2|^2 = \ldots |h_{M_R}|^2 = 1$$

then capacity when the channel is unknown at the transmitter, is

$$C = \log_2 \det \left(1 + M_R \frac{E_s}{N_0} \right) \tag{2.35}$$

The system achieves a diversity gain of M_R relative to the SISO case. For $M_R = 4$ and SNR = 10 dB, the SIMO capacity is 5.258 bit/s/Hz. The addition of receive antennas yields a logarithmic increase in capacity in SIMO channels. Knowledge of the channel at the transmitter in this case provides no additional benefit.

2.7.2 MISO Channel Capacity

In MISO channels, $M_R = 1$ and there are M_T transmit antennas. In this case, since $M_T > M_R$, we use (2.16) as it is. The channel is represented by the row matrix

$$H = \begin{pmatrix} h_1 & h_2 & \dots & h_{M_T} \end{pmatrix}$$

As $\mathbf{HH}^H = \sum_{j=1}^{M_T} |h_j|^2$, from (2.16) we obtain

$$C = \log_2 \left(1 + \sum_{j=1}^{M_T} |h_j|^2 \frac{E_s}{M_T N_0} \right) \tag{2.36}$$

If the channel coefficients are equal and normalized as $\sum_{j=1}^{M_T} |h_j|^2 = M_T$, then the capacity for the MISO case becomes

$$C = \log_2 \left(1 + \frac{E_s}{N_0} \right) \tag{2.37}$$

We note that (2.37) is the same as for a SISO case (i.e., the capacity did not increase with the number of antennas). This is for the case when the channel is unknown at the transmitter. The reason for this result is that there is no array gain at the transmitter because the transmitter has no knowledge of the channel parameters. Array gain is the average increase in the SNR at the receiver that arises from the coherent combining effect of multiple antennas at the receiver or transmitter or both. If the channel is known to the transmitter, the transmitter will weight the transmission with weights depending on the channel coefficients, so that there is coherent combining at the receiver (MISO case). If we take the case when the channel is known at the transmitter, we apply (2.31). Since the channel matrix has rank 1, there is only one term in the sum in (2.31) and only one nonzero eigenvalue given by

$$\lambda = \sum_{j=1}^{M_T} |h_j|^2$$

Hence, capacity is

$$C = \log_2 \left(1 + \sum_{j=1}^{M_T} |h_j|^2 \frac{E_s}{N_0} \right) \tag{2.38}$$

If the channel coefficients are equal and normalized as $\sum_{j=1}^{M_T} |h_j|^2 = M_T$, the capacity becomes

$$C = \log_2 \left(1 + M_T \frac{E_s}{N_0} \right) \quad (2.39)$$

For $M_T = 4$ and SNR = 10 dB, the MISO capacity is 5.258 bit/s/Hz. *This is with channel knowledge at the transmitter.* In both cases of SIMO and MISO there is only one spatial data pipe (i.e., the rank of the channel matrix is one). Basically, the channel matrix is a $M_R \times M_T$ matrix. In a MISO case, $M_R = 1$ and in a SIMO case, $M_T = 1$. In either case, the channel matrix has only one eigenvalue and its rank is 1. Physically, this means that there is only one route from transmitter to receiver for the signals to pass through. Hence, we have one data pipe. If we had $M_T = M_R = 2$, then we would have a MIMO case with a channel matrix of rank 2 and having two eigenvalues, hence, two routes from transmitter to receiver (i.e., we have two data pipes and so on).

2.8 Random Channels

We have until now discussed MIMO capacity when the channel is a deterministic channel. We now consider the case when **H** is chosen randomly according to a Rayleigh distribution in a quasi-static channel. This is a real-life situation encountered, for example, in wireless LANs with high data rates and low fade rates. We assume that the receiver has perfect knowledge of the channel and the transmitter has no knowledge of the channel. Since the channel is random, the information rate associated with it is also random. The cumulative distribution function (CDF) of the information rate of a flat fading MIMO channel is shown in Figure 2.6 for a 2×2 system. The SNR is 10 dB and the channel is unknown to the transmitter.

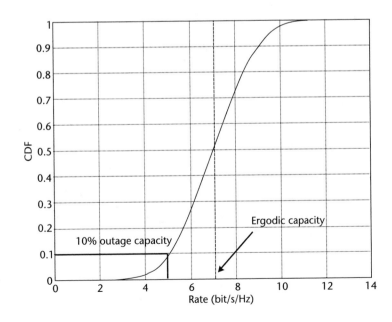

Figure 2.6 CDF of information rate for i.i.d. channel matrix with a 2×2 system and SNR = 10 dB.

2.8.1 Ergodic Capacity

The ergodic capacity of a MIMO channel is the ensemble average of the information rate over the distribution of the elements of the channel matrix $\mathbf{H}$ [7]. It is the capacity of the channel when every channel matrix $\mathbf{H}$ is an independent realization [i.e., it has no relationship to the previous matrix but is typically representative of it class (ergodic)]. This implies that it is a result of infinitely long measurements. Since the process model is ergodic, this implies that the coding is performed over an infinitely long interval. Hence, it is the Shannon capacity of the channel. Based on (2.20) the ergodic capacity is expressed as

$$ C = \epsilon \left\{ \sum_{i=1}^{r} \log_2 \left(1 + \frac{\rho}{M_T} \lambda_i \right) \right\} \qquad (2.40) $$

where $\rho = E_s / N_0$. The expectation operator applies in this case because the channel is random. Since $\mathbf{H}$ is random, the information rate associated with it is also random. The CDF of the information rate is depicted in Figure 2.6.

The ergodic capacity is the median of the CDF curve. In this case it is 7.0467 bit/s/Hz. Figure 2.7 shows the ergodic capacity over different system configurations as a function of ρ. We note that ergodic capacity increases with increasing ρ and with increasing M_T and M_R.

Ergodic capacity when the channel is known to the transmitter is based on the water-filling algorithm and is given from (2.25)

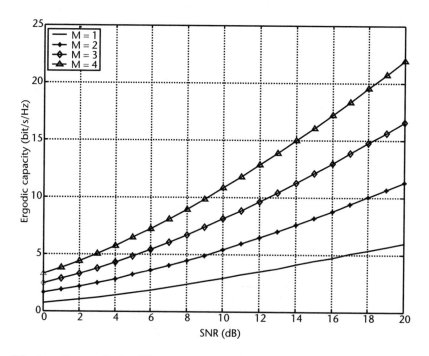

Figure 2.7 Ergodic capacity for different antenna configurations with $M_T = M_R = M$.

$$C = \epsilon \left\{ \sum_{i=1}^{r} \log_2 \left(1 + \frac{E_s \gamma_i}{M_T N_0} \lambda_i \right) \right\} \qquad (2.41)$$

Equation (2.41) is the ensemble average of the capacity achieved when the water-filling optimization is performed for each realization of **H**. Figure 2.8 shows the performance comparison of ergodic capacity of a MIMO channel with $M_T = M_R = 4$ when the channel is unknown to the transmitter and also when known and the channel is Rayleigh i.i.d.

The ergodic capacity when the channel is known to the transmitter is always higher then when it is unknown. This advantage reduces at high SNRs. This is because at high SNRs (2.41) tends to (2.40). Another way of looking at this situation is to appreciate the fact that at high SNRs, all eigenchannels perform equally well (i.e., there is no difference in quality between them). Hence, all the channels will perform to their capacities, making both cases nearly identical.

2.8.2 Outage Capacity

In reality, the block lengths are finite. The common example is speech transmission. In such cases, we speak of outage capacity. Outage capacity is the capacity that is guaranteed with a certain level of reliability. We define $p\%$ outage capacity as the information rate that is guaranteed for $(100-p)\%$ of the channel realizations, that is, $P(C \leq C_{\text{out}}) = p\%$ [12]. We show 10% outage capacity in Figure 2.9.

Figure 2.9 shows the 10% outage capacity for several MIMO cases, when the channel is i.i.d. and unknown at the transmitter. We note that as the SNR increases, the capacity increases and as the number of antennas increases, so does the capacity. From (2.16) for the case when $M_T = M_R = M$ and the channel is i.i.d.,

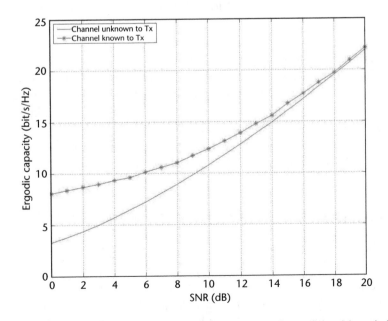

Figure 2.8 Ergodic capacity of an $M = 4$ channel with and without channel knowledge at the transmitter. The difference in ergodic capacities decreases with SNR.

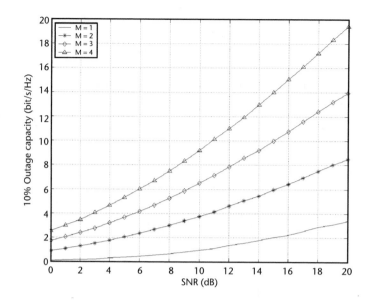

Figure 2.9   10% outage capacity for various antenna configurations. Outage capacity improves with rising $M_T = M_R = M$.

$$\frac{1}{M}\mathbf{H}_\omega\mathbf{H}_\omega^H \to \mathbf{I}_M \text{ as } M \to \infty$$

Therefore,

$$C \to M \log_2(1 + \rho) \text{ where } \rho \text{ is the SNR} \tag{2.42}$$

Asymptotically in M, the capacity in spatially white MIMO channel becomes deterministic and increases linearly with M for a fixed SNR. Also for every 3-dB increase in SNR, we get M bit/s/Hz increase in capacity for a MIMO channel, compared with 1 bit/s/Hz in a SISO channel. The outage capacity curves substantiate this conclusion. If the channel is known at the transmitter, Figure 2.10 shows that water-filling is a superior solution.

The same arguments for convergence of the curves at high SNRs apply to Figure 2.10 as for Figure 2.8 but in the context of outage capacities.

2.9   Influence of Fading Correlation on MIMO Capacity

In reality, the channel is not ideally Rayleigh i.i.d. There are various factors that cause it to deviate from this and, as a result, the performance of MIMO systems deteriorate. One of these is correlation. Correlation problems arise because of the separation distance between *antenna elements* in a base station. Usually this separation distance is in the order of a few centimeters, whereas the separation between the mobile and the base station is in the order of a few kilometers! Hence,

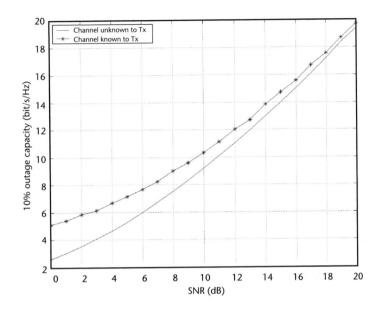

Figure 2.10   10% outage capacity of an $M = 4$ channel with and without channel knowledge at the transmitter. The difference in outage capacities decreases with SNR.

the signals arriving at the base station from a receiver will necessarily be very close together, giving rise to correlation between them. This occurs because all the antenna elements receive the same signal, due to the geometry of the phenomenon. The degree of "sameness" determines the correlation coefficient with 1 as maximum correlation and 0 as no correlation. This is overcome in a base station by:

- Using independent dipole antennas separated by a distance D that exceeds the coherent distance for that channel.
- Using two separate antenna arrays separated by a distance D that exceeds the coherent distance for that channel.

These cases are illustrated in Figure 2.11.

In Figure 2.11, both cases (a) and (b) are feasible. The separation distance D is usually of the order of 10 to 16 wavelengths for a base station, because it is on a high vantage point and far from the mobile receivers. The problem is not so severe for mobile phones because they are invariably located in a high scattering environment. In such cases the separation distance is usually 2 to 3 wavelengths. In the event of correlation, the elements of the channel matrix are correlated and may be modeled as [7]

$$\text{vec}(\mathbf{H}) = \mathbf{R}^{1/2} \, \text{vec}(\mathbf{H}_\omega) \tag{2.43}$$

where $\mathbf{H}_\omega$ is a Rayleigh i.i.d. spatially white MIMO channel matrix of size $M_R \times M_T$ and $\mathbf{R}$ is a $M_T M_R \times M_T M_R$ covariance matrix defined as

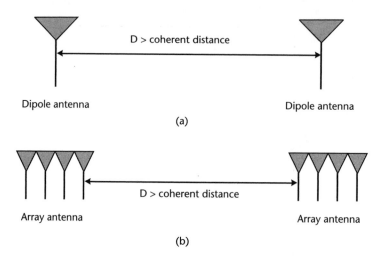

Figure 2.11 The correlation problem. We can deploy (a) separate dipole antennas well separated from each other or (b) antenna arrays well separated from each other.

$$\mathbf{R} = \epsilon \left\{ vec\,(\mathbf{H})\, vec\,(\mathbf{H})^H \right\} \qquad (2.44)$$

$\mathbf{R}$ is a positive semidefinite Hermitian matrix. If $\mathbf{R}$ is full rank (i.e., $\mathbf{R} = \mathbf{I}_{M_T M_R}$), then in such a case $\mathbf{H} = \mathbf{H}_\omega$. The idea of such a model is to efficiently portray the correlation effects in the channel. This approach is elaborated by using a more generalized model given by

$$\mathbf{H} = \mathbf{R}_r^{1/2} \mathbf{H}_\omega \mathbf{R}_t^{1/2} \qquad (2.45)$$

where $\mathbf{R}_t$ is the $M_T \times M_T$ transmit covariance matrix and $\mathbf{R}_r$ is the $M_R \times M_R$ receive covariance matrix. Both $\mathbf{R}_t$ and $\mathbf{R}_r$ are positive semidefinite Hermitian matrixes. Equation (2.45) is explained as follows:

- The transmitted signal, when it reaches the receiver, is correlated by virtue of the geometry at the receiver ($\mathbf{R}_r$). The channel *per se* has been portrayed as Rayleigh i.i.d. ($\mathbf{H}_\omega$).
- The transmitted signal is correlated at the transmitter itself due to the geometry at the transmitter ($\mathbf{R}_t$) or due to a low angle of spread.
- $\mathbf{R}$, $\mathbf{R}_t$ and $\mathbf{R}_r$ are related by $\mathbf{R} = \mathbf{R}_t^T \otimes \mathbf{R}_r$ where $\otimes$ denotes Kronecker product.

We note that $\mathbf{H}_\omega$ is full rank *per se*, but the effective rank of $\mathbf{H}$ gets reduced due to correlation at the transmitter or at the receiver or both and this effective rank is expressed as $\min\,(r(\mathbf{R}_r), r(\mathbf{R}_t))$ where $r(\mathbf{A})$ denotes rank of A.

If we assume that both the matrixes $\mathbf{R}_r$ and $\mathbf{R}_t$ are normalized so that they have unity values along their diagonals, this yields $\epsilon\{|h_{i,j}|^2\} = 1$. The capacity of the MIMO channel in the presence of spatial fading correlation without channel knowledge at the transmitter follows from (2.16) as

$$C = \log_2 \det\left(\mathbf{I}_{M_R} + \frac{\rho}{M_T}\mathbf{R}_r^{1/2}\mathbf{H}_\omega\mathbf{R}_t\mathbf{H}_\omega^H\mathbf{R}_r^{H/2}\right) \qquad (2.46)$$

Assume $M_R = M_T = M$ and that the receive and transmit correlation matrixes are full rank. Then, at high SNR, the capacity can be approximated as

$$C \approx \log_2 \det\left(\frac{\rho}{M}\mathbf{H}_\omega\mathbf{H}_\omega^H\right) + \log_2 \det(\mathbf{R}_r) + \log_2 \det(\mathbf{R}_t) \qquad (2.47)$$

We note from (2.47) that both correlation matrixes have the same impact on the channel capacity. We now examine the conditions on $\mathbf{R}_r$ that maximize capacity. The same arguments apply to $\mathbf{R}_t$.

$$\det(\mathbf{R}_r) = \prod_{i=1}^{M} \lambda_i(\mathbf{R}_r) \leq 1 \qquad (2.48)$$

Remember that there is a power constraint in that $\sum_{i=1}^{M}\lambda_i(\mathbf{R}_r) = M$. This means that $\log_2 \det(\mathbf{R}_r) \leq 0$. It can only equal zero if all eigenvalues of $\mathbf{R}_r$ are equal (i.e., $\mathbf{R}_r = \mathbf{I}_M$). Therefore, fading signal correlation does reduce the number of eigenvalues and thereby reduces the MIMO channel capacity. This loss in ergodic or outage capacity is given by $(\log_2 \det(\mathbf{R}_r) + \log_2 \det(\mathbf{R}_t))$ bit/s/Hz.

If we assume an orthogonal channel where $M_T = M_R = 2$ and further assume that there is correlation only at the receiver, then we choose a receive correlation matrix as

$$\mathbf{R}_r = \begin{bmatrix} 1 & \sigma \\ \sigma & 1 \end{bmatrix} \qquad (2.49)$$

We take a correlation coefficient of 0.8.

We note from Figure 2.12 that there is a loss of 2.47 bit/s/Hz at high SNR compared with the case with no correlation. This is the loss expected from the $\log_2 \det(\mathbf{R}_r)$ component. If the correlation coefficient of either or both of $\mathbf{R}_r$ and $\mathbf{R}_t$ is unity, then the $\mathbf{H}$ matrix will also become rank 1 (i.e., it becomes an SISO channel). Hence, correlation is not a good thing!

2.10 Influence of LOS on MIMO Capacity

We now examine another aspect, which makes a channel deviate from Rayleigh i.i.d. Until now we have only considered a Rayleigh i.i.d. channel. This is far removed from reality. It is better to depict the real-world channel as

$$\mathbf{H} = \mathbf{H}_{Ric} + \mathbf{R}_r^{1/2}\mathbf{H}_\omega\mathbf{R}_t^{1/2} \qquad (2.50)$$

where $\mathbf{H}_{Ric}$ is the Rician or line-of-sight (LOS) component. The other terms were discussed earlier. The LOS is a component that exists by virtue of a direct path

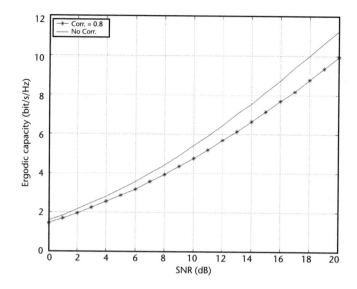

Figure 2.12   Ergodic capacity with high and low correlation. The loss in ergodic capacity is about 2.47 bit/s/Hz when $\sigma = 0.8$.

between the transmitter and the receiver, which are so located as to be within line of sight of each other. The LOS in (2.50) can also be shown as a sum of a fixed component and a scattered component as follows [13]

$$\mathbf{H} = \sqrt{\frac{K}{1 + K}}\,\overline{\mathbf{H}} + \sqrt{\frac{1}{1 + K}}\,\mathbf{H}_\omega \qquad (2.51)$$

where $\sqrt{K/(1 + K)}\,\overline{\mathbf{H}} = \epsilon(\mathbf{H})$ is the LOS component of the channel and $\sqrt{1/(1 + K)}\,\mathbf{H}_\omega$ is the fading component that assumes uncorrelated fading. The elements of $\overline{\mathbf{H}}$ are assumed to have unit power. K in (2.51) is the Rician K-factor of the system and is essentially the ratio of the power in the LOS component of the channel to the power in the fading component. $K = 0$ corresponds to a pure Rayleigh i.i.d. channel, whereas $K = \infty$ corresponds to a nonfading channel. The LOS component manifests itself in the following two cases:

- The separation distance between antennas as previously discussed.
- LOS component created due to a poor scattering environment. This is shown in Figure 2.13.

In Figure 2.13, we discuss two indoor wireless environment cases, like a WLAN environment. We have a laptop with two receiving antennas. In a poor scattering environment we are likely to encounter a situation as shown in the left half of the figure. Due to colocated antennas, we have a LOS component. If the scattering is rich enough, the antennas do not appear colocated, as shown in the right half of the figure. This sort of environment is close to Rayleigh i.i.d. and is desirable. The

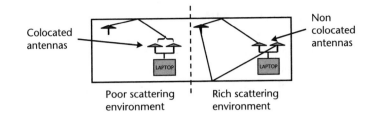

Figure 2.13   Colocation problem in a fixed WLAN environment.

former gives rise to a LOS component. Hence, we note that the LOS phenomenon can occur both in indoor as well as outdoor environments.

In either case, the end result is the same regarding correlation. We take a $\overline{\mathbf{H}}$ matrix of

$$\overline{\mathbf{H}} = \begin{bmatrix} 1 & 0.8 \\ 0.8 & 1 \end{bmatrix} \tag{2.52}$$

Equation (2.52) pertains to a correlation coefficient of 0.8, similar to the correlation effect in the example in Section 2.9. In Figure 2.14, we have plotted ergodic capacity using this channel matrix with varying K-factor.

We note from Figure 2.14 that rising K-factor is detrimental to capacity. Hence, we must be careful to minimize the LOS component. This is one of the major engineering hurdles in MIMO technology.

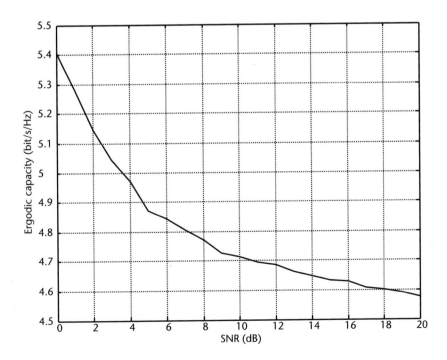

Figure 2.14   Ergodic capacity versus K-factor for a MIMO channel. Capacity declines with rising K-factor.

2.11   Influence of XPD on MIMO Capacity

The channel models discussed so far assume that the antennas at the base station and at the receivers have identical polarizations. The use of antennas with orthogonal polarizations at the transmitter and receiver leads to a gain (or power) and correlation imbalance between the elements of **H** [14]. These polarizations are usually ±45° or horizontal/vertical (0°/90°). Basically they require being orthogonal to each other. This ideally ensures zero coupling between the antennas. Therefore, signals with vertical polarizations, for example, are transmitted by one set of antennas and received by another set of vertical polarized antennas at the receiver. The same is the case with the horizontal polarized antennas. In view of the fact that these polarizations are orthogonal to each other, the signals do not "see" each other (i.e., they are independent). This is the ideal case. The reality is quite different. A certain amount of each signal "leaks" into the other signal and vice versa. We introduce the terms XPD and cross-polarization coupling (XPC). The former tells us as to how well one antenna discriminates its polarization from the other antenna. The latter term refers to the coupling between these polarizations during their propagation through the channel and is caused due to the rich scattering nature of the environment [15]. These phenomena are collectively defined by a constant α ($0 \le \alpha \le 1$), where *in the absence of XPC*, 0 implies that we have good XPD (i.e., the antennas discriminate between each other's polarizations extremely well (no interference) and 1 implies no XPD, meaning that the antennas cannot discriminate at all between each other's signals). It was found that, typically, at distances of 2.6 Kms and above, $\alpha = 1$, due to the rich scattering nature of the environment [16]. If we assume the power in the individual channel elements to be

$$\epsilon\{|h_{1,1}|^2\} = \epsilon\{|h_{2,2}|^2\} = 1 \qquad (2.53)$$

$$\epsilon\{|h_{1,2}|^2\} = \epsilon\{|h_{2,1}|^2\} = \alpha \qquad (2.54)$$

Assuming a Rayleigh i.i.d. channel, the channel **H** with cross-polarized antennas may be modeled approximately as

$$\mathbf{H} = \boldsymbol{\beta} \odot \left(\mathbf{R}_r^{1/2}\mathbf{H}_\omega\mathbf{R}_t^{1/2}\right) \qquad (2.55)$$

where

$$\boldsymbol{\beta} = \begin{bmatrix} 1 & \sqrt{\alpha} \\ \sqrt{\alpha} & 1 \end{bmatrix} \qquad (2.56)$$

and $\odot$ stands for the Hadamard product (if $\mathbf{A} = \mathbf{B} \odot \mathbf{C}$ then $[\mathbf{A}]_{i,j} = [\mathbf{B}]_{i,j}[\mathbf{C}]_{i,j}$). The covariance matrixes $\mathbf{R}_r$ and $\mathbf{R}_t$ are already well known to us as portraying the correlations extant at the receiver and the transmitter, respectively, and also include XPD, XPC, and antenna spacing as factors influencing their structure. The XPC phenomenon occurs in a scattering environment. If the environment through which the signal propagates is nonscattering, then $\mathbf{H} = \boldsymbol{\beta}$ (i.e., the right half of (2.55) vanishes).

If we assume the environment as nonscattering (i.e., deterministic), then the capacity for a 2×2 system is given by [from (2.21)]

$$C_{\alpha=0} = 2 \log_2 \left(1 + \frac{\rho}{2} \right) \tag{2.57}$$

and

$$C_{\alpha=1} = \log_2 (1 + 2\rho) \tag{2.58}$$

where ρ is SNR. For (2.58), the $\mathbf{H}$ matrix is all ones, yielding eigenvalues of 0 and 4.

At very low SNR ($\rho \ll 1$), using $\log_2 (1 + x) \approx x \log_2 e$ for $x \ll 1$,

$$C_{\alpha=0} \approx \rho \log_2 e \tag{2.59}$$

and

$$C_{\alpha=1} \approx 2\rho \log_2 e \tag{2.60}$$

Hence, good XPD is detrimental to capacity at low SNR.
In high SNR conditions ($\rho \gg 1$),

$$C_{\alpha=0} \approx 2 \log_2 \left(\frac{\rho}{2} \right) \tag{2.61}$$

$$C_{\alpha=1} \approx \log_2 (2\rho) \tag{2.62}$$

Hence, in high SNR conditions, good XPD ($\alpha = 0$) performs better than poor XPD, which is exactly the reverse of the case at low SNRs. Figure 2.15 confirms this performance for a 2×2 channel.

2.12 Keyhole Effect: Degenerate Channels

We consider a system with two transmit and two receive antennas surrounded by scatterers. These antennas are uncorrelated. If the channel were a Rayleigh i.i.d. channel, this would have yielded a channel matrix of full rank and size as $M_R \times M_T$. Now suppose, as shown in Figure 2.16, a screen separates these two sets of antennas with a small hole in it. This gives rise to a propagation condition called "keyhole." The only way for the transmitted signals to propagate is to pass through the keyhole. The transmitted signal vector is given by

$$\mathbf{S} = [s_1 \quad s_2]^T \tag{2.63}$$

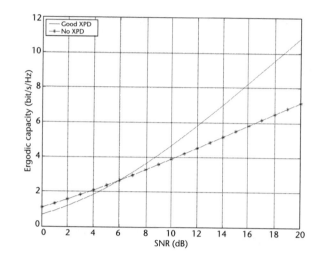

Figure 2.15   Ergodic capacity of a MIMO channel with good XPD ($\alpha = 0$) and no XPD ($\alpha = 1$).

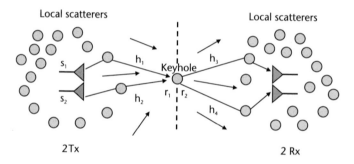

Figure 2.16   The keyhole effect.

where s_1 and s_2 are the signals transmitted from the first and second antennas, respectively. The signal incident at the keyhole is given by

$$\mathbf{r}_1 = \mathbf{H}_t \mathbf{S} \tag{2.64}$$

where $\mathbf{H}_t$ is given by

$$\mathbf{H}_t = [h_1 \quad h_2] \tag{2.65}$$

where h_1 and h_2 are the channel coefficients corresponding to transmitted signals s_1 and s_2 respectively. h_1 and h_2 are independent complex Gaussian variables. The signal across the keyhole is given by

$$\mathbf{r}_2 = \vartheta \mathbf{r}_1 \tag{2.66}$$

where ϑ is the keyhole attenuation.

The signal vector at the receive antenna across the keyhole, denoted by $\mathbf{r}_3$, is given by

$$\mathbf{r}_3 = \mathbf{H}_r \mathbf{r}_2 \tag{2.67}$$

where $\mathbf{H}_r$ is the channel matrix describing the propagation on the right-hand side of the keyhole and is given by

$$\mathbf{H}_r = [h_3 \quad h_4]^T \tag{2.68}$$

where h_3 and h_4 are the channel coefficients corresponding to the first and second receive antennas, respectively.

The effective channel $\mathbf{H}$ is given by

$$\mathbf{H} = \mathbf{H}_r \mathbf{H}_t^T \tag{2.69}$$

The received signal vector at the right-hand side of the keyhole is then given by

$$\mathbf{r}_3 = \vartheta \mathbf{H} \mathbf{S} \tag{2.70}$$

The channel matrix from (2.69) is

$$\mathbf{H} = \begin{bmatrix} h_1 h_3 & h_2 h_3 \\ h_1 h_4 & h_2 h_4 \end{bmatrix} \tag{2.71}$$

Since $\mathbf{H}$ is constructed from the product of two vectors, every realization of the channel $\mathbf{H}$ is rank-deficient with a rank of 2. The distribution of $\mathbf{H}$ is double Rayleigh [17] and is given by

$$f(x) = \int_0^\infty \frac{x}{\omega \sigma_r^4} e^{-\frac{\omega^4 + x^2}{2\omega^2 \sigma_r^2}} \, d\omega, \, x \geq 0 \tag{2.72}$$

where the amplitude distribution is the product of two independent Rayleigh distributions, each with the power of $2\sigma_r^2$. There is only one channel or data-pipe between the transmitter and the receiver. The corresponding channel capacity is given by [from (2.20)]

$$C = \log_2 \left(1 + \frac{\rho}{2} \lambda \right) \tag{2.73}$$

where ρ is SNR and λ is the solitary eigenvalue. The capacity with increasing SNR increases logarithmically, although the underlying channel is a MIMO channel. Figure 2.17 shows the performance of a degenerate channel. The drop in capacity compared with a regular channel is evident.

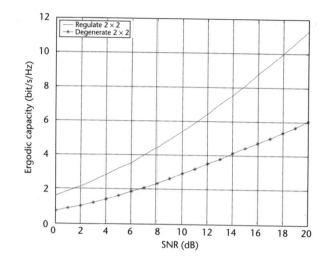

Figure 2.17 Performance of a degenerate channel for a 2 × 2 system.

Keyhole effects occur sometimes when the transmitted wavefront arrives with no angle spread (e.g., signals penetrating into buildings through small windows and also in narrow streets).

2.13 Capacity of Frequency Selective MIMO Channels

We now consider a real-life situation wherein the channel is not narrowband but frequency selective. Intuitively, subdividing the wideband channel into N narrowband ones, and then summing the capacities of these N frequency flat channels can achieve this. The bandwidth of each of these subchannels will be B/N Hz where B is the overall channel bandwidth. This is provided the coherent bandwidth of the channel permits this (i.e., it is more than or equal to B/N Hz), as otherwise the subchannels will not be frequency flat.

We take the ith subchannel. The input-output relationship is defined as [from (2.6)],

$$\mathbf{r}_i = \mathbf{H}_i \mathbf{s}_i + \mathbf{n}_i \tag{2.74}$$

where $\mathbf{r}_i$ is the $M_R \times 1$ received signal vector, $\mathbf{s}_i$ is the $M_T \times 1$ transmitted signal vector and $\mathbf{n}_i$ is the $M_R \times 1$ noise vector for the ith subchannel.

Hence, for the overall wideband channel we deal with block matrixes as

$$\mathcal{R} = \mathcal{H}\mathcal{S} + \mathcal{N} \tag{2.75}$$

where $\mathcal{R} = \left[\mathbf{r}_1^T \, \mathbf{r}_2^T \ldots \mathbf{r}_N^T\right]^T$ is $M_R N \times 1$, $\mathcal{S} = \left[\mathbf{s}_1^T \, \mathbf{s}_2^T \ldots \mathbf{s}_N^T\right]^T$ is $M_T N \times 1$, $\mathcal{N} = \left[\mathbf{n}_1^T \, \mathbf{n}_2^T \ldots \mathbf{n}_N^T\right]^T$ is $M_R N \times 1$ and $\mathcal{H}$ is an $M_R N \times M_T N$ block diagonal matrix with $\mathbf{H}_i$ as block diagonal elements. $\mathbf{R}_{ss} = \epsilon\left\{\mathcal{S}\mathcal{S}^H\right\}$ is the covariance matrix of S, constrained so that $\mathrm{Tr}(\mathbf{R}_{ss}) = NM_T$. This constrains the total average transmit power to E_s. From (2.17), the capacity of such a channel is given by

$$C_{FS} = \frac{B}{N} \max_{\mathrm{Tr}(\mathbf{R}_{ss})=NM_T} \log_2 \det\left(\mathbf{I}_{M_R N} + \frac{E_s}{M_T N_0} \mathcal{H}\mathbf{R}_{ss}\mathcal{H}^H\right) \text{ bps/Hz} \quad (2.76)$$

We now examine the two usual cases of when the channel is unknown to the transmitter and when it is known to the transmitter.

2.13.1 Channel Unknown to the Transmitter

In this case, we should choose $\mathbf{R}_{ss} = \mathbf{I}_{M_T N}$, which implies that the covariance matrix is of full rank (no correlation) and this in turn means that transmit power is allocated evenly across space (transmit antennas) and frequency (subchannels). This yields a deterministic capacity of [from (2.16)]

$$C_{FS} \approx \frac{B}{N} \sum_{i=1}^{N} \log_2 \det\left(\mathbf{I}_{M_R} + \frac{E_s}{M_T N_0} \mathbf{H}_i \mathbf{H}_i^H\right) \text{ bps/Hz} \quad (2.77)$$

If the frequency response of the channel is flat (we are talking about the *entire* channel being narrowband), [i.e., $\mathbf{H}_i = \mathbf{H}$ ($i = 1, 2, \ldots, N$)], then

$$C_{FS} = \log_2 \det\left(\mathbf{I}_{M_R} + \frac{E_s}{M_T N_0} \mathbf{H}\mathbf{H}^H\right) \quad (2.78)$$

which is the same as (2.16), the capacity of a frequency flat MIMO channel.

Further if all $\mathbf{H}_i$ are i.i.d. (i.e., the coherence bandwidth is B/N Hz), then as

$$N \to \infty, \ C_{FS} \to C_{FS}^{\infty} \quad (2.79)$$

(i.e., the capacity of such a frequency selective channel approaches a fixed quantity).

If the channel is random, we then have the usual two cases of ergodic and outage capacities. The ergodic capacity is given by

$$\overline{C}_{FS} \approx \epsilon\left\{\frac{B}{N} \sum_{i=1}^{N} \log_2 \det\left(\mathbf{I}_{M_R} + \frac{E_s}{M_T N_0} \mathbf{H}_i \mathbf{H}_i^H\right)\right\} \text{ bps/Hz} \quad (2.80)$$

The outage capacity is similarly defined. However, this outage capacity will be much better (higher) than for the earlier examined cases of frequency flat channels (at low outage rates). This is due to the high amount of frequency diversity present in the frequency selective channel. This is manifest in Figure 2.18.

In Figure 2.18, as the number of narrowband channels increases, with increasing frequency selectivity, the outage capacity also rises proportionately because of rising frequency diversity. Hence, the more the frequency selectivity, the higher the outage capacity. Note also the tendency of the curve to flatten with rising frequency

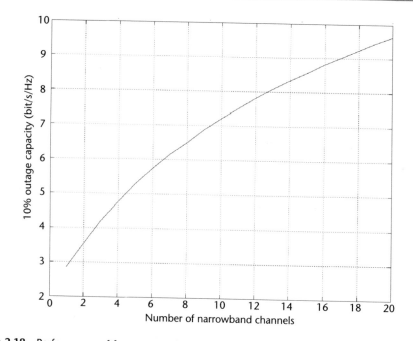

Figure 2.18 Performance of frequency selectivity versus 10% outage capacity.

selectivity and rising N. This bears out the statement in (2.79) that as $N \to \infty$, the capacity tends to a fixed value. This means that asymptotically (in N), the outage capacity of a sample realization of the frequency selective MIMO channel equals its ergodic capacity (because $N \to \infty$). The influence of multiple physical parameters such as delay spread, cluster angle spread and total angle spread on ergodic and outage capacity of frequency selective MIMO channels will be studied in Chapter 9.

2.13.2 Channel Known to the Transmitter

The treatment regarding this case is similar as was done earlier for frequency flat channels. In this case, we need to distribute the energy or power across space (antennas) and frequency (subchannels) so as to maximize spectral efficiency. This is called space-frequency water-filling [18]. Since water-filling is applicable only to purely orthogonal channels, it becomes necessary to achieve orthogonal channels by using OFDM techniques (see Chapter 7) to convert a frequency select channel into a set of parallel frequency flat channels, which are orthogonal to each other.

In such an event, if the composite channel $\mathcal{H}$ is known to the transmitter, the channel may be decomposed into $r(\mathcal{H})$ space-frequency modes, where $r(\mathbf{A})$ of matrix $\mathbf{A}$ stands for rank. The capacity is then given by [using (2.31)]

$$C_{FS} = \frac{B}{N} \max_{\sum_{i=1}^{r(\mathcal{H})} \gamma_i = NM_T} \sum_{i=1}^{r(\mathcal{H})} \log_2 \left(1 + \frac{E_s \gamma_i}{M_T N_0} \lambda_i(\mathcal{H}\mathcal{H}^H) \right), \text{ bps/Hz} \quad (2.81)$$

where $\lambda_i(\mathcal{H}\mathcal{H}^H)$ $(i = 1, 2, \ldots, r(\mathcal{H}))$ are the positive eigenvalues of $\mathcal{H}\mathcal{H}^H$ and γ_i is the energy allocated to the ith space-frequency subchannel. We can define ergodic and outage capacities of such channels, as was done earlier for frequency flat channels.

References

[1]   Rappaport, T. S., *Wireless Communications: Principles and Practice,* Upper Saddle River, NJ: Prentice Hall, 1996.

[2]   Alamouti, S. M., "A Simple Transmit Diversity Technique for Wireless Communications," *IEEE Journal Select. Areas Commun.,* Vol. 16, No. 8, October 1998, pp. 1451–1458.

[3]   Bölcskei, H., and A. J. Paulraj, "Multiple-Input Multiple-Output (MIMO) Wireless Systems," Chapter in *The Communications Handbook,* 2nd edition, J. Gibson (ed.), CRC Press, 2002, pp. 90.1–90.14.

[4]   Telatar, E., "Capacity of Multi-Antenna Gaussian Channels," *European Transactions on Telecommunications,* Vol. 10, No. 6, November/December 1999, pp. 585–595.

[5]   Foschini, G. J., and M. J. Gans, "On Limits of Wireless Communications in a Fading Environment When Using Multiple Antennas," *Wireless Personal Communications,* Vol. 6, 1998, pp. 311–335.

[6]   Vucetic, Branka, and Jinhong Yuan, *Space-Time Coding,* Chichester, UK: John Wiley & Sons Ltd., 2003.

[7]   Paulraj, Arogyaswami, Rohit Nabar, and Dhananjay Gore, *Introduction to Space-Time Wireless Communications,* Cambridge, UK: Cambridge University Press, 2003.

[8]   Golub, G., and C. Van Loan, *Matrix Computations,* Baltimore, MD: John Hopkins University Press, 2nd edition, 1989.

[9]   Cover, T., and J. Thomas, *Elements of Information Theory,* New York: John Wiley & Sons, 1992.

[10] Marzetta, T., and B. Hochwald, "Capacity of a Mobile Multiple-Antenna Communication Link in Rayleigh Flat Fading," *IEEE Trans. Inf. Theory,* 45(1), January 1999, pp. 139–157.

[11] Hassibi, B., and T. Marzetta, "Multiple Antennas and Isotropically Random Unitary Inputs: The Received Signal Density in Closed Form," *IEEE Trans. Inf. Theory,* 48(6), June 2002, pp. 1473–1484.

[12] Biglieri, E., J. Proakis, and S. Shamai, "Fading Channels: Information Theoretic and Communications Aspects," *IEEE Trans. Inf. Theory,* 44(6), October 1998, pp. 2619–2692.

[13] Rashid-Farrokhi, F., et al., "Spectral Efficiency of Wireless Systems with Multiple-Transmit and Receive Antennas," *PIMRC,* London, UK, September 2000, pp. 373–377.

[14] Nabar, R., et al., "Performance of Multi-Antenna Signaling Techniques in the Presence of Polarization Diversity," *IEEE Trans. Sig. Proc.,* 50(10), October 2002, pp. 2553–2562.

[15] Vaughan, R., "Polarization Diversity in Mobile Communications," *IEEE Trans. Veh. Tech.,* 39(3), August 1990, pp. 177–186.

[16] Baum, D. S., et al., "Measurement and Characterization Broadband MIMO Fixed Wireless Channels at 2.5 GHz," *Proc. IEEE Int. Conf. Pers. Wireless Comm.,* Hyderabad, India, December 2000, pp. 203–206.

[17] Erceg, V., et al., "Comparison of the WISE Propagation Tool Prediction With Experimental Data Collected in Urban Microcellular Environments," *IEEE J. Sel. Areas Comm.,* 15(4), May 1997, pp. 677–684.

[18] Raleigh, G., and J. Cioffi, "Spatio-Temporal Coding for Wireless Communication," *IEEE Trans. Comm.* 46(3), March 1998, pp. 357–366.

Channel Propagation, Fading, and Link Budget Analysis

3.1 Introduction

In Chapter 2 we examined the mutual information capacity of wireless communication based on MIMO channels. We found that this capacity grows linearly with the number of antennas in flat fading channels, due to the increase in the number of spatial data pipes. All this is accomplished *without* increasing the bandwidth or power.

In this chapter, we examine channel fading and propagation issues. We will also discuss a few channel propagation models and carry out link budget analysis. Finally we examine certain diversity combining techniques like selection diversity, maximal ratio combining, and equal gain combining.

3.2 Radio Wave Propagation

The mobile radio channel experiences a lot of limitations on the performance of wireless systems. The transmission path can vary from line-of-sight to one severely obstructed by buildings and foliage. Unlike wired channels, radio channels are extremely random and do not offer easy analysis. The speed of motion, for example, impacts on how the signal level fades as the mobile terminal moves in space. This modeling is therefore based more on statistics and requires specific measurements for an intended communication system.

Broadly the mechanics of electromagnetic wave propagation are confined to reflection, diffraction, and scattering.

Reflection, diffraction, and scattering are the three basic propagation mechanisms for radio waves. Received power (or its reciprocal, path loss) is generally the most important parameter predicted by large-scale propagation models and is based on these three phenomena. This is also applicable to small-scale fading and multipath propagation, which will be discussed later in this chapter.

3.2.1 Reflection

This occurs when electromagnetic waves bounce off objects whose dimensions are large compared with the wavelength of the propagating wave. They usually occur from the surface of the earth and off buildings and walls. A radio wave, when propagating in one medium, impinges on another medium with different electrical properties and is partially transmitted and reflected. The plane wave is incident on a perfect dielectric, part of the energy is transmitted into the second medium and part is reflected back into the first medium without any loss of energy in absorption. If the second medium is a perfect conductor then all of the incident energy is reflected back into the first medium without loss of energy. The electric field intensity of the reflected and transmitted waves may be related to the incident wave in the medium of origin through the *Fresnel reflection coefficient*. This reflection coefficient is a function of the material properties and generally depends on the wave polarization, angle of incidence, and frequency of the propagating wave.

3.2.2   Diffraction

Diffraction allows radio signals to propagate around the curved surface of the earth, beyond the horizon, and behind obstructions. The received field strength decreases rapidly as a receiver moves deeper into the obstructed (shadowed) region and is usually of enough intensity so as to produce a discernable signal. The phenomenon of diffraction can be explained by the Huygens principle, which states that all points on a wavefront can be considered as point sources for the production of secondary wavelets and that these secondary wavelets combine to produce a new wavefront in the direction of propagation. Diffraction is caused by the propagation of secondary wavelets into a shadowed region. The field strength of a diffracted wave in the shadowed region is the vector sum of the electric field components of all the secondary wavelets in the space around the obstacle.

3.2.3   Scattering

The actual received signal in a mobile radio environment is often stronger than what is predicted by reflection and diffraction models alone. This occurs because when a radio wave impinges on a rough surface, the reflected energy is spread out (diffused) in all directions due to scattering. Objects such as lampposts and trees tend to scatter energy in all directions, thereby providing additional radio energy at the receiver.

Sometimes reflection, diffraction and scattering are collectively referred to as scattering. Further discussions on these phenomena are beyond the scope of this book and the interested reader is referred to [1–4].

Cellular systems usually operate in urban areas, where there is no direct line-of-sight path between the transmitter and receiver and where high-rise buildings cause severe diffraction loss. Multiple reflections from various objects cause the electromagnetic waves to travel along different paths of varying lengths. The interaction between these waves causes multipath fading at a given location, because their phases are such that sometimes they add and sometimes they subtract (fade). The strengths of these waves slowly reduce with distance from the transmitter.

Propagation models based on average-received signal strength at a given distance from the transmitter are useful to estimate a radio coverage area and are called *large-scale propagation models* or *macroscopic fading models*. They are characterized by a large separation—usually a few kilometers—between the transmitter and receiver. On the other hand, propagation models that characterize the rapid fluctuations of the received signal strength over very short distances (a few wavelengths) or short time durations (on the order of seconds) are called *small-scale fading models* or *microscopic fading models*. The latter gives rise to rapid fluctuations as the mobile moves over short distances and the received power sometimes varies as much as 30 to 40 dB when the receiver moves only a fraction of a wavelength. Large-scale fading is manifest when the mobile moves over larger distances, causing the local average signal level to gradually decrease. It is this local average signal level that is predicted by large-scale propagation models. We shall examine a few of these models in this chapter.

Typically, the local average-received power is measured by averaging signal measurements over a measurement track of 5λ to 40λ. For cellular frequencies in the 1-to 2-GHz band, this works out to movements of 1 to 10m [1]. Figure 3.1 shows an example of small-scale fading and large-scale fading. Small-scale fading movements are rapid fluctuations, whereas large-scale fading movements are much slower average changes in signal strength [1]. The statistical distribution of this mean is influenced by parameters like frequency, antenna heights, environments and so on. However, it has been observed that the received power averaged over microscopic fading approaches a normal distribution when plotted on a logarithmic scale (i.e., in decibels) and is called log-normal distribution [5]. It is given by

$$f(x) = \frac{1}{\sqrt{2\pi}\sigma} e^{-\frac{(x-\mu)^2}{2\sigma^2}} \qquad (3.1)$$

In (3.1) x is in decibels and is a random variable representing the long-term signal power level fluctuation; μ and σ are, respectively, the mean and standard

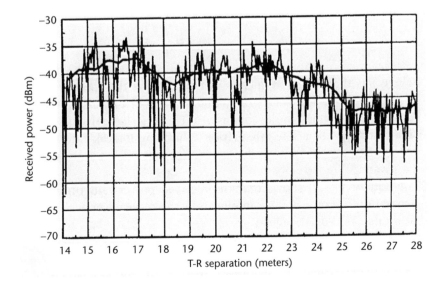

Figure 3.1 Small-scale and large-scale fading. (*From:* [1]. © 2002. Reprinted by permission of Pearson Education Inc., Upper Saddle River, NJ.)

deviation of x expressed in decibels. μ is the path loss described earlier. A typical value for σ is 8 dB [6].

3.3 Large-Scale Fading or Macroscopic Fading

3.3.1 Free-Space Propagation Model

If there is a clear unobstructed line-of-sight path between the transmitter and receiver, then we resort to the free-space propagation model. Satellite communication systems and microwave line-of-sight radio links undergo free-space propagation. In this model, the power is presumed to decay with distance from the transmitter according to some power law, usually as square of the distance from the transmitter. The free-space power received by an antenna at a distance d from the transmitter is given by [1],

$$P_r(d) = \frac{P_t G_1 G_2 \lambda^2}{(4\pi)^2 d^2 L} \tag{3.2}$$

where P_t is the transmitted power, $P_r(d)$ is the received power as a function of the separation distance d in meters, G_1 is the transmit antenna gain, G_2 is the receive antenna gain, L is the system loss not related to propagation ($L \geq 1$) and λ is the wavelength in meters. The gain of an antenna is related to its effective aperture by

$$G = \frac{4\pi A_e}{\lambda^2} \tag{3.3}$$

λ is related to the carrier frequency by

$$\lambda = \frac{c}{f} \tag{3.4}$$

where f is the carrier frequency in Hz and c is the speed of light in meters/sec (3×10^8 m/sec). The values of P_t and P_r must be expressed in identical units and G_t and G_r are dimensionless quantities. The miscellaneous losses ($L \geq 1$) are usually due to transmission line attenuation (plumbing losses), filter losses, and antenna losses in the communication system. $L = 1$ indicates no losses in the system hardware.

Equation (3.2) shows that the received power falls off as the square of the separation distance d. This implies that the received power decays with distance at a rate of 20 dB/decade.

We define an *isotropic* radiator as an ideal antenna that radiates power with unit gain uniformly in all directions and is often used as a reference antenna gain in wireless systems. The *effective isotropic radiated power* (EIRP) is defined as

$$EIRP = P_t G_t \tag{3.5}$$

and represents the maximum radiated power available from a transmitter in the direction of maximum antenna gain compared with an isotropic radiator. In practice, antenna gains are given in units of dBi (dB gain with respect to an isotropic antenna).

The path loss, which is the amount of attenuation suffered by the signal in dBs, is defined as the difference (in dB) between the transmitted power and the received power and is given by

$$PL \ (dB) = 10 \log \frac{P_t}{P_r} = -10 \log \left[\frac{G_t G_r \lambda^2}{(4\pi)^2 d^2} \right] \tag{3.6}$$

It is important to note that the free-space model is only applicable in the so-called far-field region of the transmitting antenna or in the *Fraunhofer* region and is defined as

$$d_f = \frac{2D^2}{\lambda} \tag{3.7}$$

where D is the largest physical linear dimension of the antenna (e.g., the length of a rectangular array antenna).

From (3.2), we note that the equation does not hold for $d = 0$. Hence, large-scale propagation models use a close-in distance, d_0, as a known received power reference point. The received power at any distance $d > d_0$ may then be related to P_r and d_0. The value $P_r(d_0)$ may be predicted from (3.2) by extrapolation or may be measured in the radio environment by taking the average received power at many points located at a close-in radial distance d_0 from the transmitter. The reference distance must be so chosen that it lies in the far-field (i.e., $d_0 \geq d_f$) and d_0 is chosen to be smaller than any practical distance used in the mobile communication system. Thus from (3.2), the received power in free space at a distance greater than d_0 is given by

$$P_r(d) = P_r(d_0) \left(\frac{d_0}{d} \right)^2, \ d \geq d_0 \geq d_f \tag{3.8}$$

In mobile radio systems, P_r changes by many orders of magnitude over a typical coverage area of several square kilometers. In view of this very large dynamic range of received power levels, dBm or dBW units are used to express received power levels. dBm is the power in dBs referred to one milliwatt. dBW is the power in dBs referred to one watt. For example,

$$P_r(d) \ dBm = 10 \log \left[\frac{P_r(d_0)}{0.001W} \right] + 20 \log \left(\frac{d_0}{d} \right), \ d \geq d_0 \geq d_f \tag{3.9}$$

where $P_r(d_0)$ is in watts.

The reference distance d_0 for practical systems using low-gain antennas in the 1–2 GHz region is typically 1m in indoor environments and 100m or 1 Km in outdoor environments, so that the numerator in (3.8) and (3.9) is a multiple of 10. This makes loss computations easy in dB units [1].

We now illustrate what we have learned through examples [1].

Example 1

Find the far-field distance for an antenna with maximum dimension of 2m and operating frequency of 900 MHz.

Solution

Given:

Largest dimension of antenna, $D = 2$m.

Operating frequency $f_c = 900$ MHz, $\lambda = c/f = \dfrac{3 \times 10^8 \text{ m/s}}{900 \times 10^6 \text{ Hz}} = 0.33$m

Using (3.7),

$$d_f = \frac{2D^2}{\lambda} = \frac{2(2)^2}{0.33} = 24.24\text{m}$$

Example 2

If a transmitter produces 50W of power, express the transmit power in units of (a) dBm and (b) dBW. If 50W is applied to an antenna of gain 1, find the received power in dBm at a free-space distance of 100m from the antenna. What is P_r (10 km)? Assume a gain of 2 for the receiver antenna and no system losses.

Solution

Given:

Transmitter power, $P_t = 50$W
Carrier frequency, $f_c = 900$ MHz
Using (3.9)
(a) Transmitter power,

$$P_t \text{ (dBm)} = 10 \log [P_t \text{ (mW)}/(1 \text{ mW})]$$

$$= 10 \log [50 \times 10^3] = 47 \text{ dBm}$$

(b) Transmitter power,

$$P_t \text{ (dBW)} = 10 \log [P_t \text{ (W)}/(1\text{W})]$$

$$= 10 \log [50] = 17 \text{ dBW}$$

The received power can be determined using (3.2)

$$P_r = \frac{P_t G_t G_r \lambda^2}{(4\pi)^2 d^2 L} = \frac{50(1)(2)(0.33)^2}{(4\pi)^2 (100)^2 (1)} = 6.9 \times 10^{-3} \text{ mW}$$

$$P_r \text{ (dBm)} = 10 \log P_r \text{ (mW)} = 10 \log (6.9 \times 10^{-3} \text{ mW}) = -21.6 \text{ dBm}$$

The received power at 10 Km can be expressed in terms of dBm using (3.9) where $d_0 = 100$m and $d = 10$ Km

$$P_r \,(10 \text{ Km}) = P_r \,(100) + 20 \log \left[\frac{100}{10,000} \right] = -21.6 - 40 \text{ dB}$$

$$= -61.6 \text{ dBm}$$

3.3.2   Outdoor Propagation Models

Free-space propagation is rarely encountered in real-life situations. In reality, we need to take into account the terrain profile in a particular area for estimating path loss. The terrain may vary from a simple curved earth profile to a highly mountainous profile. The presence of trees, buildings, and other obstacles must be taken into account. A number of propagation models are available to predict path loss over irregular terrain. These models differ in their ability to predict signal strength at a particular receiving point or in a specific local area (called a sector) because their approach is different and their results vary in terms of accuracy and complexity. These models are based on iterative experiments conducted over a period of time by measuring data in a specific area. We discuss two such well-known models.

3.3.2.1   Okumura Model

This is a widely used model for signal prediction in an urban area. It is applicable for frequencies in the range of 150 to 1,920 MHz and can be extrapolated up to 3 GHz and distances of 1 to 100 Km. It can be used for base station antenna heights ranging from 30 to 1,000m.

Okumura [7] developed a set of curves giving the median attenuation relative to free space (A_{mu}) in an urban area over a quasi-smooth terrain with a base station effective antenna height (h_{te}) of 200m and a mobile antenna height (h_{re}) of 3m. These curves were developed from extensive measurements using vertical omni-directional antennas at both base and mobile and are plotted as a function of frequency in the range 100 to 1,920 MHz and as a function of distance from the base station in the range of 1 to 100 Km. To use these curves, we first determine the free-space path loss between the points of interest and then the value of $A_{mu}(f, d)$ (as read from the curves) is added to it along with correction factors to account for the type of terrain. The model is expressed as

$$L_{50} \text{ (dB)} = L_F + A_{mu}(f, d) - G(h_{te}) - G(h_{re}) - G_{\text{AREA}} \qquad (3.10)$$

where L_{50} is the 50th percentile (i.e., median) value of propagation path loss, L_F is the free-space propagation loss, A_{mu} is the median attenuation relative to free space, $G(h_{te})$ is the base station antenna height gain factor, $G(h_{re})$ is the mobile antenna height gain factor, and G_{AREA} is the gain due to the type of environment. The antenna height gains are strictly a function of height and have nothing to do with the antenna patterns.

Plots of $A_{mu}(f, d)$ and G_{AREA} for a wide range of frequencies are shown in Figures 3.2 and 3.3 [1]. Moreover, Okumura determined that $G(h_{te})$ varies at a rate of 20 dB/decade and $G(h_{re})$ varies at a rate of 10 dB/decade for heights of less than 3m.

$$G(h_{te}) = 20 \log\left(\frac{h_{te}}{200}\right), \ 1{,}000\text{m} > h_{te} > 30\text{m} \tag{3.11}$$

$$G(h_{re}) = 10 \log\left(\frac{h_{re}}{3}\right), \ h_{re} \leq 3\text{m} \tag{3.12}$$

$$G(h_{re}) = 20 \log\left(\frac{h_{re}}{3}\right), \ 10\text{m} > h_{re} > 3\text{m} \tag{3.13}$$

Other corrections may also be applied to Okumura's model. Some of these are terrain undulation height (Δh), isolated ridge height, average slope of the terrain, and the mixed land-sea parameter. Once the terrain-related parameters are calculated, the necessary correction factors can be added or subtracted as required. All these correction factors are also available as Okumura curves [7].

Okumura's model is completely based on measured data and there is no analysis to justify it. All extrapolations to these curves for other conditions are highly

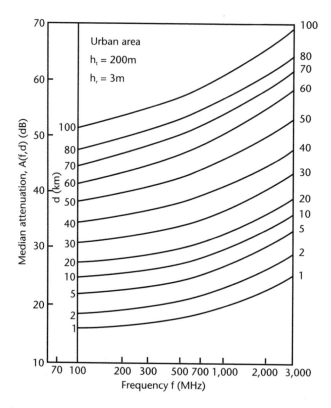

Figure 3.2 Median attenuation relative to free space ($A_{mu}(f, d)$) over a quasi-smooth terrain. (*From:* [7]. © 1968, IEEE.)

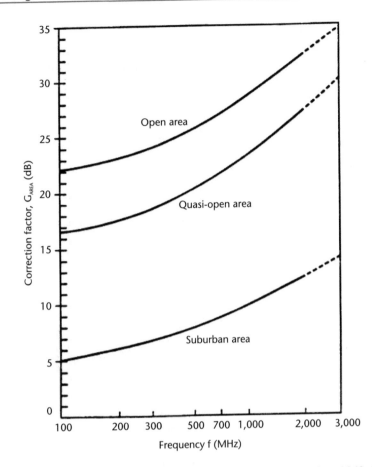

Figure 3.3   Correction factor G_{AREA} for different types of terrain. (*From:* [7]. © 1968, IEEE.)

subjective. Yet it is considered the simplest and best in terms of accuracy in path loss prediction for cellular systems in a cluttered environment. It has become a standard in Japan. The major disadvantage is its slow response to rapid changes in terrain. Hence, it is not so good in rural areas. Common standard deviations between predicted and measured path loss values are around 10 dB to 14 dB.

We now work out an example by way of illustration [1].

Example 3
Find the median path loss using Okumura's model for d = 50 Km, h_{te} = 100m, h_{re} = 10m in a suburban environment. If the base station transmitter radiates an EIRP of 1 kW at a carrier frequency of 900 MHz, find the power at the receiver (assume a gain of 2 at the receiving antenna).

Solution
The free-space path loss can be calculated using (3.2) as

$$L_F = 10 \log \left[\frac{\lambda^2}{(4\pi)^2 d^2} \right] = 10 \log \left[\frac{0.33^2}{(4\pi)^2 \times (50 \times 10^3)} \right] = 125.5 \text{ dB}$$

From the Okumura curves

$$A_{mu}(900 \text{ MHz } (50 \text{ km})) = 43 \text{ dB}$$

and

$$G_{AREA} = 9 \text{ dB}$$

Using (3.11) and (3.13)

$$G(h_{te}) = 20 \log \left(\frac{h_{te}}{200} \right) = 20 \log \left(\frac{100}{200} \right) = -6 \text{ dB}$$

$$G(h_{re}) = 20 \log \left(\frac{h_{re}}{3} \right) = 20 \log \left(\frac{10}{3} \right) = 10.46 \text{ dB}$$

Using

$$
\begin{aligned}
L_{50} \text{ (dB)} &= L_F + A_{mu}(f, d) - G(h_{te}) - G(h_{re}) - G_{AREA} \\
&= 125.5 \text{ dB} + 43 \text{ dB} - (-6 \text{ dB}) - 10.46 \text{ dB} - 9 \text{ dB} \\
&= 155.04 \text{ dB}
\end{aligned}
$$

Therefore, the median received power is

$$
\begin{aligned}
P_r(d) &= EIRP \text{ (dBm)} - L_{50} \text{ (dB)} + G_r \text{ (dB)} \\
&= 60 \text{ dBm} - 155.04 \text{ dB} + 3 \text{ dB} = -92.04 \text{ dBm}
\end{aligned}
$$

3.3.2.2 Hata Model

The Hata model is an empirical formulation of the graphical path loss data provided by Okumura and is valid from 150 to 1,500 MHz. Hata presented the loss as a standard formula and supplied correction equations for application to other situations. The standard formula for median path loss in urban areas is given by [1]

$$L_{50} (urban) \text{ (dB)} = 69.55 + 26.16 \log f_c - 13.82 \log h_{te} - a(h_{re}) \quad (3.14)$$
$$+ (44.9 - 6.55 \log h_{te}) \log d$$

where f_c is the frequency in MHz from 150 to 1,500 MHz, h_{te} is the effective transmitter (base station) antenna height (in meters) ranging from 30 to 200m, h_{re} is the effective receiver (mobile) antenna height (in meters) ranging from 1 to 10m, d is the T-R separation distance (in Km), and $a(h_{re})$ is the correction factor for effective mobile antenna height, which is a function of the size of the coverage area. For a small to medium-sized city, the correction factor is given by

$$a(h_{re}) = (1.1 \log f_c - 0.7)h_{re} - (1.56 \log f_c - 0.8) \text{ dB} \quad (3.15)$$

and for a large city,

$$a(h_{re}) = 8.29(\log 1.54h_{re})^2 - 1.1 \text{ dB for } f_c \le 300 \text{ MHz} \qquad (3.16)$$

$$a(h_{re}) = 3.2(\log 11.75h_{re})^2 - 4.97 \text{ dB for } f_c \ge 300 \text{ MHz} \qquad (3.17)$$

To obtain the path loss in a suburban area, the standard Hata formula in (3.13) is modified as

$$L_{50} \text{ (dB)} = L_{50} \text{ (urban)} - 2[\log(f_c/28)]^2 - 5.4 \qquad (3.18)$$

and for path loss in open rural areas, the formula is modified as

$$L_{50} \text{ (dB)} = L_{50} \text{ (urban)} - 4.78(\log f_c)^2 + 18.33 \log f_c - 40.94 \qquad (3.19)$$

The predictions of Hata's model compare very closely with the original Okumura model, if d exceeds 1 Km. This model is well-suited to large cell mobile systems.

This concludes our discussions on outdoor propagation models. The interested reader is referred to [1] and the references listed therein.

3.4   Small-Scale Fading

Small-scale fading or simply *fading* is used to describe the rapid fluctuations of the amplitude, phases, or multipath delays of a radio signal over a short period of time or travel distance, *so that large-scale path loss effects may be ignored.* Fading is caused by a number of signals (two or more) arriving at the reception point through different paths, giving rise to constructive (strengthening) vectorial summing of the signal or destructive (weakening) vectorial subtraction of the signals, depending on their phase and amplitude values. These different signals other than the main signal are called *multipath* waves.

Multipath in a radio channel creates small-scale fading effects. These effects are commonly characterized as causing:

- Rapid changes in signal strength over a small travel distance or time interval.
- Random frequency modulation due to varying Doppler shifts on different multipaths.
- Time dispersion (echoes) caused by multipath propagation delays.

Even when a line-of-sight exists, multipath still occurs due to reflections from the ground or surrounding structures. Assume that there is no moving object in the channel. In such a case, fading is purely a spatial phenomenon. The signals add or subtract, creating standing waves in the area where the mobile is located. In such a case, as the mobile moves, it encounters temporal fading as it moves through the multipath field. In a more serious case, the mobile may stop at a particular point at which the received signal is in deep fade. Maintaining good communication in that case becomes very difficult. It can only be countered using diversity techniques, as discussed in Chapter 2. Figure 2.2 is one such example.

3.4.1　Microscopic Fading

Microscopic fading refers to the rapid fluctuations of the received signal in space, time, and frequency and is caused by the signal scattering off objects between the transmitter and receiver. Since this fading is a superposition of a large number of independent scattered components, then by the central limit theorem, the components of the received signal can be assumed to be independent zero mean Gaussian processes. The envelope of the received signal is consequently Rayleigh distributed and is given by

$$f(x) = \frac{2x}{\Omega} e^{-\frac{x^2}{\Omega}} u(x) \tag{3.20}$$

where Ω is the average received power and $u(x)$ is the unit step function defined as

$$u(x) = \begin{cases} 1 & x \geq 0 \\ 0 & x < 0 \end{cases} \tag{3.21}$$

If there is a direct LOS path between the transmitter and receiver, the signal envelope is no longer Rayleigh and the distribution of the signal is Ricean. The Ricean distribution is often defined in terms of the Ricean factor, K, which is the ratio of the power in the mean component of the channel to the power in the scattered component. The Ricean PDF of the envelope is given by

$$f(x) = \frac{2(K + 1)}{\Omega} e^{\left(-K - \frac{(K + 1)\mu^2}{\Omega}\right)} I_0 \left(2x \sqrt{\frac{K(K + 1)}{\Omega}}\right) \mu(x) \tag{3.22}$$

where I_0 is the zero-order modified Bessel function of the first kind defined as

$$I_0(x) = \frac{1}{2\pi} \int_0^{2\pi} e^{-x \cos \theta} d\theta \tag{3.23}$$

In the absence of a direct path, $K = 0$ and the Ricean PDF reduces to Rayleigh PDF, since $I_0(0) = 1$. Figure 3.4 shows the combined effects of path loss and macroscopic and microscopic fading on received power in a wireless channel [6].

We note from Figure 3.4 that the mean propagation loss increases monotonically with range. Local deviations from this mean occur due to macroscopic and microscopic fading.

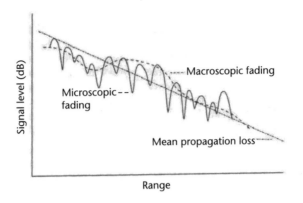

Figure 3.4 Signal power fluctuation versus range in wireless channels. (*From:* [6]. Reprinted with the permission of Cambridge University Press.)

There are three types of microscopic fading [6]:

- Doppler spread-time selective fading;
- Delay spread-frequency selective fading;
- Angle spread-space selective fading.

3.4.1.1 Doppler Spread-Time Selective Fading

Time varying fading due to the motion of a scatterer or the motion of a transmitter or receiver or both results in Doppler spread. The term spread is used to denote the fact that a pure tone frequency f_c in hertz spreads across a finite bandwidth ($f_c \pm f_{max}$). There is a direct relationship between the autocorrelation function of a signal and its spectrum and is defined by the Wiener-Khinchin equations [8]. The Fourier transform of the time autocorrelation of the channel response to a continuous wave (CW) tone is defined as Doppler power spectrum $\psi_{Do}(f)$ with $f_c - f_{max}$ $\leq f \leq f + f_{max}$. Figure 3.5(a) shows the Doppler spectrum [6].

The Doppler power spectrum has a classical U-shaped form and is approximated by Jakes model [5]. The Doppler shift of the received signal denoted by f_d is given by

$$f_d = \frac{v f_c}{c} \cos \theta \qquad (3.24)$$

where v is the velocity of the moving object (or vehicle speed, if we are talking about static scatterers and a moving vehicle), θ is the relative angle between the moving object and the point of reception of the Doppler signal, and c is the speed of light. Obviously, the maximum Doppler will be received at a relative angle of 0° (i.e., when the moving object is ahead or astern).

The root mean square (RMS) bandwidth of $\psi_{Do}(f)$ is called the Doppler spread and is given by

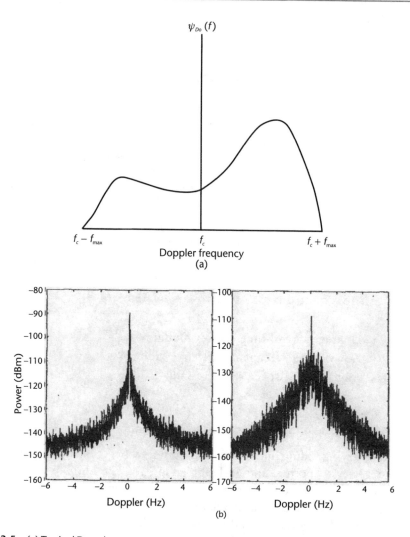

Figure 3.5 (a) Typical Doppler power spectrum. (*From:* [6]. Reprinted with permission of Cambridge University Press.) (b) Fixed wireless Doppler spectra. (© 2000, IEEE.)

$$f_{RMS} = \sqrt{\frac{\int (f - \bar{f})^2 \psi_{Do}(f)\, df}{\int \psi_{Do}(f)\, df}} \tag{3.25}$$

where $\bar{f}$ is the average frequency of the Doppler spectrum and is given by

$$\bar{f} = \frac{\int f \psi_{Do}(f)\, df}{\int \psi_{Do}(f)\, df} \tag{3.26}$$

In LOS cases the spectrum is modified by an additional discrete frequency component given by f_d. We define coherence time of the channel as

$$T_c \approx \frac{1}{f_{RMS}} \tag{3.27}$$

where T_c is defined as the time lag for which the signal autocorrelation coefficient reduces to 0.7. It serves as a measure of how fast the channel changes in time, implying that the larger the coherence time, the slower the channel fluctuation. The Doppler spectrum shown in Figure 3.5(a) pertains to a mobile receiver moving at constant speed (e.g., in a car). However, in a fixed wireless channel, the receiver is static but there is movement in the environment (e.g., trees and foliage moving in a random manner due to wind). In such cases, the Doppler spectrum is as shown in Figure 3.5(b) [9].

In Figure 3.5(b), the left-hand graph pertains to a low Doppler spread and the right-hand one pertains to a higher Doppler spread. The curves level off at high Doppler shifts due to the prevailing noise levels. If there is moving traffic around the mobile, Doppler components can occur at much higher frequencies, but the shape of the Doppler spectra will be similar.

3.4.1.2   Delay Spread-Frequency Selective Fading

The small-scale variations of a mobile radio signal can be directly related to the impulse response of the mobile radio channel. This stems from the fact that a mobile radio channel may be modeled as a linear filter with a time varying impulse response, where the time variation is due to receiver motion in space. The filtering nature of the channel is caused by the summation of amplitudes and delays of the multiple arriving waves at any instant of time. Therefore, the impulse response is a useful characterization of the channel because it can be used to predict and compare the performance of many different mobile communication systems and transmission bandwidths for a particular channel condition.

To compare different multipath channels and develop some general design guidelines for wireless systems, certain parameters were decided on as benchmarks to quantify the multipath channel. These parameters are the mean excess delay, RMS delay spread, and excess delay spread and they can be determined from the power delay profile. These are shown in Figure 3.6 [1].

We define [1]:

- *Mean excess delay:* The mean excess delay is the first moment of the power delay profile and is defined as

$$\overline{\tau} = \frac{\sum\limits_{k} P(\tau_k)\tau_k}{\sum\limits_{k} P(\tau_k)} \tag{3.28}$$

- *RMS delay spread:* The RMS delay spread is the square root of the second central moment of the power delay profile and is defined as

$$\sigma_\tau = \sqrt{\overline{\tau^2} - (\overline{\tau})^2} \tag{3.29}$$

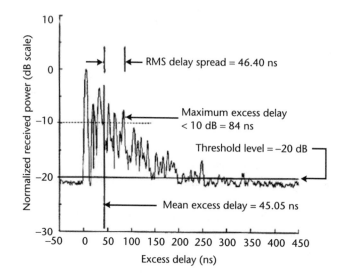

Figure 3.6   Example of an indoor power delay profile. (*From:* [1]. © 2002. Reprinted by permission of Pearson Education Inc., Upper Saddle River, NJ.)

where

$$\overline{\tau^2} = \frac{\sum_k P(\tau_k)\tau_k^2}{\sum_k P(\tau_k)} \tag{3.30}$$

These delays are measured relative to the first detectable signal arriving at the receiver at $\tau_0 = 0$. Equations (3.28) to (3.30) do not rely on the absolute power level of $P(\tau)$, but only on the relative amplitudes of the multipath components within $P(\tau)$. Typical values of RMS delay spread are on the order of microseconds in outdoor mobile radio channels and on the order of nanoseconds in indoor radio channels.

- *Maximum excess delay* (X dB): This is defined to be the time delay during which multipath energy falls to X dB below the maximum. This implies that maximum excess delay is defined as $\tau_x - \tau_0$, where τ_0 is the first arriving signal and τ_k is the maximum delay at which a multipath component is within X dB of the strongest arriving multipath signal (which does not necessarily arrive at τ_0). Figure 3.6 illustrates the computation of maximum excess delay for multipath components within 10 dB of the maximum. The maximum excess delay tells us how long a multipath exists above a given threshold. This value τ_k must be specified with a threshold that relates the multipath noise floor to the maximum received multipath component. This is also sometimes called *excess delay spread*.

In practice, the values for $\overline{\tau}$, $\overline{\tau^2}$ and σ_λ depend on the choice of noise threshold used to process $P(\tau)$. The noise threshold is used to differentiate between received multipath components and thermal noise. If the noise threshold is too low, then noise will be processed as multipath, thus giving rise to values of $\bar{\tau}$, $\overline{\tau^2}$ and σ which are artificially high [1].

Delay spread causes frequency selective fading as the channel acts like a tapped delay line filter. Frequency selective fading can be characterized in terms of coherence bandwidth, B_c, which is the frequency lag for which the channel's autocorrelation function reduces to 0.7 (remember that the autocorrelation function and the spectrum are connected by the Wiener-Khinchin equations [8]). We define coherence bandwidth as

$$B_c \approx \frac{1}{\sigma_\tau} \qquad (3.31)$$

When the coherence bandwidth is comparable with or less than the signal bandwidth, the channel is said to be frequency selective. Otherwise it is frequency flat or non-selective. A "flat" channel passes all spectral components with approximately equal gain and linear phase. It is not possible to provide an exact relationship between coherence bandwidth and RMS delay spread, as it is a function of specific channel impulse response and applied signals. We need to resort to spectral analysis to determine the exact impact that time varying multipath has on a transmitted signal. Hence, accurate multipath modeling is essential.

3.4.1.3   Rician K-Factor Measurement

There are many techniques to measure Rician K-factor from the power profile. The moment-method estimation of K-factor has found popular appeal [10]. The details are beyond the scope of this book.

We now solve an example to firm up our ideas!

Example 4
Calculate the mean excess delay, RMS delay spread, and maximum excess delay (10 dB) for the multipath profile given in Figure 3.7. Estimate the coherence bandwidth of the channel.

Solution
Using definition of maximum excess delay (10 dB), we determine that $\tau_{10\,\text{dB}} =$ 3 μsec.

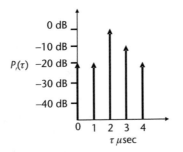

Figure 3.7   Multipath profile for Example 4.

The mean excess delay is [using (3.28)]

$$\bar{\tau} = \frac{(0.01)(0) + (0.01)(1) + (1)(2) + (0.1)(3) + (0.01)(4)}{[0.01 + 0.01 + 1 + 0.1 + 0.01]} = \frac{2.35}{1.13} = 2.08 \; \mu\text{sec}$$

The second moment for the given power delay profile is (using 3.30)

$$\overline{\tau^2} = \frac{(0.01)(0)^2 + (0.01)(1)^2 + (1)(2)^2 + (0.1)(3)^2 + (0.01)(4)^2}{[0.01 + 0.01 + 1 + 0.1 + 0.01]} = \frac{5.07}{1.13} = 4.49 \; \mu\text{sec}$$

Therefore the RMS delay spread is (using 3.29)

$$\sigma_\tau = \sqrt{4.49 - (2.08)^2} = 0.4 \; \mu\text{sec}$$

The coherence bandwidth is (using 3.31)

$$B_c = \frac{1}{\sigma_\tau} = \frac{1}{0.4 \; \mu\text{sec}} = 2.5 \text{ MHz}$$

Hence, communications systems operating within a bandwidth of 2.5 MHz need not use equalizers.

3.4.1.4 Angle Spread-Space Selective Fading

Angle spread at the receiver refers to the angle of arrival (AOA) of the multipath components at the receive antenna. Similarly, the angle of departure (AOD) from the transmitter of the multipath that reaches the receiver is called the angle spread at the transmitter.

We denote AOA by θ and the rest of the analysis is as was done for delay spread, the only difference being that instead of τ we substitute θ. The terminology now becomes RMS angle spread and is given by

$$\sigma_\theta = \sqrt{\overline{\theta^2} - (\bar{\theta})^2} \tag{3.32}$$

where

$$\overline{\theta^2} = \frac{\sum_k P(\theta_k)\theta_k^2}{\sum_k P(\theta_k)} \tag{3.33}$$

A typical angle (power) spectrum is shown in Figure 3.8.

The RMS angle spread is measured similar to the RMS delay spread. These angles are measured relative to the first detectable signal arriving at the receiver at $\theta_0 = 0$.

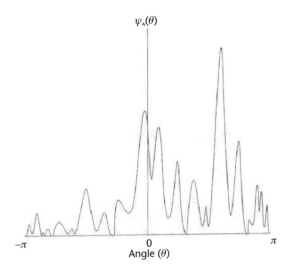

Figure 3.8 Typical angle (power) spectrum $\psi_A(\theta)$. (*From:* [6]. © 2003. Reprinted with the permission of Cambridge University Press.)

Equations (3.32) and (3.33) do not rely on the absolute power level of $P(\theta)$, but only on the relative amplitudes of the multipath components within $P(\theta)$. Once again, instead of coherence bandwidth in the delay case, we have coherent distance in this case. Angle spread causes space selective fading, which means that signal amplitude depends on the spatial location of the antenna. Space selective fading is characterized by coherent distance, D_c, which is the spatial separation for which the autocorrelation coefficient of the spatial fading drops to 0.7. It is inversely proportional to angle spread and is given by

$$D_c \propto \frac{1}{\theta_{RMS}} \tag{3.34}$$

The value of D_c varies from typically 10 to 16 wavelengths on a base station to 3 to 5 wavelengths at the mobile. This is explained by the fact that at the base station, which is located at a height, the signals received tend to be closer spaced in azimuth, hence θ_{RMS} is small. The case at the mobile terminal is the reverse, in view of the comparatively rich scattering environment.

3.5 Microscopic Fading Measurements

We have by now become well acquainted with the importance of channel measurements in wireless communications and the need to determine the microscopic fading parameters in a radio channel. We will now examine briefly three important channel sounding techniques to achieve this. These are:

- Direct pulse measurements.
- Spread-spectrum sliding correlator measurements.
- Swept frequency measurements.

These are just the basic approaches to the problem. The actual equipments used may vary in detail.

3.5.1 Direct Pulse Measurements

This technique enables engineers to rapidly determine the power delay profile of the channel [11, 12]. Basically we generate a pulse train of narrowband pulses of width T_p. These pulses are then received by a receiver that has a bandpass filter at its input of bandwidth $BW = 2/T_p$. The signal is then amplified, envelope detected, and given to a storage oscilloscope. This gives an immediate measurement of the square of the channel impulse response (envelope detection) convolved with the probing pulse. If the oscilloscope is set on averaging mode, we obtain the average power delay profile of the channel. The advantage here is that the system is not complex. The minimum resolvable delay between multipath components is equal to the probing pulse width T_p. Due to the wideband input filter, the system is subject to a lot on noise. Also, the pulse system relies on the ability to trigger the oscilloscope on the first arriving signal. If this signal is in deep fade, the system may not trigger properly. In addition, the phase of the multipath components is lost due to the envelope detector. This problem can be solved by using a coherent detector.

3.5.2 Spread-Spectrum Sliding Correlator Channel Sounding

Figure 3.9 describes a spread-spectrum channel impulse response measurement system [1, 13]. In the previous effort, we saw that if the first trigger is not available

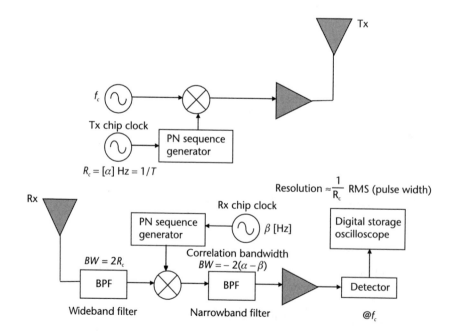

Figure 3.9   Spread-spectrum channel impulse response measurement system. (*From:* [1]. © 2002. Reprinted by permission of Pearson Education Inc., Upper Saddle River, NJ.)

due to deep fades, the system fails. The problem is further compounded by the fact that the input filter, being wideband, lets noise into the system. To counter this, the spread-spectrum system was developed. The idea here is to "spread" the carrier signal over a wide bandwidth by mixing it with a binary pseudonoise (PN) sequence having chip duration T_c and a chip-rate R_c equal to $1/T_c$ Hz. The power spectrum envelope of the transmitted signal is given by [13]

$$S(f) = \left[\frac{\sin \pi(f - f_c)T_c}{\pi(f - f_c)T_c} \right]^2 \qquad (3.35)$$

and the null-to-null radio frequency (RF) bandwidth is

$$BW = 2R_c \qquad (3.36)$$

The signal is then transmitted and at the receiver the reverse operation takes place (i.e., it is "despread" using the same PN sequence). However, there is a nuance here. The transmitted PN sequence is at a slightly higher rate than the PN sequence at the receiver. This causes the window to "slide" at the receiver at the difference frequency given by

$$\gamma = \frac{\alpha}{\alpha - \beta} \qquad (3.37)$$

where

α = transmitter chip clock rate (Hz)

β = receiver chip clock rate (Hz)

Mixing the chip sequence in this fashion gives rise to a "sliding correlator." Therefore, as the delayed multipaths arrive one after the other, they are reflected as peaks on the power delay profile. The PN sequences are selected to have good autocorrelation and cross-correlation properties. This implies that "own" sequence (the true signal) will give rise to a maximum peak signal and the rest will be treated as "noise" and be spread throughout the bandwidth. In this way the narrowband filter that follows the correlator can reject almost the entire incoming signal power. Hence, the wideband input filter problem, as was noted in the previous method, is absent. This type of processing results in processing gain and is given by

$$PG = \frac{2R_c}{T_p} = \frac{2T_p}{T_c} = \frac{SNR_{out}}{SNR_{in}} \qquad (3.38)$$

where $T_p = 1/R_p$ is the period of the baseband information and SNR is signal-to-noise ratio.

Since the incoming spread-spectrum signal is mixed with a receiver PN sequence that is slower than the transmitter PN sequence, the signal is essentially down-converted to a low-frequency narrowband signal. Hence, the relative rates of the two codes slipping past each other is the rate of information transferred to the oscilloscope. This narrowband signal allows narrowband processing, eliminating much of the passband noise and interference. The processing gain is then realized using a narrowband filter $(BW = 2(\alpha - \beta))$.

The equivalent time measurements refer to the relative times of multipath components as they are displayed on the oscilloscope. The observed time scale on the oscilloscope using a sliding correlator is related to the actual propagation time scale by

$$\text{Actual Propagation Time} = \frac{\text{Observed Time}}{\gamma} \qquad (3.39)$$

This effect is due to the relative rate of information transfer to the sliding correlator and must be kept in mind when measuring. This effect is known as *time dilation*. The length of the PN sequence must be greater than the longest multipath propagation delay; otherwise, these delays will be missed out.

The advantages of this system are:

- Passband noise is rejected.
- Transmitter and receiver synchronization problem is eliminated by the sliding correlator.
- Sensitivity is adjustable by changing the sliding factor and the postcorrelator filter bandwidth.
- Required transmitter powers can be considerably lower than comparable direct pulse systems due to the inherent "processing gain" of the spread-spectrum systems.

The disadvantages are:

- The measurements are not made in real time, unlike in direct pulse systems, because they are compiled as the PN codes slide past each other.
- Time taken to measure the channel is very high.
- Phase measurement is not possible because the detector is noncoherent. Even if we use a coherent detector, the sweep time of a spread-spectrum signal induces delay such that the phases of individual multipath components with different time delays would be measured at substantially different times, during which the channel might change.

3.5.3　Frequency Domain Channel Sounding

Figure 3.10 shows the frequency domain channel impulse response measurement system [14, 15]. This method exploits the dual relationship between time domain and frequency domain. In this case we measure the channel in the frequency domain and then convert it into time domain impulse response by taking its inverse discrete Fourier transform (IDFT). A vector network analyzer controls a swept frequency synthesizer. An S-parameter test set is used to monitor the frequency response of

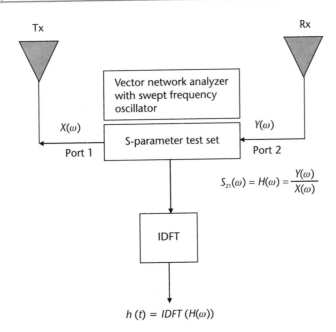

Figure 3.10 Frequency domain channel impulse response measurement system. (*From:* [1]. © 2002. Reprinted by permission of Pearson Education Inc., Upper Saddle River, NJ.)

the channel. The sweeper scans a particular frequency band, centered on the carrier, by stepping through discrete frequencies. The number and spacing of the frequency step impacts the time resolution of the impulse response measurement. For each frequency step, the S-parameter test set transmits a known signal level at port 1 and monitors the received signal at port 2. These signals allow the analyzer to measure the complex response, $S_{21}(\omega)$, of the channel over the measured frequency range. The $S_{21}(\omega)$ measure is the measure of the signal flow from transmitter antenna to receiver antenna (i.e., the channel).

This technique works well and indirectly provides amplitude and phase information in the time domain. However, it requires careful calibration and hard-wired synchronization between the transmitter and receiver, making it suitable only for indoor channel measurements. This system is also nonreal-time. Hence, it is not suitable for time-varying channels unless the sweep times are fast enough.

This ends our review of channel measurement techniques. The interested reader is referred to [1] and the references listed at the end of this chapter.

3.6 Antenna Diversity

3.6.1 Diversity Combining Methods

In Chapter 2, we examined the advantages of diversity and how it helps to combat fading. In this chapter we examine diversity combining techniques for SIMO channels. There are three principal methods—selection combining, maximal ratio combining, and equal gain combining.

3.6.1.1 Selection Combining

This is the simplest combining method. Consider a M_R receiver system. In this method, we select the signal coming into each of the M_R antennas that has the highest instantaneous SNR at every symbol interval. This makes the output of the combiner equal to that of the best incoming signal. The advantage here is that this method does not require any additional RF receiver chain. In other words, all receive antennas share a single RF receiver chain. This keeps the cost down. In practice the strongest signals are selected [i.e., signals with the highest $(S + N)/N$ ratio] because it is difficult to measure SNR alone.

Consider M_R independent Rayleigh fading channels available at the receiver. Each channel is called a diversity branch. Assume that each branch has an average SNR, η, given by

$$\eta = \frac{\overline{E_s}}{N_0} = E\left(h_i^2\right)\frac{E_s}{N_0} \tag{3.40}$$

If each branch has an instantaneous $SNR = \gamma_i$, then

$$\gamma_i = h_i^2 \frac{E_s}{N_0}, \ i = 1, 2, \ldots, M_R$$

The probability that the SNR for the ith receive antenna is lower than a threshold ν is given by

$$P(\gamma_i \leq \nu) = \int_0^\nu f_{\gamma_i} \tag{3.41}$$

where $f_{\gamma_i}(\alpha)$ denotes the probability density function of γ_i, which is assumed to be the same for all antennas. If we have M_R independent receive antennas, the probability that all of them have an SNR below the threshold ν is given by

$$P(\gamma_i \leq \nu, \ldots, \gamma_{M_R} \leq \nu) = [P(\gamma_i \leq \nu)]^{M_R} \tag{3.42}$$

and this decreases as M_R increases. This is also the CDF of the random variable

$$\overline{\gamma} = \max\{\gamma_1, \ldots, \gamma_{M_R}\} \tag{3.43}$$

Hence, $\overline{\gamma} < \nu$, iff $\gamma_1, \ldots, \gamma_{M_R}$ are all less than ν. Therefore the PDF follows directly from the derivative of the CDF with respect to ν.

We explain this through an example [1].

Example 5

Assume a four-branch diversity, where each branch receives an independent Rayleigh fading signal. If the average SNR is 20 dB, determine the probability that the

SNR will drop below 10 dB. Compare this with the case of a single receiver without diversity.

Solution
For a Rayleigh fading channel, the fading amplitude α has a Rayleigh distribution, so the fading power α^2 and therefore, X, have a chi-square distribution with two degrees of freedom. Then

$$p(X) = \frac{1}{\Gamma} e^{-\frac{X}{\Gamma}}, \quad X \geq 0$$

where $\Gamma = (\overline{E_s}/N_0)$ is the average value of the SNR. In such a case, if each branch has an instantaneous $SNR = E_s$, then the PDF of γ_i is

$$p(\gamma_i) = \frac{1}{\Gamma} e^{-\frac{\gamma_i}{\Gamma}}, \quad \gamma_i \geq 0$$

where Γ is the mean SNR in each branch. The probability that a single branch has an instantaneous SNR less than some threshold ν is

$$P[\gamma_i \leq \nu] = \int_0^\nu p(\gamma_i) \, d\gamma_i = \int_0^\nu \frac{1}{\Gamma} e^{-\frac{\gamma_i}{\Gamma}} \, d\gamma_i = 1 - e^{-\frac{\nu}{\Gamma}}$$

Therefore,

$$P[\gamma_1, \ldots, \gamma_{M_R} \leq \nu] = \left(1 - e^{-\nu/\Gamma}\right)^{M_R}$$

Armed with this math,

$$P_4(10 \text{ dB}) = \left(1 - e^{-0.1}\right)^4 = 0.000082$$

where $\nu = 10$ dB, $\Gamma = 20$ dB and $\nu/\Gamma = 0.1$.
    For a SISO channel, $M_R = 1$. Hence,

$$P_1(10 \text{ dB}) = \left(1 - e^{-0.1}\right)^1 = 0.095$$

The advantage of selection diversity becomes evident!

3.6.1.2 Maximal Ratio Combining

In maximal ratio combining (MRC), the signals from all of the M_R branches are weighted according to their individual SNRs and then summed. Here the individual signals need to be brought into phase alignment before summing. This implies individual RF receiver tracts. If the signals are r_i from each branch, and each branch has a gain G_i, then

$$r_{M_R} = \sum_{i=1}^{M_R} G_i r_i \qquad (3.44)$$

where $r_i = h_i s_i + v_i$, $s_i = 2E_s$ being the transmitted signal, v_i is the noise in each branch with a power spectral density of $2N_0$ and h_i is the channel coefficient.

Therefore, from (3.44)

$$r_{M_R} = \sum_{i=1}^{M_R} G_i h_i s_i + \sum_{i=1}^{M_R} G_i v_i \qquad (3.45)$$

The power spectral density of the noise after MRC is given by

$$S_v = 2N_0 \sum_{i=1}^{M_R} |G_i|^2 \qquad (3.46)$$

The instantaneous signal energy is

$$2E_s \left| \sum_{i=1}^{M_R} |G_i h_i|^2 \right| \qquad (3.47)$$

This results in the SNR applied to the detector as

$$\gamma_{M_R} = \frac{E_s \left| \sum_{i=1}^{M_R} |G_i h_i|^2 \right|}{N_0 \sum_{i=1}^{M_R} |G_i|^2} \qquad (3.48)$$

From Cauchy-Schwartz inequality defined as

$$\left| \sum_{i=1}^{M_R} |a_i b_i|^2 \right| \leq \sum_{i=1}^{M_R} |a_i|^2 \sum_{i=1}^{M_R} |b_i|^2 \qquad (3.49)$$

We obtain, if $G_i = h_i$ for all i (perfect channel knowledge)

$$\gamma_{M_R} = \frac{E_s}{N_0} \sum_{i=1}^{M_R} |G_i|^2 \qquad (3.50)$$

Note that $E_s |G_i|^2 / N_0$ is the SNR per antenna, (3.50) is nothing more than the sum of the SNRs of each antenna, which means that γ_{M_R} can be large even if the individual SNRs are small. The performance of MRC is shown in Figure 3.11.

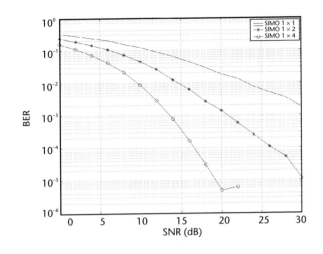

Figure 3.11 Error rate performance for MRC in Rayleigh fading.

This is for a Rayleigh fading channel. The modulation is 16 quadrature amplitude modulation (QAM).

It can be seen from Figure 3.11 that MRC is a powerful technique. It is most common in SIMO channels. In fact the software provided with this book uses MRC for SIMO channels. However, best results are obtained only with perfect channel knowledge, as that is the assumption in obtaining (3.50).

3.6.1.3 Equal Gain Combining

It is the same as MRC but with equal weighting for all branches. Hence, in this sense it is suboptimal. The performance is marginally inferior to MRC, but the complexities of implementation are much less.

3.6.2 MIMO Channels

Until now we have examined SIMO channels where there is only one transmit antenna and multiple receive antennas. What if there were multiple transmit and multiple receive antennas (MIMO channels) or multiple transmit and one receive antenna (MISO channels)? This is the subject of discussion in Chapters 4, 5, and 6.

References

[1] Rappaport, T. S., *Wireless Communications: Principles and Practice*, Upper Saddle River, NJ: Prentice Hall, 1996.

[2] Ramo, S., J. R. Whinney, and T. Van Duzer, *Fields and Waves in Communication Electronics*, New York: John Wiley & Sons, 1965.

[3] K. Bullington, "Radio Propagation at Frequencies Above 30 Megacycles," *Proc. of IEEE*, Vol. 35, 1947, pp. 1122–1136.

[4]   Landron, O., M. J. Feuerstein, and T. S. Rappaport, "A Comparison of Theoretical and Empirical Reflection Coefficients for Typical Exterior Wall Surfaces in a Mobile Radio Environment," *IEEE Trans. on Antennas and Propagation*, Vol. 44, No. 3, March 1996, pp. 341–351.

[5]   Jakes, W., *Microwave Mobile Communications*, New York: John Wiley & Sons, 1974.

[6]   Paulraj, A., R. Nabar, and D. Gore, *Introduction to Space-Time Wireless Communications*, Cambridge, UK: Cambridge University Press, 2003.

[7]   Okumura, T., E. Ohmori, and K. Fuluda, "Field Strength and Its Variability in VHF and UHF Land Mobile Service," *Review Electrical Communication Laboratory*, Vol. 16, No. 9–10, September-October 1968, pp. 825–873.

[8]   Proakis, J., et al., *Advanced Digital Signal Processing*, Singapore: MacMillan, January 1992.

[9]   Baum, D.S., et al., "Measurement and Characterization Broadband MIMO Fixed Wireless Channels at 2.5 GHz," *Proc. IEEE Int. Conf. Pers. Wireless Comm.*, Hyderabad, India, December 2000, pp. 203–206.

[10] Greenstein, L. J., D. G. Michelson, and V. Erceg, "Moment-Method Estimation of the Rician K-factor," *IEEE Commn, Letters*, Vol. 3, No. 6, June 1999.

[11] Rappaport, T. S., "Characterization of UHF Multipath Radio Channels in Factory Buildings," *IEEE Trans. on Antennas and Propagation*, Vol. 37, No. 8, August 1989, pp. 1058–1069.

[12] Rappaport, T. S., S. Y. Seidel, and R. Singh, "900 MHz Multipath Propagation Measurements for U.S. Digital Cellular Radiotelephone," *IEEE Trans. on Veh. Tech.*, May 1990, pp. 132–139.

[13] Dixon, R.C., *Spread Spectrum Systems*, 2nd edition, New York: John Wiley & Sons, 1984.

[14] Zaghloul, H., G. Morrison, and M. Fattouche, "Frequency Response and Path Loss Measurements of Indoor Channels," *Electronic Letters*, Vol. 27, No. 12, June 1991, pp. 1021–1022.

[15] Zaghloul, H., G. Morrison, and M. Fattouche, "Comparison of Indoor Propagation Channel Characteristics at Different Frequencies," *Electronic Letters*, Vol. 27, No. 22, October 1991, pp. 2077–2079.

CHAPTER 4

Space-Time Block Coding

4.1 Introduction

Space-time block coding is a simple yet ingenious transmit diversity technique in MIMO technology. We shall now discuss *space-time block codes* (STBC) and evaluate their performance in MIMO fading channels. We shall first examine the Alamouti code [1], which started it all. Basically Alamouti proposed a simple scheme for a 2×2 system that achieves a full diversity gain with a simple maximum likelihood decoding algorithm. We shall then examine higher-order diversity systems involving a large number of antennas, but whose basic approach is derived from the method proposed by Alamouti [1]. The premise in all of these approaches has been that we have perfect knowledge of the channel at the receiver and that the data streams are independent. In reality, however, this is not true. Therefore, we shall then go on to examine the behavior of these space-time codes in the presence of imperfect channel estimates and correlated slow Rayleigh fading channels.

4.2 Delay Diversity Scheme

Early attempts to obtain transmit diversity were based on the so-called delay diversity scheme [2]. Suppose we assume $M_T = 2$ and $M_R = 1$ (this will be a MISO channel). Initially let us examine what will happen if we transmit the same signal simultaneously from both antennas. If we assume a flat fading environment, where the channel signatures corresponding to the transmit antennas are given by h_1 and h_2, the received signal r may be expressed as

$$r = \sqrt{\frac{E_s}{2}}(h_1 + h_2)s + n \tag{4.1}$$

where E_s is the average energy available at the transmitter over a symbol period and is evenly divided between the transmit antennas and n is the ZMCSCG noise, representing additive white Gaussian noise sample at the receiver. We know from probability theory that the sum of two complex Gaussian random variables is also complex Gaussian. Hence, we determine that $\frac{1}{\sqrt{2}}(h_1 + h_2)$ is also ZMCSCG with unit variance. Hence,

$$r = \sqrt{E_s}\,hs + n \tag{4.2}$$

where h is ZMCSCG with $E\{|h|^2\} = 1$. Therefore, we can readily infer that this technique does not impart diversity.

In the delay diversity scheme, however, this approach is implemented, but with a major difference. We do not transmit the same symbol simultaneously from both antennas, but with a delay between the transmissions (i.e., we transmit the data signal from the first antenna and a delayed replica of the same signal from the second antenna after an interval).

If we assume that this delay is one symbol interval, the effective channel as "seen" by the receiver now becomes two channels as given by,

$$h[i] = h_1\,\delta[i] + h_2\,\delta[i-1], \; i = 0, 1, 2, \ldots \tag{4.3}$$

where h_1 and h_2 are the channel gains between the two transmit antennas and the single receive antenna, respectively. We assume that h_1 and h_2 are i.i.d. ZMCSCG random variables with unit variance. From the point of view of the receiver, such a channel looks exactly like a two-path channel with independent path fading and equal average path energy. If we now employ a maximum likelihood (ML) detector at the receiver, we can capture full second-order diversity at the receiver. The negative side to this approach is that the method introduces interference between symbols and the complexity of the ML detectors rises exponentially with the number of transmit antennas. Hence, there was a need to look for an alternate approach. This need was fulfilled by Alamouti [1].

4.3 Alamouti Space-Time Code

The approach as outlined by Alamouti is shown in Figure 4.1.

The information bits are first modulated using an M-ary modulation scheme. The encoder then takes a block of two modulated symbols s_1 and s_2 in each encoding operation and gives it to the transmit antennas according to the code matrix,

$$S = \begin{bmatrix} s_1 & -s_2^* \\ s_2 & s_1^* \end{bmatrix} \tag{4.4}$$

In (4.4), the first column represents the first transmission period and the second column the second transmission period. The first row corresponds to the symbols

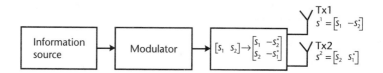

Figure 4.1 A block diagram of the Alamouti space-time encoder.

transmitted from the first antenna and the second row corresponds to the symbols transmitted from the second antenna. Elaborating further, during the first symbol period, the first antenna transmits s_1 and the second antenna transmits s_2. During the second symbol period, the first antenna transmits $-s_2^*$ and the second antenna transmits s_1^* being the complex conjugate of s_1.

This implies that we are transmitting both in space (across two antennas) and time (two transmission intervals). This is space-time coding. Looking at the equations,

$$s_1 = \begin{bmatrix} s_1, & -s_2^* \end{bmatrix} \tag{4.5}$$
$$s_2 = \begin{bmatrix} s_2, & s_1^* \end{bmatrix}$$

where s_1 is the information sequence from the first antenna and s_2 is the information sequence from the second antenna.

A close examination of (4.5) reveals that the sequences are orthogonal (i.e., the inner product of s_1 and s_2 is zero). This inner product is given by,

$$s_1 s_2 = s_1 s_2^* - s_2^* s_1 = 0 \tag{4.6}$$

If we assume one antenna at the receiver, the receiver signals are defined as follows, based on the scheme at Figure 4.2.

The fading coefficients from antennas 1 and 2 are defined by $h_1(t)$ and $h_2(t)$, respectively, at time t. If we assume that these coefficients are constant across two consecutive symbol transmission periods, we obtain,

$$h_1(t) = h_1(t + T) = h_1 = |h_1| e^{j\theta_1} \tag{4.7}$$
$$h_2(t) = h_2(t + T) = h_2 = |h_2| e^{j\theta_2}$$

where $|h_i|$ and θ_i, $i = 1, 2$ are the amplitude gain and phase shift for the path from transmit antenna i to the receive antenna and T is the symbol duration.

At the receiver the signals after passing through the channel can be expressed as,

$$r_1 = h_1 s_1 + h_2 s_2 + n_1 \tag{4.8}$$
$$r_2 = -h_1 s_2^* + h_2 s_1^* + n_2$$

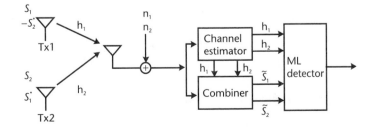

Figure 4.2   Alamouti's two-antenna transmit diversity scheme.

where n_1 and n_2 are independent complex variables with zero mean and unit variance, representing additive white Gaussian noise samples at time t and $t + T$, respectively.

4.3.1   Maximum Likelihood Decoding

We now assume that the channel coefficients h_1 and h_2 can be recovered perfectly at the receiver. We use these coefficients as the CSI. The combiner combines the received signal as follows:

$$\tilde{s}_1 = h_1^* r_1 + h_2 r_2^* = \left(\alpha_1^2 + \alpha_2^2\right)s_1 + h_1^* n_1 + h_2 n_2^* \tag{4.9}$$

$$\tilde{s}_2 = h_2^* r_1 - h_1 r_2^* = \left(\alpha_1^2 + \alpha_2^2\right)s_2 - h_1 n_2^* + h_2^* n_1$$

and sends them to the maximum likelihood detector, which minimizes the following decision metric

$$\left| r_1 - h_1 s_1 - h_2 s_2 \right|^2 + \left| r_2 + h_1 s_2^* - h_2 s_1^* \right|^2 \tag{4.10}$$

over all possible values of s_1 and s_2. Expanding this and deleting terms that are independent of the code words, the above minimization reduces to separately minimizing

$$\left| r_1 h_1^* + r_2^* h_2 - s_1 \right|^2 + \left(\alpha_1^2 + \alpha_2^2 - 1\right)\left| s_1 \right|^2$$

for detecting s_1 and

$$\left| r_1 h_2^* - r_2^* h_1 - s_2 \right|^2 + \left(\alpha_1^2 + \alpha_2^2 - 1\right)\left| s_2 \right|^2$$

for decoding s_2.

Equivalently, if we use the notation $d^2(x, y) = (x - y)(x^* - y^*) = \left| x - y \right|^2$, the decision rule for each combined signal $\tilde{s}_j$, $j = 1, 2$ becomes: Pick s_i, *iff*

$$\left(\alpha_1^2 + \alpha_2^2 - 1\right)\left| s_i \right|^2 + d^2(\tilde{s}_j, s_i) \le \left(\alpha_1^2 + \alpha_2^2 - 1\right)\left| s_k \right|^2 + d^2(\tilde{s}_j, s_k) \quad \forall i \ne k \tag{4.11}$$

For PSK signals (equal energy constellations), this simplifies to

$$d^2(\tilde{s}_j, s_i) \le d^2(\tilde{s}_j, s_k) \quad \forall i \ne k \tag{4.12}$$

4.3.2   Maximum Ratio Combining

In the case of maximum ratio combining (see Figure 4.3), the resulting received signals are

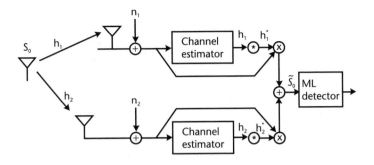

Figure 4.3 Maximum ratio combining with 1 Tx and 2 Rx.

$$r_1 = h_1 s_0 + n_1$$

$$r_2 = -h_2 s_0 + n_2$$

and the combined signal is

$$\tilde{s}_0 = h_1^* s_0 + h_2^* r_2 \tag{4.13}$$

$$= \left(\alpha_1^2 + \alpha_2^2\right) s_0 + h_1^* n_1 + h_2^* n_2$$

The maximum likelihood detector decides signal s_i using exactly the same decision rule in (4.11) or (4.12) for PSK signals.

Note that the MRC signal $\tilde{s}_0$ in (4.13) is equivalent to the resulting combined signals of the transmit diversity scheme in (4.11), except for a phase difference in the noise components that do not affect the effective SNR. This shows that the diversity order from Alamouti's two-antenna transmit diversity (with one receive antenna) is the same as that of the two-branch MRC.

4.3.3 Transmit Diversity

The transmissions in the Alamouti scheme are orthogonal. This implies that the receiver antenna "sees" two completely orthogonal streams. Hence, we obtain a transmit diversity of two. Consider two distinct code sequences S and $\hat{S}$ generated by the inputs (s_1, s_2) and $(\hat{s}_1, \hat{s}_2)$, respectively, where $(s_1, s_2) \neq (\hat{s}_1, \hat{s}_2)$. The code word difference matrix is given by

$$B(S, \hat{S}) = \begin{bmatrix} s_1 - \hat{s}_1 & -s_2^* + \hat{s}_2^* \\ s_2 - \hat{s}_2 & s_1^* - \hat{s}_1^* \end{bmatrix} \tag{4.14}$$

Since the rows of the code matrix are orthogonal, the rows of the code word difference matrix are orthogonal as well. The code word distance matrix is given by

$$A(S, \hat{S}) = B(S, \hat{S}) B^H(S, \hat{S}) \tag{4.15}$$

$$= \begin{bmatrix} |s_1 - \hat{s}_1|^2 + |s_2 - \hat{s}_2|^2 & 0 \\ 0 & |s_1 - \hat{s}_1|^2 + |s_2 - \hat{s}_2|^2 \end{bmatrix}$$

Since $(s_1, s_2) \neq (\hat{s}_1, \hat{s}_2)$, very obviously the distance matrixes of any two distinct code words have a full rank of two. In other words, the Alamouti scheme gives us a transmit diversity of $M_T = 2$. The determinant of matrix $A(S, \hat{S})$ is given by

$$\det(A(S, \hat{S})) = \left(\left| s_1 - \hat{s}_1 \right|^2 + \left| s_2 - \hat{s}_2 \right|^2 \right)^2 \qquad (4.16)$$

The code word distance matrix at (4.15) has two identical eigenvalues. The minimum eigenvalue is equal to the minimum squared Euclidian distance in the signal constellation. Hence, the minimum distance between any two transmitted code sequences remains the same as in the uncoded system. This implies that the coding gain is 1. This is the disadvantage of the Alamouti scheme in that, unlike space-time trellis codes (to be discussed in Chapter 5), the scheme achieves full transmit diversity gain *without* CSI at the transmitter but has no coding gain unless it is given the CSI.

4.3.4 Summary of Alamouti's Scheme

Alamouti further extended this scheme to the case of two transmit antennas and M_R receive antennas and showed that the scheme provided a diversity order of $2M_R$. Characteristics of this scheme include:

- No feedback from receiver to transmitter is required for CSI to obtain full transmit diversity.
- No bandwidth expansion (as redundancy is applied in space across multiple antennas, not in time or frequency).
- Low complexity decoders.
- Identical performance as MRC if the total radiated power is doubled from that used in MRC. This is because, if the transmit power is kept constant, this scheme suffers a 3-dB penalty in performance since the transmit power is divided in half across two transmit antennas.
- No need for complete redesign of existing systems to incorporate this diversity scheme. Hence, it is very popular as a candidate for improving link quality based on dual transmit antenna techniques, without any drastic system modifications.

4.4 Space-Time Block Codes

The Alamouti scheme brought in a revolution of sorts in multiantenna systems by providing full diversity of two without CSI at the transmitter and a very simple maximum likelihood decoding system at the receiver. Maximum likelihood decoders provide full diversity gain of M_R at the receiver. Hence, such a system provides a guaranteed overall diversity gain of $2M_R$, without CSI at the transmitter. This is achieved by the key feature of orthogonality between the sequences generated by the two transmit antennas. Due to these reasons, the scheme was generalized to an arbitrary number of transmit antennas by applying the theory of *orthogonal designs*. The generalized schemes are referred to as space-time block codes (STBCs) [3]. These codes can achieve the full transmit diversity of $M_T M_R$, while allowing a very simple maximum likelihood decoding algorithm, based only on linear processing of received signals [4].

Let M_T represent the number of transmit antennas and p represent the number of time periods for transmission of one block of coded symbols. Let us also assume that the signal constellation consists of 2^m points. Then each encoding operation maps a block of km information bits into the signal constellation to select k modulated signals $s_1, s_2, \ldots, s_k$, where each group of m bits selects a constellation signal. These k modulated signals are then encoded in a space-time block encoder to generate M_T parallel signal sequences of length p, as shown in Figure 4.4. This gives rise to a transmission matrix S of size $M_T \times p$. These sequences are transmitted through M_T transmit antennas simultaneously in p time periods. Therefore, the number of symbols the encoder takes as its input in each encoding operation is k. The number of transmission periods required to transmit the entire S matrix is p. The *rate* of the space-time block code is defined as the ratio between the number of symbols the encoder takes as its input and the number of space-time coded symbols transmitted from each antenna. It is given by

$$R = \frac{k}{p} \tag{4.17}$$

The spectral efficiency of the space-time block code is given by

$$\eta = \frac{r_b}{B} = \frac{r_s m R}{r_s} = \frac{km}{p} \text{ bits/s/Hz} \tag{4.18}$$

where r_b and r_s are the bit and symbol rate, respectively, and B is the bandwidth.

The entries of the transmission matrix S are so chosen that they are linear combinations of the k modulated symbols $s_1, s_2, \ldots, s_k$ and their conjugates $s_1^*, s_2^*, \ldots, s_k^*$. The matrix itself is so constructed based on orthogonal designs such that [3]

$$S.S^H = c\left(|s_1|^2 + |s_2|^2 + \ldots |s_k|^2\right)I_{M_T} \tag{4.19}$$

where c is a constant, M_T is the number of transmit antennas, S^H is the Hermitian of S, and I_{M_T} is an $M_T \times M_T$ identity matrix. This approach yields a diversity of M_T. These code transmission matrixes are cleverly constructed such that the rows and columns of each matrix are orthogonal to each other (i.e., the dot product of

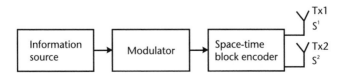

Figure 4.4 Encoder for STBC.

each row with another row is zero). If this condition is satisfied, (4.19) will be satisfied, yielding the full transmit diversity of M_T. Another way of looking at this problem is recalling from linear algebra, that if the rows of a matrix are orthogonal (i.e., their dot product is zero), then the rows of that matrix are deemed independent. This implies that each row contributes an eigenvalue (i.e., the matrix is of full rank). Hence, full transmit diversity s is achieved as each transmit antenna contributes to one row in that matrix. The code rates will, however, vary depending on how the matrix is constructed. Based on (4.17), we can have $R = 1$, which is a full rate. This implies that there is no bandwidth expansion involved, whereas a code with rate $R < 1$ implies a bandwidth expansion factor of $1/R$. It will be shown in this chapter that code rates of unity (i.e., full rates) are relatively easily achievable if the matrix is real, but the choice for full-rate codes is more restricted if the matrix is complex. Using (4.19), the orthogonality achieved in all cases enables us to achieve full transmit diversity, *irrespective of the code rate* and additionally allows the receiver to decouple the signals transmitted from different antennas. Consequently, a simple maximum likelihood decoding, based only on linear processing of the received signals, can be employed at the receiver.

4.4.1 STBC for Real Signal Constellations

In the preceding paragraph we mentioned that based on the type of signal constellation, space-time block codes can be classified into STBC with real signals or STBC with complex signals. We shall now examine the generation of real transmission matrixes.

At the outset, it should be noted that (4.19) is crucial to our design. Let us consider square transmission matrixes. Such matrixes exist *if* the number of transmit antennas $M_T = 2$, 4, or 8 [4]. These codes are full rate, since the matrix is square, and also full transmit diversity of M_T. The transmission matrixes are given by

$$S_2 = \begin{bmatrix} s_1 & -s_2 \\ s_2 & s_1 \end{bmatrix} \tag{4.20}$$

for $M_T = 2$ transmit antennas. The reader can verify that for this matrix, (4.19) is satisfied.

$$S_4 = \begin{bmatrix} s_1 & -s_2 & -s_3 & -s_4 \\ s_2 & s_1 & s_4 & -s_3 \\ s_3 & -s_4 & s_1 & s_2 \\ s_4 & s_3 & -s_2 & s_1 \end{bmatrix} \tag{4.21}$$

for $M_T = 4$ transmit antennas, and

$$
S_8 = \begin{bmatrix}
s_1 & -s_2 & -s_3 & -s_4 & -s_5 & -s_6 & -s_7 & -s_8 \\
s_2 & s_1 & -s_4 & s_3 & -s_6 & s_5 & s_8 & -s_7 \\
s_3 & s_4 & s_1 & -s_2 & -s_7 & -s_8 & s_5 & s_6 \\
s_4 & -s_3 & s_2 & s_1 & -s_8 & s_7 & -s_6 & s_5 \\
s_5 & s_6 & s_7 & s_8 & s_1 & -s_2 & -s_3 & -s_4 \\
s_6 & -s_5 & s_8 & -s_7 & s_2 & s_1 & s_4 & -s_3 \\
s_7 & -s_8 & -s_5 & s_6 & s_3 & -s_4 & s_1 & s_2 \\
s_8 & s_7 & -s_6 & -s_5 & s_4 & s_3 & -s_2 & s_1
\end{bmatrix} \tag{4.22}
$$

for $M_T = 8$ transmit antennas.

The reader can verify that all the preceding matrixes have independent rows in that their dot product is zero for any real constellation, such as M-ASK. This automatically satisfies (4.19). We can also verify by inspection that the code rate for all these matrixes is unity. For example, if we consider (4.21), we note that there are four transmit antennas (i.e., we are dealing with a space-time block code of size 4, corresponding to four rows). There are also four transmission periods p corresponding to each column of the matrix. There are also four symbols (i.e., $k = 4$, s_1, s_2, s_3, and s_4). Hence, during the first transmission interval, s_1, s_2, s_3, and s_4 are transmitted, wherein s_1 is transmitted from the first antenna, s_2 from the second antenna and so on. During the next transmission interval, $-s_2$, s_1, $-s_4$, and s_3 are transmitted, wherein $-s_2$ is transmitted from the first antenna, s_1 from the second antenna and so on. This gives us,

$$
R = \frac{k}{p} = \frac{4}{4} = 1 \tag{4.23}
$$

(i.e., a code rate of unity).

If we now desire to construct full code rate $R = 1$ transmission schemes for any number of antennas, since full code rates are desirable and are bandwidth efficient, we can do so by following another set of rules applicable to both square and nonsquare matrixes. This rule [3] says that for M_T transmit antennas, the minimum value of transmission periods p to achieve the full rate is given by

$$
p = \min\left(2^{4c+d}\right) \tag{4.24}
$$

where the minimization is taken over the set

$$
c, d \,\big|\, 0 \le c, \; 0 \le d \le 4, \text{ and } 8c + 2^d \ge M_T \tag{4.25}
$$

Based on (4.24) and (4.25), we construct nonsquare matrixes of size 3, 5, 6, and 7 for real numbers, yielding full diversity and full rate. These are as follows [3]:

$$S_3 = \begin{bmatrix} s_1 & -s_2 & -s_3 & -s_4 \\ s_2 & s_1 & s_4 & -s_3 \\ s_3 & -s_4 & s_1 & s_2 \end{bmatrix} \tag{4.26}$$

$$S_5 = \begin{bmatrix} s_1 & -s_2 & -s_3 & -s_4 & -s_5 & -s_6 & -s_7 & -s_8 \\ s_2 & s_1 & -s_4 & s_3 & -s_6 & s_5 & s_8 & -s_7 \\ s_3 & s_4 & s_1 & -s_2 & -s_7 & -s_8 & s_5 & s_6 \\ s_4 & -s_3 & s_2 & s_1 & -s_8 & s_7 & -s_6 & s_5 \\ s_5 & s_6 & s_7 & s_8 & s_1 & -s_2 & -s_3 & -s_4 \end{bmatrix} \tag{4.27}$$

$$S_6 = \begin{bmatrix} s_1 & -s_2 & -s_3 & -s_4 & -s_5 & -s_6 & -s_7 & -s_8 \\ s_2 & s_1 & -s_4 & s_3 & -s_6 & s_5 & s_8 & -s_7 \\ s_3 & s_4 & s_1 & -s_2 & -s_7 & -s_8 & s_5 & s_6 \\ s_4 & -s_3 & s_2 & s_1 & -s_8 & s_7 & -s_6 & s_5 \\ s_5 & s_6 & s_7 & s_8 & s_1 & -s_2 & -s_3 & -s_4 \\ s_6 & -s_5 & s_8 & -s_7 & s_2 & s_1 & s_4 & -s_3 \end{bmatrix} \tag{4.28}$$

$$S_6 = \begin{bmatrix} s_1 & -s_2 & -s_3 & -s_4 & -s_5 & -s_6 & -s_7 & -s_8 \\ s_2 & s_1 & -s_4 & s_3 & -s_6 & s_5 & s_8 & -s_7 \\ s_3 & s_4 & s_1 & -s_2 & -s_7 & -s_8 & s_5 & s_6 \\ s_4 & -s_3 & s_2 & s_1 & -s_8 & s_7 & -s_6 & s_5 \\ s_5 & s_6 & s_7 & s_8 & s_1 & -s_2 & -s_3 & -s_4 \\ s_6 & -s_5 & s_8 & -s_7 & s_2 & s_1 & s_4 & -s_3 \\ s_7 & -s_8 & -s_5 & s_6 & s_3 & -s_4 & s_1 & s_2 \end{bmatrix} \tag{4.29}$$

Once again, we take as an example, the transmission matrix S_7. In this example, $k = 8$, as there are eight symbols involved, $s_1, s_2, \ldots, s_8$, with eight transmission periods, $p = 8$. We transmit these eight symbols over eight transmission periods from *seven antennas* as before. This yields the full diversity of $M_T = 8$ and a code rate of

$$R = \frac{k}{p} = \frac{8}{8} = 1 \text{ (i.e., unity)} \tag{4.30}$$

4.4.2 STBC for Complex Signal Constellations

Complex orthogonal design matrixes are defined as matrixes of size $M_T \times p$ with complex entries of $s_1, s_2, \ldots, s_k$ and their conjugates and satisfying (4.19). Such matrixes provide the full transmit diversity of M_T with a code rate of k/p.

The Alamouti scheme is itself one such matrix with complex entries for two transmit antennas. This is represented by

$$G_2 = \begin{bmatrix} s_1 & -s_2^* \\ s_2 & s_1^* \end{bmatrix}$$

(4.31)

This scheme provides the full diversity of 2 with a full code rate of 1. The Alamouti scheme is unique in that for complex entries, it is the only such matrix with a code rate of unity. Hence, for higher order modulations other than binary phase shift keying (BPSK), this has found wide application.

The design rules for this class of transmission matrixes are identical to those already discussed for real entries (i.e., we design for full diversity by satisfying (4.19) and we minimize the value of p to minimize the decoding delay). We present the following complex transmission matrixes of size $M_T = 3$ and $M_T = 4$ incorporating a code rate of 1/2 [4].

$$G_3 = \begin{bmatrix} s_1 & -s_2 & -s_3 & -s_4 & s_1^* & -s_2^* & -s_3^* & -s_4^* \\ s_2 & s_1 & s_4 & -s_3 & s_2^* & s_1^* & s_4^* & -s_3^* \\ s_3 & -s_4 & s_1 & s_2 & s_3^* & -s_4^* & s_1^* & s_2^* \end{bmatrix}$$

(4.32)

$$G_4 = \begin{bmatrix} s_1 & -s_2 & -s_3 & -s_4 & s_1^* & -s_2^* & -s_3^* & -s_4^* \\ s_2 & s_1 & s_4 & -s_3 & s_2^* & s_1^* & s_4^* & -s_3^* \\ s_3 & -s_4 & s_1 & s_2 & s_3^* & -s_4^* & s_1^* & s_2^* \\ s_4 & s_3 & -s_2 & s_1 & s_4^* & s_3^* & -s_2^* & s_1^* \end{bmatrix}$$

(4.33)

The reader can easily verify that the inner product of any two rows of these matrixes is zero. This proves that the matrix is orthogonal and of full rank yielding full diversity of $M_T = 3$ and $M_T = 4$, respectively. In the case of G_3, for example, we note that there are four symbols, s_1, s_2, s_3, and s_4 and their complex conjugates, yielding $k = 4$, and there are eight transmission periods, yielding $p = 8$. This gives us a code rate of $R = k/p = 4/8 = 1/2$. Similarly, G_4 has a code rate of $R = k/p = 4/8 = 1/2$, but with a diversity of $M_T = 4$.

The desire for higher code rates leads us to more complex linear processing. The following are size 3 and 4 codes with rate 3/4 [4]:

$$\mathcal{H}_3 = \begin{bmatrix} s_1 & -s_2^* & \dfrac{s_3^*}{\sqrt{2}} & \dfrac{s_3^*}{\sqrt{2}} \\ s_2 & s_1^* & \dfrac{s_3^*}{\sqrt{2}} & \dfrac{-s_3^*}{\sqrt{2}} \\ \dfrac{s_3}{\sqrt{2}} & \dfrac{s_3}{\sqrt{2}} & \dfrac{\left(-s_1 - s_1^* + s_2 - s_2^*\right)}{2} & \dfrac{\left(s_2 + s_2^* + s_1 - s_1^*\right)}{2} \end{bmatrix}$$

(4.34)

$$\mathcal{H}_4 = \begin{bmatrix} s_1 & -s_2 & \dfrac{s_3^*}{\sqrt{2}} & \dfrac{s_3^*}{\sqrt{2}} \\[2mm] s_2 & s_1 & \dfrac{s_3^*}{\sqrt{2}} & \dfrac{-s_3^*}{\sqrt{2}} \\[2mm] \dfrac{s_3}{\sqrt{2}} & \dfrac{s_3}{\sqrt{2}} & \dfrac{\left(-s_1 - s_1^* + s_2 - s_2^*\right)}{2} & \dfrac{\left(s_2 + s_2^* + s_1 - s_1^*\right)}{2} \\[2mm] \dfrac{s_3}{\sqrt{2}} & \dfrac{-s_3}{\sqrt{2}} & \dfrac{\left(-s_2 - s_2^* + s_1 - s_1^*\right)}{2} & \dfrac{-\left(s_1 + s_1^* + s_2 - s_2^*\right)}{2} \end{bmatrix} \tag{4.35}$$

The search is still on for codes with rates greater than 0.5. This is still an open field for further research.

4.5 Decoding of STBC

The decoding of these codes is similar to the one proposed for Alamouti's scheme. We present the formula for decoding G_3 and G_4. The reader is advised to refer to [4] for the decoding procedures for the other codes.

The decoder for G_3 minimizes the decision metric

$$\left| \left[\sum_{j=1}^{m} \left(r_1^j \alpha_{1,j}^* + r_2^j \alpha_{2,j}^* + r_3^j \alpha_{3,j}^* + \left(r_5^j\right)^* \alpha_{1,j} + \left(r_6^j\right)^* \alpha_{2,j} + \left(r_7^j\right)^* \alpha_{3,j} \right) \right] - s_1 \right|^2$$
$$+ \left(-1 + 2 \sum_{j=1}^{m} \sum_{i=1}^{3} |\alpha_{i,j}|^2 \right) |s_1|^2$$

for decoding s_1, the decision metric

$$\left| \left[\sum_{j=1}^{m} \left(r_1^j \alpha_{2,j}^* - r_2^j \alpha_{1,j}^* + r_4^j \alpha_{3,j}^* + \left(r_5^j\right)^* \alpha_{2,j} - \left(r_6^j\right)^* \alpha_{1,j} + \left(r_8^j\right)^* \alpha_{3,j} \right) \right] - s_2 \right|^2$$
$$+ \left(-1 + 2 \sum_{j=1}^{m} \sum_{i=1}^{3} |\alpha_{i,j}|^2 \right) |s_2|^2$$

for decoding s_2, the decision metric

$$\left| \left[\sum_{j=1}^{m} \left(r_1^j \alpha_{3,j}^* - r_3^j \alpha_{1,j}^* - r_4^j \alpha_{2,j}^* + \left(r_5^j\right)^* \alpha_{3,j} - \left(r_7^j\right)^* \alpha_{1,j} - \left(r_8^j\right)^* \alpha_{2,j} \right) \right] - s_3 \right|^2$$
$$+ \left(-1 + 2 \sum_{j=1}^{m} \sum_{i=1}^{3} |\alpha_{i,j}|^2 \right) |s_3|^2$$

for decoding s_3, the decision metric

$$\left| \left[\sum_{j=1}^{m} \left(-r_2^j \alpha_{3,j}^* + r_3^j \alpha_{2,j}^* - r_4^j \alpha_{1,j}^* - \left(r_6^j\right)^* \alpha_{3,j} + \left(r_7^j\right)^* \alpha_{2,j} - \left(r_8^j\right)^* \alpha_{1,j} \right) \right] - s_4 \right|^2$$

$$+ \left(-1 + 2 \sum_{j=1}^{m} \sum_{i=1}^{3} |\alpha_{i,j}|^2 \right) |s_4|^2$$

for decoding s_4.

The decoder for G_4 minimizes the decision metric

$$\left| \left[\sum_{j=1}^{m} \left(r_1^j \alpha_{1,j}^* + r_2^j \alpha_{2,j}^* + r_3^j \alpha_{3,j}^* + r_4^j \alpha_{4,j}^* + \left(r_5^j\right)^* \alpha_{1,j} + \left(r_6^j\right)^* \alpha_{2,j} + \left(r_7^j\right)^* \alpha_{3,j} \right. \right. \right.$$

$$\left. \left. \left. + \left(r_8^j\right)^* \alpha_{4,j} \right) \right] - s_1 \right|^2 + \left(-1 + 2 \sum_{j=1}^{m} \sum_{i=1}^{3} |\alpha_{i,j}|^2 \right) |s_1|^2$$

for decoding s_1, the decision metric

$$\left| \left[\sum_{j=1}^{m} \left(r_1^j \alpha_{2,j}^* - r_2^j \alpha_{1,j}^* - r_3^j \alpha_{4,j}^* + r_4^j \alpha_{3,j}^* + \left(r_5^j\right)^* \alpha_{2,j} - \left(r_6^j\right)^* \alpha_{1,j} - \left(r_7^j\right)^* \alpha_{4,j} \right. \right. \right.$$

$$\left. \left. \left. + \left(r_8^j\right)^* \alpha_{3,j} \right) \right] - s_2 \right|^2 + \left(-1 + 2 \sum_{j=1}^{m} \sum_{i=1}^{3} |\alpha_{i,j}|^2 \right) |s_2|^2$$

for decoding s_2, the decision metric

$$\left| \left[\sum_{j=1}^{m} \left(r_1^j \alpha_{3,j}^* + r_2^j \alpha_{4,j}^* - r_3^j \alpha_{1,j}^* - r_4^j \alpha_{2,j}^* + \left(r_5^j\right)^* \alpha_{3,j} + \left(r_6^j\right)^* \alpha_{4,j} - \left(r_7^j\right)^* \alpha_{1,j} \right. \right. \right.$$

$$\left. \left. \left. - \left(r_8^j\right)^* \alpha_{2,j} \right) \right] - s_3 \right|^2 + \left(-1 + 2 \sum_{j=1}^{m} \sum_{i=1}^{3} |\alpha_{i,j}|^2 \right) |s_3|^2$$

for decoding s_3, the decision metric

$$\left| \left[\sum_{j=1}^{m} \left(r_1^j \alpha_{4,j}^* - r_2^j \alpha_{3,j}^* + r_3^j \alpha_{2,j}^* - r_4^j \alpha_{1,j}^* + \left(r_5^j \right)^* \alpha_{4,j} - \left(r_6^j \right)^* \alpha_{3,j} + \left(r_7^j \right)^* \alpha_{2,j} \right. \right. \right.$$

$$\left. \left. \left. - \left(r_8^j \right)^* \alpha_{1,j} \right) \right] - s_4 \right|^2 + \left(-1 + 2 \sum_{j=1}^{m} \sum_{i=1}^{3} |\alpha_{i,j}|^2 \right) |s_4|^2$$

for decoding s_4.

4.6   Simulation Results

In this section, we show simulation results pertaining to the performance of STBC on Rayleigh fading channels. In the simulations, it is assumed that the receiver has perfect CSI and that the fading between transmit and receive antennas is mutually independent.

We further assume that the total transmitted power level from two antennas for the Alamouti scheme is the same as the transmit power from the single transmit antenna for the MRC receiver diversity scheme and that it is normalized to one. The SISO channel is shown for comparison. A 2×1 Alamouti scheme and a 1×2 MRC scheme have the same diversity of order 2. This is evident from Figure 4.5, as the slopes of both the curves are the same. However, though the curves enjoy the same diversity, the 2×1 Alamouti scheme appears to have less gain than a 1×2 MRC. This is because the power from the Alamouti scheme is divided equally between the antennas (i.e., 3 dB less per antenna than the power from the MRC scheme, which has only one antenna). The 2×2 Alamouti scheme, on the other hand, shows a better performance than either of these curves because the order of

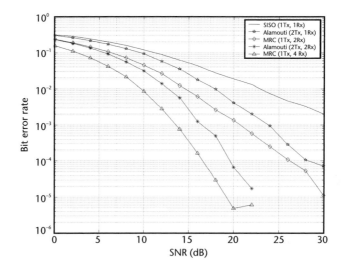

Figure 4.5   The BER performance of 16 QAM Alamouti scheme with different antenna combinations in slow Rayleigh fading channel.

diversity in this case is 4 ($M_T M_R = 2 \times 2 = 4$). Recall that the Alamouti scheme in this simulation does not possess CSI of the channel. Hence, it cannot provide array gain at the transmitter but it does provide diversity gain at the transmitter because the data streams are orthogonal. The MRC scheme, on the other hand, does not possess diversity gain at the transmitter because there is only one antenna, but it does possess array gain at the receiver because it possesses CSI at the receiver. Extending this logic further, it is to be expected that a 2×2 Alamouti scheme will be 3 dB inferior to a 1×4 MRC scheme, since both have the same diversity order, but there is a 3 dB power loss at the transmitter of the Alamouti scheme due to equal division of power between the antennas. These conclusions are borne out by the curves. To summarize, there are three types of gains that we are discussing:

- *Coding gain:* This is the gain provided by temporal coding, like convolutional coding or block coding.
- *Array gain:* This refers to the average increase in the SNR at the receiver that arises from the coherent combining effect of multiple antennas at the receiver or transmitter or both. This is the gain provided by the knowledge of the channel state (CSI), which is realized by judicious weighting of the signals from each antenna based on the knowledge of the channel. MRC at the receiver is one such example. Dominant eigenmode transmission (see Section 4.9) is one such technique for providing array gain at the transmitter.
- *Diversity gain:* This is the gain provided by spatial diversity across channels, either at the transmitter or receiver or both and is necessary for combating fading. The stipulation in this case is that the gaps across the antenna bank under question (either at the transmitter or receiver or both) should be separated by distances exceeding the coherent distance for that channel. This ensures that the channel between each transmit-receive antenna pair fades independently. Normally, systems require CSI to achieve diversity gain, but Alamouti's scheme does not require CSI to achieve transmit diversity because it is based on orthogonal data streams. The diversity order is equal to the product of the number of transmit and receive antennas.

Note that when CSI is available either at the transmitter or receiver, we obtain both array gain and diversity gain at the respective positions. The former directly increases the SNR whereas the latter reduces deep fades, improving quality of reception and thereby *also* increasing SNR. Hence, the net increase in SNR is a contribution of both array gain as well as diversity gain. In antenna theory, the elements of the antenna array are required to be spaced at a distance of half a wavelength apart (i.e., $\lambda/2$). If the distance is greater than $\lambda/2$ we experience spatial aliasing, giving rise to spurious lobes called "grating lobes." These grating lobes cause a lot of interference because they can transmit/receive interfering signals at wide angles. However, in a base station, if we require spatial diversity then the elements will need to be spaced around 16λ apart due to the geometry at the base station. This aspect was shown in Figure 2.11 in Chapter 2. Therefore, for base stations, Figure 2.11(b) is the better configuration. This will provide both array gain at each antenna array as well as diversity between the antenna arrays. On the other hand, in a mobile, due to the rich scattering environment found there, the coherent distance is quite often as little as 0.25λ. In such cases we can acquire both array gain as well as diversity gain. The array gain is calculated based on the antenna array geometry and is given in any standard textbook on antennas. The array gain can also be expressed in dBs using the relationship $10 \log_{10}(X)$. Hence, for example, at the receiver when it has CSI, the net SNR improvement is the sum of contributions from both array gain (antenna configuration permitting) and diversity gain. Like array gains, diversity gains can also be expressed in dBs. For example, in a SIMO system, if $M_R = 2$, then the diversity gain is 2 or $10 \log_{10}(2)$ = 3 dB. The nature of coding and diversity gains is different. Diversity gain manifests itself in increasing the magnitude of the slope of the curves, whereas coding gain shifts the error rate curve to the left. We must be careful when looking at the amount of coding gain from the curves, because it also includes array gain. The SNR advantage due to diversity gains increases with diversity order and lower target error rates. Coding gain, on the other hand, is typically constant at high SNRs. This is because at high SNRs the signals are already strong enough to be correctly received and therefore the presence/absence of temporal coding does not affect the situation and the coding becomes constant.

We now examine the performance of STBC using variable numbers of transmit/receive antennas on Rayleigh fading channels. During these simulations, we shall vary the type of modulation and the code sizes, with a view to maintaining constant spectral efficiency so as to ensure a fair comparison. We also assume that the receiver knows the perfect CSI.

The BER for STBC with 3 bit/s/Hz and a variable number of transmit antennas is shown in Figure 4.6 [4]. The performance of an uncoded 8 PSK is plotted for comparison. An 8 PSK modulation with a rate one code (full-rate code) will yield a spectral efficiency of 3 bit/s/Hz. This is shown for a 2×1 system. Three and four antenna systems employ rate 3/4 codes using $\mathcal{H}_3$ and $\mathcal{H}_4$ codes, respectively. This, with 16 QAM (4 bits) modulation, yields a spectral efficiency of 3 bit/s/Hz. Hence, in all cases we employ the same transmission rates (i.e., 3 bit/s/Hz). From

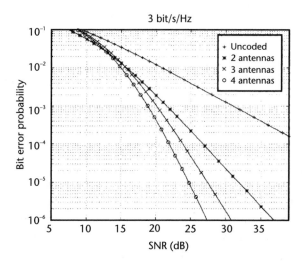

Figure 4.6 BER performance for STBC of 3 bit/s/Hz on Rayleigh fading channels with one receive antenna. (*From:* [4]. © 1999, IEEE.)

Figure 4.6 we observe that at the BER of 10^{-5}, the code G_4 is better by about 7 dB and 2.5 dB than the code G_2 and G_3, respectively.

We now examine the performance of a situation yielding a spectral efficiency of 2 bit/s/Hz with two, three, and four transmit antennas and one receive antenna on a Rayleigh fading channel. Once again, if we deploy two antennas, we need to use a rate one code G_2, but with QPSK modulation this time around to ensure a spectral efficiency of 2 bit/s/Hz. Similarly with rate 1/2 codes G_3 and G_4, we need to employ 16 QAM modulation to achieve the same transmission rate. We note in Figure 4.7 that at a BER of 10^{-3}, the code with four transmit antennas gains about 1 dB relative to the codes with two and three antennas, respectively. At higher BER this difference is even higher. This is purely because of higher transmit diversity.

Finally, we examine the performance yielding a spectral efficiency of 1 bit/s/Hz. This is once again shown for the cases of two, three, and four antennas. As usual we need to employ rate one G_2 code for two antennas and rate 1/2 codes G_3 and G_4 for three and four antennas, respectively. This implies that in order to achieve the desired spectral efficiency, we need to use BPSK modulation with G_2 code and QPSK modulation with G_3 and G_4 codes. From Figure 4.8 we note that at the same BER of 10^{-3}, the code with four antennas is superior by about 4.5 dB and 1.5 dB to the codes with two and three antennas, respectively. Once again, this is due to the higher transmit diversity of the four-antenna system.

We note that an increase in transmit diversity improves the performance. This is a very important inference from a commercial point of view, because handheld mobiles always pose a lot of problems in achieving antenna diversity at the receiver. Hence, transmit diversity at the base station holds a lot of promise. This type of diversity is easier to implement and, consequently, a lot of importance is being attached to attaining this type of diversity. The other advantage of STBC that has become abundantly clear is that we can easily increase the size of the code from two to three and to four, with a very little increase in decoding complexity, due to the fact that only linear processing is required for decoding.

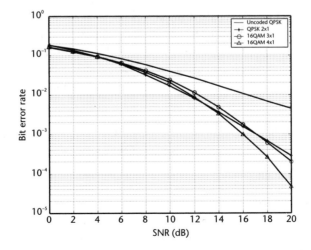

Figure 4.7   BER performance for STBC of 2 bit/s/Hz on Rayleigh fading channels with one receive antenna.

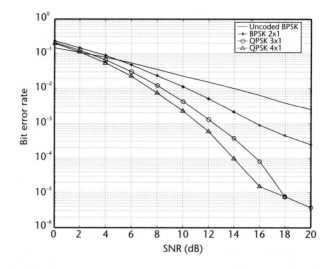

Figure 4.8 BER performance for STBC of 1 bit/s/Hz on Rayleigh fading channels with one receive antenna.

These curves were obtained using the accompanying software except for curves in Figure 4.6. The reader is advised to try to recover these graphs using this software. The reader is also encouraged to carefully study the coding with a view to finding out "how it is done." Toward this end, the software has been deliberately made unprofessional and simple to enable a novice to follow the coding procedure. There are also a lot of comments to aid the reader, as well as a complete explanation in Appendix B.

4.7   Imperfect Channel Estimation: A Performance Analysis

In the preceding paragraphs, the basic assumption was that the receiver has perfect channel knowledge. This, however, does not apply to real situations. In such cases, it becomes necessary to estimate the channel parameters. Such estimates are at best estimates and are thus called imperfect channel estimations. In this section, we will study the effects of imperfect channel estimation on the code performance. We start with a brief description of the popular methods available to achieve this. It is assumed that the channel is constant over the duration of a frame and independent between frames. Broadly, there are two principle techniques, based on the manner of structuring the training sequences [5–7], used in channel estimation:

- *Preamble structure:* In this method, we append a packet of training symbols that are strictly pilot symbols. Obviously, the more the symbols, the better the estimate.
- *Pilot structure:* In this method, the packet consists of both pilots and information symbols. Once again, the more the pilots, the better the estimate. However, these pilots infringe on the quantum of information data. Hence, there is a need to compromise between the number of pilots essential for correct estimation and the number of information data necessary to maintain the desired data throughput.

The advantage of using the preamble structure over the pilot structure is that the larger the number of pilots symbols that can be employed in this method the better the channel estimates. However, this method is effective only in slow fading channels, as the channel is required to remain static for the duration of the preamble *and* the data symbols. The pilot structure, however, allows for tracking a fast moving channel but is less accurate. Since we are dealing with a multiantenna system, it becomes necessary to discriminate between the pilots (i.e., we need to know at the receiver as to which pilots are coming from which antenna). This knowledge will give us a better estimate of the channel between a particular antenna at the transmitter and a particular antenna at the receiver. Recall that we are dealing with a narrowband system in that there is only one path between the antenna at the transmitter and the antenna at the receiver (i.e., there are no multipaths). We achieve this by making the training sequences from the individual antennas orthogonal to each other. Toward this end, we can have three types of orthogonal schemes:

- *Orthogonal in time:* In this case, we transmit from each antenna by turns. This ensures that there is no interference at the receiver from other antennas.
- *Orthogonal in frequency:* In this case, each antenna will use one tone, different from the tones from other antennas. We then carry out frequency discrimination at the receiver. This will, of course, be very expensive on spectrum.
- *Orthogonal in coding:* In this case, we choose training symbols for each antenna, such that they are orthogonal to each other (i.e., their dot product is zero).

Regardless of the way we obtain the training symbols at the receiver, there are numerous ways of processing the information [5, 6]. We shall examine three of the most popular ones.

4.7.1   Least Squares Estimation

The channel estimates can be obtained by minimizing the error matrix E_k^2 per symbol where E_k^2 is given by

$$E_k^2 = (R_k - S_k \eta_k)^H (R_k - S_k \eta_k) \tag{4.36}$$

where R_k is the received vector with the kth symbol, S_k is the kth transmitted sample, and η_k is the channel coefficient for the kth symbol.

The least squares solution for the channel estimates is obtained as

$$\eta_{k,LS} = \left(S_k^H S_k\right)^{-1} S_k^H R_k \tag{4.37}$$

where $(\ )^H$ denotes the Hermitian operator.

In case the preambles are designed such that they are orthogonal, then the matrix S is unitary. In such a case, (4.37) simplifies to

$$\eta_k = S_k^H R_k = S_k^{-1} R_k = \eta_k + \tilde{W}_k \qquad (4.38)$$

where W is the additive white Gaussian noise matrix $\tilde{W}_k = S_k^{-1} W_k$.

4.7.2 Minimum Mean Squares Estimation

This is the method used in [8]. The channel fading coefficients are estimated by inserting pilot sequences in the transmitted signals. In general, with M_T transmitting antennas we need M_T different pilot sequences $P_1, P_2, \ldots, P_{M_T}$. These pilot sequences are transmitted as a preamble of k symbols.

$$P_i = (P_{i,1}, P_{i,2}, \ldots, P_{i,k}) \qquad (4.39)$$

Obviously, these sequences will be linear superpositions of each other at the receiver. Hence, it is necessary that they be orthogonal to each other.

The received signal at antenna j at time t is given by

$$r_t^j = \sum_{i=1}^{M_T} h_{j,i} P_{i,t} + n_t^j \qquad (4.40)$$

where $h_{j,I}$ is the fading coefficient for the path from transmit antenna I to receive antenna j and n_t^j is the noise sample at receive antenna j at time t. The received signal and noise vectors at antenna j can be represented as

$$r^j = (r_1^j, r_2^j, \ldots, r_k^j) \qquad (4.41)$$
$$n^j = (n_1^j, n_2^j, \ldots, n_k^j)$$

The minimum mean squares estimation (MMSE) of $h_{j,I}$ is given by

$$\tilde{h}_{j,i} = \frac{r^j \cdot P_i}{\|P_i\|^2}$$

$$= h_{j,i} + \frac{n^j \cdot P_i}{\|P_i\|^2} \qquad (4.42)$$

$$= h_{j,i} + e_{j,i}$$

where $e_{j,i}$ is the estimation error due to the noise and is given by

$$e_{j,i} = \frac{n^j . P_i}{P_i \cdot P_i} \qquad (4.43)$$

Since n_t^j is a zero-mean complex Gaussian random variable with single-sided power spectral density N_0, the estimation error $e_{j,i}$ has a zero mean and single-sided power spectral density N_0/k. The advantage of this method over the previous one is that it is a "mean" estimator (i.e., it even averages out the noise). Hence, it is definitely superior to the least squares estimation technique. Figure 4.9 shows the result using the MMSE technique. We use a rate one code G_2 with QPSK modulation. The channel model is a slow-fading Rayleigh channel with constant coefficients over a frame of 130 symbols. The pilot sequence inserted in each frame has a length of 10 symbols in the preamble mode. The simulation results show that due to imperfect channel estimation, the code performance is degraded by about 3 dB compared with the ideal. This degradation also includes the degradation caused due to loss of signal energy by appending the pilot sequences. The reader is advised to try out these two options with different frame lengths and pilot lengths using the software accompanying this book. The performance degradation is also linked to the number of transmit antennas. If the number of transmit antennas increases, the sensitivity of the system to channel estimation error increases [8].

4.7.3   Channel Estimation Algorithm Using the FFT Method

Consider a MIMO system using M_T transmit and M_R receive antennas. Let $c_{m,k}$ be a pilot symbol transmitted from antenna m for symbol k. The received signal at receive antenna n can be expressed as

$$r_{n,k} = \sum_{i=1}^{m} h_{n,m,k} \, c_{m,k} + \eta_{n,k} \tag{4.44}$$

where $h_{n,m,k}$ is the frequency response of the channel experienced by symbol k between transmit antenna m and receive antenna n and $\eta_{n,k}$ represents additive white Gaussian noise with zero mean and variance $\sigma_n^2/2$ per dimension. Hence, the received symbol at each receive antenna is a linear combination of transmitted

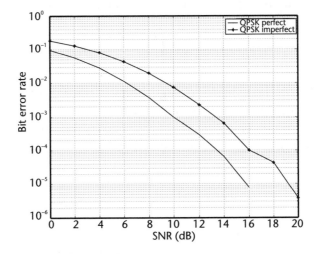

Figure 4.9 Performance of STBC with 2 bit/s/Hz using imperfect channel estimation and two receive antennas.

symbols that are modified by the channel gains and noise. We note that (4.44) can be reduced to

$$r_{n,k} = h_{n,m,k} c_{m,k} + \eta_{n,k} \tag{4.45}$$

if only one antenna is transmitting at a time. Then $h_{n,m,k}$ can be estimated by simply dividing $r_{n,k}$ by the known training symbol $c_{m,k}$. Thus, $h_{n,m,k} = r_{n,k}/c_{m,k}$ where $h_{n,m,k}$ is the channel estimate for symbol k between transmit antenna m and receive antenna n. To achieve (4.45), it is necessary to ensure that the pilot symbols from each antenna are mutually orthogonal, as previously discussed. Once again, as discussed earlier, we can employ a pilot structure or a preamble structure.

Using (4.38),

$$\eta_k = S_k^H R_k = S_k^{-1} R_k = \eta_k + \tilde{W}_k$$

we note that the channel estimate η_k for each symbol k is given by multiplying the received vector R_k by the inverse of the pilot symbol S_k. Extending this concept to a set of symbols, we obtain for N symbols (giving rise to a block matrix $M_T \times M_R$)

$$\tilde{\eta}_k = S_k^{-1} R_k = \eta_k + \tilde{W}_k \quad k = 0, 1, \ldots, N - 1 \tag{4.46}$$

where $\tilde{W}_k = S_k^{-1} W_k$, assuming that all S_k's are nonsingular (i.e., they can be inverted). In addition, if each of the S_k for $k = 1, 2, \ldots, N - 1$ are unitary, which is the case in our assumptions, since the pilot symbols are orthogonal, the variance of the noise terms in the channel estimates would remain the same. We call these channel estimates coarse channel estimates.

The channel estimates are further improved by first taking N-point inverse fast Fourier transform (IFFT) [5, 7] of the $M_T \times M_R$ coarse channel estimate vectors to convert them to the time domain. These $M_T \times M_R$ length-N vectors $\{g_m\}_{m=0}^{N-1}$ are then passed through a rectangular window such that

$$\bar{h}_m = \begin{cases} g_m & 0 \leq m \leq (G - 1) \\ 0 & m \geq G \end{cases} \tag{4.47}$$

where G is the maximum tap length of the channel. In OFDM systems (see Chapter 7) this is equal to the length of the cyclic prefix.

The time domain fine channel estimates $\bar{h}_m$ are then converted using fast Fourier transform (FFT) to fine channel estimates $\{\eta_{q,l,k}\}_{k=0}^{N-1}$ in frequency domain such that

$$\bar{\eta}_{q,l} = FFT\{\bar{h}_m\} \quad 1 \leq q \leq M_T, \, 1 \leq l \leq M_R \tag{4.48}$$
$$= \tilde{\eta}_{q,l} + \tilde{W}_{q,l}$$

Because of the windowing operation, the noise variance in the fine channel estimates is reduced to $\sigma^2 \dfrac{G}{N}$.

This method was applied to a frame of 64 symbols with a preamble of 64 pilots in a slow Rayleigh fading channel. Two curves were plotted, one using LSE and the other with FFT. The sequences were orthogonal in time. The result is shown in Figure 4.10 for a modulation of quadrature phase shift keying (QPSK) for a rate one code G_2.

We note from Figure 4.10 that at a BER of 10^{-4} there is a performance gap of 3 dB from the ideal if we use the least squares estimation method. However, this is improved to 0.7 dB from the ideal if we use the FFT method. The performance will be even better if we use MMSE instead of LSE in the first step. The curves in Figure 4.10 were plotted using the software provided with Chapter 9. We shall examine the FFT method in more detail in that chapter.

4.8   Effect of Antenna Correlation on Performance

Until now, in all our simulations, we have assumed that the data streams are uncorrelated and independent. In reality, this is difficult to achieve. We show that there is performance degradation due to correlation in slow fading channels. In Figure 4.11 we note this result for the G_2 case employing two transmit and two receive antennas.

We assume that the transmit antennas are not correlated but the receive antennas are correlated. The receive correlation matrix is given by

$$\vartheta_R = \begin{bmatrix} 1 & \alpha \\ \alpha & 1 \end{bmatrix} \tag{4.49}$$

where α is the correlation factor between the receive antennas. In the simulation, the correlation factors chosen were 0.25, 0.5, 0.75, and 1. It can be noted that

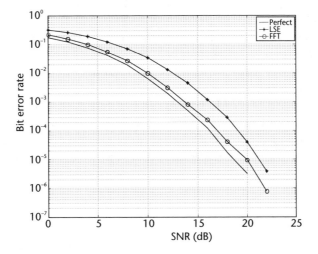

Figure 4.10   Performance of STBC with 2 bit/s/Hz using imperfect channel estimation using FFT method and two receive antennas.

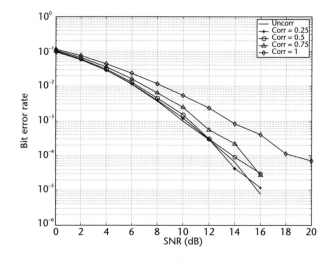

Figure 4.11   Performance of STBC with 2 bit/s/Hz on correlated slow Rayleigh fading channel with two transmit and two receive antennas.

with a correlation factor of 0.25 there is a slight degradation relative to the ideal case when there is no correlation. Thereafter, the correlation progressively degrades the performance by 0.5 dB and 1.2 dB at a BER of 10^{-3} for a correlation factor of 0.5 and 0.75, respectively. When the channels are fully correlated, the penalty on the code performance is about 4 dB at the same BER. The reader is encouraged to recover these curves for a slow fading Rayleigh channel using the accompanying software.

4.9   Dominant Eigenmode Transmission

Thus far, in all our analysis in this chapter we have assumed a common thread of a complete lack of knowledge of the channel at the transmitter. Hence, we could only obtain diversity gain due to transmit diversity, but there was no array gain at the transmitter. However, the possession of complete channel knowledge at the transmitter leads us to the very interesting concept of dominant eigenmode transmission [9].

Consider a $M_T \times M_R$ system. Such a system obtains knowledge of the channel through various techniques [9]. Once such is channel estimation using feedback. In this approach, the knowledge of the channel is obtained at the receiver and sent to the base station on the reverse link. This feedback will involve some delay (or lag), δ_{lag}. Since wireless channels are time-varying, we need

$$\delta_{lag} \ll T_c \tag{4.50}$$

where T_c is the coherence time. Therefore, δ_{lag}/T_c determines channel accuracy at the transmitter. In a fast-changing channel, we need more frequent channel updates. This results in a significant overhead.

We now transmit the same signal from all antennas in the transmit array with weight vector $\mathbf{w}$. The received signal vector is then given by

$$y = \sqrt{\frac{E_s}{M_T}} \mathbf{H}\mathbf{w}s + \mathbf{n} \tag{4.51}$$

where $\mathbf{y}$ is the $M_R \times 1$ received signal vector, $\mathbf{H}$ is the $M_R \times M_T$ channel transfer function, $\mathbf{w}$ is the $M_T \times 1$ complex weight vector, and $\mathbf{n}$ is spatially white ZMCSCG noise. Note that $\|\mathbf{w}\|^2 = M_T$ to maintain total average transmitted energy.

The weighted sum of all antenna outputs at the receiver is given by

$$z = \mathbf{g}^H \mathbf{y} \tag{4.52}$$

where $\mathbf{g}$ is a $M_R \times 1$ vector of complex weights. The SNR at the receiver, η, is then given by

$$\eta = \frac{\|\mathbf{g}^H \mathbf{H}\mathbf{w}\|_F^2}{M_T \|\mathbf{g}\|_F^2} \rho \tag{4.53}$$

where $\|\cdot\|_F$ is the Frobenious norm of a matrix, which is the sum of the norms of all the matrix elements and ρ is the SNR at the receiver for an SISO channel.

Hence, maximizing the SNR at the receiver is equivalent to maximizing $\|\mathbf{g}^H \mathbf{H}\mathbf{w}\|_F^2 / \|\mathbf{g}\|_F^2$. The singular value decomposition (SVD) [10] of $\mathbf{H}$ is given by

$$\mathbf{H} = \mathbf{U}\mathbf{\Sigma}\mathbf{V}^H \tag{4.54}$$

From (4.53) we note that η is maximized when $\mathbf{w}/\sqrt{M_T}$ and $\mathbf{g}$ are the input and output singular vectors, respectively, corresponding to the maximum singular value $\sigma_{\max}$ of $\mathbf{H}$. Using these values, the effective input-output relation for the channel reduces to

$$z = \sqrt{E_s} \sigma_{\max} s + n \tag{4.55}$$

where n is ZMCSCG noise with variance N_0. We know that singular values of $\mathbf{H}$ are the square root of eigenvalues of $\mathbf{H}\mathbf{H}^H$. Therefore, $\sigma_{\max}^2 = \lambda_{\max}$, where $\lambda_{\max}$ is the maximum eigenvalue of $\mathbf{H}\mathbf{H}^H$, the SNR at the receiver is given by

$$\eta = \lambda_{\max} \rho \tag{4.56}$$

Therefore, the array gain of dominant eigenmode transmission is given by $\epsilon\{\lambda_{\max}\}$, where ϵ is the expectation operator.

Inspection of (4.56) tells us that the SNR at the receiver of such a system is enhanced by a factor $\lambda_{\max}$, which is the maximum eigenvalue of $\mathbf{H}\mathbf{H}^H$. Hence, this higher *effective* SNR causes the performance curve to be better than a scheme, which has no channel knowledge at the transmitter, by a factor that is equal to the array gain. The probability of symbol error for such a system is given by [9]

$$\overline{N}_e \left(\frac{\rho d_{\min}^2}{4 \min (M_T, M_R)} \right)^{-M_T M_R} \geq \overline{P}_e \geq \overline{N}_e \left(\frac{\rho d_{\min}^2}{4} \right)^{-M_T M_R} \tag{4.57}$$

where $\overline{N}_e$ and $d_{\min}$ are the number of nearest neighbors and minimum distance of separation of the underlying scalar constellation, respectively. Inspection of (4.57) tells us that the symbol error rate (SER) must maintain a slope of magnitude $M_T M_R$, as a function of SNR (on a log-log scale). Hence, we conclude that dominant eigenmode transmission for a system extracts a full diversity order of $M_T M_R$ [9].

The reader should note that we have not resorted to any STBC scheme like Alamouti's scheme, during dominant eigenmode transmission. We achieve all this by simple weighted transmission based on channel knowledge at the transmitter, just as we normally do at the receiver. The transmitter weights, $\mathbf{w}$, therefore assign each signal stream to a separate eigenmode of the channel matrix, $\mathbf{H}$, thereby maintaining orthogonality among the substreams at the receiver since the eigenvalues are themselves orthogonal. The receiver weights, $\mathbf{g}$, extract these substreams from the combined signal. Note that this is known to be the signal structure which maximizes the channel capacity as discussed in Chapter 2. To optimize data throughput, the data and signal power would be assigned to the eigenmodes using a "waterpouring" solution based on the eigenvalues $\lambda_j = \sigma_j^2$ in the noise free case. As an example, it is pointed out that for a 2×2 scheme, we have two *dominant* eigenmodes. These two dominant eigenmodes enable the transmission and the remaining eigenmodes of the channel matrix are not used and considered redundant. Hence, the name dominant eigenmode transmission is given to this method.

4.10 Capacity of OSTBC Channels

The SNR at the receiver is given by $(\rho/M_T)\|\mathbf{H}\|_F^2$ and the capacity is given by [11]

$$C_{OSTBC} = r_s \left(1 + \frac{\rho}{M_T} \|\mathbf{H}\|_F^2 \right) \tag{4.58}$$

where r_s is the code rate.

We know from (2.17) the capacity of a MIMO channel when the channel is unknown to the transmitter is given by

$$C = \log_2 \det \left(\mathbf{I}_{M_R} + \frac{E_s}{M_T N_0} \mathbf{H}\mathbf{H}^H \right)$$

$$= \log_2 \prod_{k=1}^{r} \left(1 + \frac{\rho}{M_T} \lambda_k \right) \tag{4.59}$$

$$= \log_2 \left(1 + \frac{\rho}{M_T} \|\mathbf{H}\|_F^2 + \frac{\rho^2}{M_T} (\cdot) + \dots \right)$$

$$\geq C_{OSTBC}$$

where λ_k are the eigenvalues of $\mathbf{HH}^H$. As was stated in Chapter 2, $\lambda_k \geq 0$, $(k = 1, 2, \ldots, r)$ and $\Sigma_{k=1}^{r} \lambda_k = \mathrm{Tr}(\mathbf{HH}^H) = \|\mathbf{H}\|_F^2$. Hence, the capacity of orthogonal space-time block code (OSTBC) channels is inferior to the channel with optimal coding except in the case of Alamouti's scheme wherein the code rate $r_s = 1$ causing $C = C_{OSTBC}$. However, the outage properties of OSTBC will be superior to the outage obtained with optimal coding for a given transmission rate, because OSTBC fundamentally improves the link.

The main conclusions from this chapter are summarized in Table 4.1 [9].

4.11 Simulation Exercises

1. The reader is advised to try out the exercises suggested in this chapter.
2. Implement a 2×2 scheme using dominant eigenmode transmission by modifying the code in the software. How does it compare with the Alamouti (G_2) scheme?

Table 4.1 Array Gain and Diversity Order for Different Multiple Antenna Configurations

Configurations	Expected Array Gain	Diversity Order
SIMO(CU)	M_R	M_R
SIMO(CK)	M_R	M_R
MISO(CU)	1	M_R
MISO(CK)	M_T	M_T
MIMO(CU)	M_R	$M_R M_T$
MIMO(CK)	$\epsilon(\lambda_{\max})$	$M_R M_T$

CU: Channel unknown to the transmitter
CK: Channel known to the transmitter
Source: [9]. Reprinted with the permission of Cambridge University Press.

References

[1] Alamouti, S. M. "A Simple Transmit Diversity Technique for Wireless Communications," *IEEE Journal Select. Areas Commun.,* Vol. 16, No. 8, October 1998, pp. 1451–1458.

[2] Seshadri, N., and J. Winters, "Two Signaling Schemes for Improving the Error Performance of Frequency-Division-Duplex (FDD) Transmission Systems Using Transmitter Antenna Diversity," *Int. J. Wireless Information Networks,* Vol. 1, January 1994, pp. 49–60.

[3] Tarokh, V., H. Jafarkhani, and A. R. Calderbank, "Space-Time Block Codes From Orthogonal Designs," *IEEE Trans. Inform. Theory,* Vol. 45, No. 5, July 1999, pp. 1456–1467.

[4] Tarokh, V., H. Jafarkhani, and A. R. Calderbank, "Space-Time Block Coding for Wireless: Performance Results," *IEEE J. Select. Areas Commun.,* Vol. 17, No. 3, March 1999, pp. 451–460.

[5] Mody, A. N., and G. L. Stuber, "Parameter Estimation for OFDM With Transmit Receive Diversity," *IEEE Vehicular Technology Conference,* Rhodes, Greece, May 2001.

[6] Mody, A. N., and G. L. Stuber, "Synchronization for MIMO-OFDM Systems," *IEEE Global Communications Conference,* San Antonio, TX, November 2001.

[7] Siew, J., et al., "A Channel Estimation Method for MIMO-OFDM Systems," *London Communications Symposium,* London, September 2002.

[8]   Tarokh, V., et al., "Space-Time Codes for High Data Rate Wireless Communication: Performance Criteria in the Presence of Channel Estimation Errors, Mobility, and Multiple Paths," *IEEE Trans. Commun.*, Vol. 47, No. 2, February 1999, pp. 199–207.

[9]   Paulraj, A., R. Nabar, and D. Gore, *Introduction to Space-Time Wireless Communications,* Cambridge, UK: Cambridge University Press, 2003.

[10]   Golub G., and C. Van Loan, *Matrix Computations,* Baltimore, MD: John Hopkins University Press, 2nd edition, 1989.

[11]   Hassibi B., and B. Hochwald, "High-Rate Codes That Are Linear in Space and Time," *IEEE Trans. Inf. Theory,* Vol. 48, No. 7, July 2002, pp. 1804–1824.

CHAPTER 5

Space-Time Trellis Codes

5.1 Introduction

In Chapter 4, we discussed space-time block codes. These codes provided maximum diversity advantage using simple decoding techniques. However, space-time block codes did not provide coding gain, and nonfullrate space-time block codes introduced bandwidth expansion. In view of this, it becomes worthwhile to consider a joint design of error control coding, modulation, transmit, and receive diversity to develop an effective signaling scheme called space-time trellis codes (STTC), which is able to combat effects of fading. This concept was first introduced by Tarokh, Seshadri, and Calderbank [1]. It became extremely popular because STTC can simultaneously offer coding gain with spectral efficiency and full diversity over fading channels. This coding gain should not be confused with the coding gain discussed in Section 4.6. This coding gain is achieved through the inherent nature of the STTC itself and is distinct from the coding gain achieved by temporal block codes and convolution codes.

In this chapter, we explore the basic theory leading to such code design using M-PSK schemes for various numbers of transmit antennas and spectral efficiencies, in both slow as well as fast fading channels. The code performance is examined with simulations and the effects of imperfect channel estimations and correlation are also considered.

5.2 Space-Time Coded Systems

We consider a baseband space-time coded system with M_T transmit antennas and M_R receive antennas, as shown in Figure 5.1.

The data to be transmitted are encoded by a space-time encoder. At each instant t, a block of m binary information symbols denoted by

$$c_t = \left(c_t^1, c_t^2, \ldots, c_t^m\right) \tag{5.1}$$

is fed into the space-time encoder. The space-time encoder maps the block of m binary input data into M_T modulation symbols from a signal set of $M = 2^m$ points. The coded data are applied to a serial-to-parallel (S/P) converter to produce a sequence of M_T parallel symbols, arranged as a $M_T \times 1$ column vector

$$s_t = \left(s_t^1, s_t^2, \ldots, s_t^{M_T}\right)^T \tag{5.2}$$

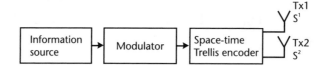

Figure 5.1   A block diagram of space-time trellis encoder.

where T means the transpose of a matrix. The M_T parallel outputs are simultaneously transmitted from all the M_T antennas, whereby symbol s_t^i, $1 \le i \le M_T$ is transmitted by antenna i and all transmitted symbols have the same duration of T sec. The vector of coded modulation symbols from different antennas, as shown in (5.2), is called a *space-time symbol*. The spectral efficiency of the system is

$$\eta = \frac{r_b}{B} = m \text{ b/s/Hz} \tag{5.3}$$

where r_b is the data rate and B is the channel bandwidth. The spectral efficiency in (5.3) is equal to the spectral efficiency of a reference uncoded system with one transmit antenna.

The multiple antennas at both the transmitter and receiver create a MIMO channel. We assume flat fading between each transmit and receive antenna and we also assume that the channel is memoryless.

The channel matrix at any given time t is given by

$$H_t = \begin{bmatrix} h_{1,1}^t & h_{1,2}^t & \cdots & h_{1,M_T}^t \\ h_{2,1}^t & h_{2,2}^t & \cdots & h_{2,M_T}^t \\ \vdots & \vdots & \vdots & \vdots \\ h_{M_R,1}^t & h_{M_R,2}^t & \cdots & h_{M_R,M_T}^t \end{bmatrix} \tag{5.4}$$

where the jith element, denoted by $h_{j,i}^t$, is the fading attenuation coefficient for the path from transmit antenna i to receive antenna j.

The coefficients in (5.4) are assumed to be i.i.d. Gaussian. There are two cases that we need examine. The first case is that we assume that the channel is a slow fading channel (i.e., the fading coefficients are constant during a frame and vary from one frame to another). This is also called quasi-static fading. The second case is a fast fading channel (i.e., the fading coefficients are constant within each symbol period and vary from one symbol to another). At the receiver, we note that the signal at each antenna is a noisy superposition of M_T transmitted signals degraded by channel fading. At time t the received signal at antenna j, $j = 1, 2, \ldots, M_R$ denoted by r_t^j is given by

$$r_t^j = \sum_{i=1}^{M_T} h_{j,t}^t s_t^i + n_t^j \tag{5.5}$$

where n_t^j is the noise component of receive antenna j at time t, which is also i.i.d. Gaussian.

We represent

$$r_t = \left(r_t^1, r_t^2, \ldots, r_t^{M_R}\right) \tag{5.6}$$

and

$$n_t = \left(n_t^1, n_t^2, \ldots, n_t^{M_R}\right) \tag{5.7}$$

Thus the received signal vector can be represented as

$$r_t = H_t s_t + n_t \tag{5.8}$$

The decoder uses a maximum likelihood algorithm to estimate the transmitted information sequence and we assume that the receiver has complete knowledge of the channel. The transmitter, however, has no knowledge of the channel. The decision metric is computed based on the squared Euclidian distance between the received sequence as measured and the actual received sequence, as

$$\sum_t \sum_{j=1}^{M_R} \left| r_t^j - \sum_{i=1}^{M_T} h_{j,i}^t s_t^i \right|^2 \tag{5.9}$$

The decoder selects a code word with the minimum decision metric as the decoded sequence. This decoder is implemented as a Viterbi decoder.

5.3 Space-Time Code Word Design Criteria

We assume that the transmitted data frame length is L symbols long for each antenna. This leads to a space-time code word matrix $M_T \times L$,

$$S = [s_1, s_2, \ldots, s_L] = \begin{bmatrix} s_1^1 & s_2^1 & \cdots & s_L^1 \\ s_1^2 & s_2^2 & \cdots & s_L^2 \\ \vdots & \vdots & \vdots & \vdots \\ s_1^{M_T} & s_2^{M_T} & \cdots & s_L^{M_T} \end{bmatrix} \tag{5.10}$$

where each row corresponds to the data sequence transmitted from each antenna and each column is the space-time symbol at time t.

The pairwise error probability (PEP) is the probability that the maximum likelihood decoder selects as its estimate a signal $e = e_1^1 e_1^2 \ldots e_1^{M_T} e_2^1 \ldots e_2^{M_T} e_L^1 \ldots e_L^{M_T}$ when in fact the signal $s = s_1^1 s_1^2 \ldots s_1^{M_T} s_2^1 \ldots s_2^{M_T} s_L^1 \ldots s_L^{M_T}$ was transmitted. This will occur if, summing over all symbols, antennas, and time periods

$$\sum_{t=1}^{L}\sum_{j=1}^{M_R}\left|r_t^j - \sum_{i=1}^{M_T} h_{i,j}^t s_t^i\right|^2 \geq \sum_{t=1}^{L}\sum_{j=1}^{M_R}\left|r_t^j - \sum_{i=1}^{M_T} h_{i,j}^t e_t^i\right|^2$$

which can be rewritten as

$$\sum_{t=1}^{L}\sum_{j=1}^{M_R} 2\,\mathrm{Re}\left\{(\eta_t^j)^* \sum_{i=1}^{M_T} h_{i,j}^t (e_t^i - s_t^i)\right\} \geq \sum_{t=1}^{L}\sum_{j=1}^{M_R}\left|\sum_{i=1}^{M_T} h_{i,j}^t (e_t^i - s_t^i)\right|^2 \quad (5.11)$$

where $\mathrm{Re}\{\cdot\}$ refers to the real part of the argument.

If we assume that the receiver has perfect knowledge of the channel, then for a given instance of channel path gains $\{h_{i,j}\}$, the term on the right of (5.11) is a constant equal to $d^2(e, s)$ and the term on the left is a zero-mean Gaussian random variable with variance $4\sigma^2 d^2(e, s)$. Hence, the PEP conditioned on knowing $\{h_{i,j}\}$ is given by

$$P(s \to e \,|\, h_{i,j}, i = 1, \ldots, M_T, j = 1, \ldots, M_R) = Q\left(\frac{d(s, e)}{2\sigma}\right)$$

$$\leq \exp\left(-d^2(s, e)\, E_s / 4N_0\right)$$

where $Q(x)$ is the complementary error function given by $Q(x) = \int_x^\infty e^{-t^2/2}\, dt$.

Now, $d^2(s, e)$ can be rewritten as $\sum_{i=1}^{M_T}\sum_{i'=1}^{M_T} h_{i,j}\, \overline{h_{i',j}}\, (s_t^i - e_t^i)\left(\overline{s_t^{i'} - e_t^{i'}}\right)$ where $\overline{x}$ is the complex conjugate of x. Thus

$$d^2(s, e) = \sum_{t=1}^{L}\sum_{j=1}^{M_R}\sum_{i=1}^{M_T}\sum_{i'=1}^{M_T} h_{i,j}\, \overline{h_{i',j}}\, (s_t^i - e_t^i)\left(\overline{s_t^{i'} - e_t^{i'}}\right) \quad (5.12)$$

If we denote $\Omega_j = (h_{1,j} h_{2,j}, \ldots h_{M_T,j})$, we can rewrite

$$d^2(s, e) \sum_{j=1}^{M_R} \Omega_j A \Omega_j^\dagger \quad (5.13)$$

where $\Omega_j^\dagger$ denotes the Hermitian transpose of Ω_j and $A = A(e, s)$ is an $M_T \times M_T$ matrix independent of time and contains entries $A_{p,q} = \sum_{t=1}^{L}(s_t^p - e_t^p)\left(\overline{s_t^q - e_t^q}\right)$. So, the PEP becomes

$$P(s \to e \,|\, h_{i,j}, i = 1, \ldots, M_T, j = 1, \ldots, M_R) \leq \prod_{j=1}^{M_R} \exp\left(-\frac{\Omega_j A \Omega_j^\dagger E_s}{4N_0}\right)$$

$$(5.14)$$

Since A is Hermitian, there exists a unitary matrix V that satisfies $VV^\dagger = I$ and a real diagonal matrix D such that

$$VAV^\dagger = D$$

The rows $\{v_1, v_2, \ldots, v_{M_T}\}$ of V are the eigenvectors of A and form a complete orthonormal basis of an M_T-dimensional vector space. Furthermore, the diagonal elements of D are the eigenvalues λ_i, $i = 1, 2, \ldots, M_T$ of A, including multiplicities, and are nonnegative real numbers since A is Hermitian. By the construction of A, the *code word difference matrix*, B where

$$B(s, e) = \begin{pmatrix} e_1^1 - s_1^1 & e_2^1 - s_2^1 & \cdots & e_L^1 - s_L^1 \\ e_1^2 - s_1^2 & e_2^2 - s_2^2 & \cdots & e_L^2 - s_L^2 \\ \vdots & \vdots & \vdots & \vdots \\ e_1^{M_T} - s_1^{M_T} & e_2^{M_T} - s_2^{M_T} & \cdots & e_L^{M_T} - s_L^{M_T} \end{pmatrix}$$

is a square root of A (i.e., $A = BB^\dagger$).

Next, we express $d^2(s, e)$ in terms of the $\{\lambda_i\}$'s. Let $(\beta_{1,j}, \beta_{2,j}, \ldots, \beta_{M_T,j}) = \Omega_j V^\dagger$, then

$$\Omega_j A \Omega_j^\dagger = \Omega_j V^\dagger D V \Omega_j^\dagger = \sum_{i=1}^{M_T} \lambda_i |\beta_{i,j}|^2 \tag{5.15}$$

$$\therefore d^2(s, e) = \sum_{j=1}^{M_R} \sum_{i=1}^{M_T} \lambda_i |\beta_{i,j}|^2$$

Since $h_{i,j}$ are samples of a complex Gaussian random variable with mean $Eh_{i,j}$, let

$$K^j = (Eh_{1,j}, Eh_{2,j}, \ldots, Eh_{M_T,j})$$

Since V is unitary, this implies $\beta_{i,j}$ are independent complex Gaussian random variables with variance 0.5 per dimension and with mean $K^j \cdot v_i$.

5.4   Design of Space-Time Trellis Codes on Slow Fading Channels

5.4.1   Error Probability on Slow Fading Channels

In the case of flat Rayleigh fading, $Eh_{i,j} = 0$ for all i and j. To obtain an upper bound on the average probability of error, we average

$$\prod_{j=1}^{M_R} \exp\left(-\left(\frac{E_s}{4N_0}\right) \sum_{i=1}^{M_T} \lambda_i |\beta_{i,j}|^2\right) \tag{5.16}$$

with respect to the independent Rayleigh distributions of $|\beta_{i,j}|$ with probability density $p(|\beta_{i,j}|) = 2|\beta_{i,j}| \exp(-|\beta_{i,j}|^2)$. Letting $c = E_s/4N_0$, we get

$$E \prod_{j=1}^{M_R} e^{-c \sum_{i=1}^{M_T} \lambda_i |\beta_{i,j}|^2} = \prod_{j=1}^{M_R} \prod_{i=1}^{M_T} E\left(e^{-c\lambda_i |\beta_{i,j}|^2}\right)$$

by independence and

$$E\left(e^{-c\lambda_i |\beta_{i,j}|^2}\right) = 2\int_0^\infty e^{-c\lambda_i w^2} w e^{-w^2} dw$$

$$= 2\int_0^\infty w e^{-w^2(1 + c\lambda_i)} dw \text{ So,}$$

$$= \frac{1}{1 + c\lambda_i} \int_0^\infty u e^{-u} du$$

$$= \frac{1}{1 + c\lambda_i}$$

So,

$$P(s \rightarrow e) \leq \left(\frac{1}{\prod_{i=1}^{M_T}\left(1 + \frac{E_s}{4N_0}\lambda_i\right)}\right)^{M_R} \tag{5.17}$$

Let r be the rank of A. Then the eigenvalue 0 has a multiplicity of $M_T - r$. Let the nonzero eigenvalues of A be $\lambda_1, \lambda_2, \ldots, \lambda_r$. Then for high enough SNR, $\left(1 + \frac{E_s}{4N_0}\lambda_i\right) \approx \frac{E_s}{4N_0}\lambda_i$ and the PEP in (5.17) becomes

$$P(s \rightarrow e) \leq \left(\prod_{i=1}^r \lambda_i\right)^{-M_R} \left(\frac{E_s}{4N_0}\right)^{-rM_R} = \left(\prod_{i=1}^r \lambda_i^{1/r}\right)^{-rM_R} \left(\frac{E_s}{4N_0}\right)^{-rM_R} \tag{5.18}$$

Thus a *diversity advantage* of $M_R r$ and a *coding advantage* of $\left(\prod_{i=1}^r \lambda_i\right)^{1/r}$ are obtained. Note that the diversity advantage is defined as the power of the SNR in the denominator of the right-hand expression of (5.18). Intuitively the coding advantage is an estimate of the gain over an uncoded system with the same diversity advantage.

5.4.2 Design Criteria for Slow Rayleigh Fading STTCs

We now define the basic design criteria for slow Rayleigh STTCs:

- *The rank criterion:* To obtain maximal diversity, we need to maximize the minimum rank r of the matrix B over all pairs of distinct code words. A diversity advantage of rM_R is achieved.
- *The determinant criterion:* Let rM_R be the target diversity advantage. Then the design goal is to maximize the minimum determinant $\prod_{i=1}^{r} \lambda_i$ of the matrix A along the pairs of distinct code words with that minimum rank.

These criteria are referred to as *rank and determinant criteria*. This is also called the TSC criteria after the authors of this paper [1]. Maximizing the minimum rank r of the matrix B implies that we need to find a space-time code that achieves the full rank of the matrix B (i.e., M_T). This is not always achievable due to the restriction of the trellis code structure. Vucetic [2–6] has given further details on these aspects. For a space-time trellis code with memory order of ν, the length of an error event, denoted by L, can be lower bounded as [2]

$$L \geq [\nu/2] + 1$$

For an error event path of length L in the trellis, $B(s, e)$ is a matrix of size $M_T \times L$, which results in the maximum achievable rank of $\min(M_T, L)$. This yields the upper bound as given by $\min(M_T, [\nu/2] + 1)$. The upper bound of the rank values for STTC with various numbers of transmit antennas and memory orders is listed in Table 5.1 [2].

We note that the full rank is achievable only for STTC with two transmit antennas. In case of three and four transmit antennas, we require a memory order of the encoder of at least four and six, respectively, to achieve full rank. Further examination of Table 5.1 tells us that there is an interaction between the maximum achievable rank, the number of transmit antennas, and the memory order of an STTC. We shall come back to this issue later in this chapter.

From (5.11), the PEP conditioned on knowing $\{h_{i,j}\}$ is given by

$$P(s \rightarrow e | h_{i,j}, i = 1, \ldots, M_T, j = 1, \ldots, M_R) = Q\left(\frac{d(s, e)}{2\sigma}\right) \qquad (5.19)$$

$$\leq \exp\left(-d^2(s, e)E_s/4N_0\right)$$

where $Q(x)$ is the complementary error function given by $Q(x) = \int_x^{\infty} e^{-t^2/2} \, dt$.

Table 5.1 Upper Bound of the Rank Values for STTC

	$M_T = 2$	$M_T = 3$	$M_T = 4$	$M_T = 5$	$M_T \geq 6$
$\nu = 2$	2	2	2	2	2
$\nu = 3$	2	2	2	2	2
$\nu = 4$	2	3	3	3	3
$\nu = 5$	2	3	3	3	3
$\nu = 6$	2	3	4	4	4

From: [2]. © 2001 IEEE.

From (5.13) and (5.15),

$$d^2(s, e) = \sum_{j=1}^{M_R} \sum_{i=1}^{M_T} \lambda_i |\beta_{j,i}|^2 \tag{5.20}$$

Substituting in (5.19) and using the inequality $Q(x) \leq \frac{1}{2} e^{-x^2/2}$, $x \geq 0$,

$$P(s \rightarrow e|H) \leq \frac{1}{2} \exp\left(-\sum_{j=1}^{M_R} \sum_{i=1}^{M_T} \lambda_i |\beta_{j,i}|^2 \frac{E_s}{4N_0}\right) \tag{5.21}$$

This is the upper bound on the conditional pairwise error probability expressed as a function of $|\beta_{j,i}|$, which is contingent upon $h_{j,i}$. We know that $|\beta_{j,i}|^2$ follow the central Chi-square distribution with the mean value and the variance given by

$$\mu_{|\beta_{j,i}|^2} = 1$$

and

$$\sigma^2_{|\beta_{j,i}|^2} = 1$$

For a large rM_R (>3) value according to the central limit theorem [7], the expression

$$\sum_{j=1}^{M_R} \sum_{i=1}^{M_T} \lambda_i |\beta_{j,i}|^2 \tag{5.22}$$

approaches a Gaussian random variable D with the mean value

$$\mu_D = M_R \sum_{i=1}^{M_T} \lambda_i$$

and the variance

$$\sigma^2_D = M_R \sum_{i=1}^{M_T} \lambda_i^2$$

Thus the unconditional pairwise error probability can be upper-bounded by

$$P(s \rightarrow e) \leq \int_{D=0}^{+\infty} \frac{1}{2} \exp\left(-\frac{E_s}{4N_0} D\right) p(D) \, dD \tag{5.23}$$

Then we arrive at

$$P(s \rightarrow e) \leq \frac{1}{2} \exp\left(\frac{1}{2}\left(\frac{E_s}{4N_0}\right)^2 \sigma_D^2 - \frac{E_s}{4N_0}\mu_D\right) Q\left(\frac{\frac{E_s}{4N_0}\sigma_D^2 - \mu_D}{\sigma_D}\right)$$

By using inequality $Q(x) \leq \frac{1}{2}e^{-x^2/2}$, $x \geq 0$
We further get

$$P(s \rightarrow e) \leq \frac{1}{4} \exp\left(-M_R \frac{E_s}{4N_0} \sum_{i=1}^{M_T} \lambda_i\right) \qquad (5.24)$$

To minimize the PEP, it follows from (5.24) that the sum of the eigenvalues of matrix $A(s, e)$ where $A(s, e) = B(s, e)B^*(s, e)$ should be maximized. For a square matrix the sum of all the eigenvalues is equal to the trace of the matrix, denoted by $tr(\nu)$. It can be written as

$$tr(\nu) = \sum_{i=1}^{M_T} \lambda_i = \sum_{i=1}^{M_T} A_{ii} \qquad (5.25)$$

where $A_{i,i}$ are the elements on the main diagonal of matrix $A(s, e)$. The trace of matrix $A(s, e)$ can be expressed as

$$tr(\nu) = \sum_{i=1}^{M_T} \sum_{j=1}^{L} \left|e_j^i - s_j^i\right|^2 \qquad (5.26)$$

As (5.26) shows, the trace of $A(s, e)$ is equivalent to the Euclidian distance between code words s and e over all the transmit antennas. In other words, the pairwise error probability is minimized if the Euclidian distance is maximized. This is consistent with the conclusions on the convergence of a fading channel to an AWGN channel for a large number of diversity branches and with the design criteria for trellis coded modulation on fading channels in [8, 9]. Note that a larger rM_R value provides faster convergence to an ideal AWGN channel.

Summarizing, if rM_R is sufficiently large (>3), the performance of STTC codes is dominated by the minimum trace of $A(s, e)$ taken over all pairs of distinct code words s and e. Thus the coding gain is maximized if the minimum trace of $A(s, e)$ over all code word pairs is maximized. We refer to this design rule as the *trace criterion*.

It is important to note that the condition of the trace criterion, $rM_R > 3$, is met for most of the combinations of the transmit and receive antenna numbers. In most cases, the condition $rM_R > 3$ can translate into $M_T M_R > 3$. When $M_T = 2$, it is important for $A(s, e)$ to have full rank of $r = 2$. When $M_T \geq 4$, the full rank requirement is not necessary to minimize the PEP. This is because, in such cases, the largest minimum trace dominates the performance of the code. We will confirm this through simulations later in this chapter.

When rM_R is small (<4), (5.22) no longer approaches a Gaussian distribution. In this case one can follow the rank and determinant criteria (TSC criteria). It is noted that Vucetic had determined the boundary value to be 4. In (5.22) this boundary is determined by the required number of random variables rM_R to satisfy the central limit theorem. This is based on the argument that for random variables with smooth PDFs, the central limit theorem can be applied if the number of random variables in the sum is larger than 4 [8]. Simulations have justified this, as it was found that as long as $rM_R \geq 4$, the best codes based on the trace criterion outperform the best codes based on the TSC criteria.

We illustrate this observation through an example [2]. Consider the codes shown in Table 5.2.

Consider three QPSK codes as shown in Table 5.2. These are 4-state codes with two transmit antennas. The three codes are denoted by A, B, and C, respectively. These codes have the same bandwidth efficiency of 2 bit/s/Hz. The minimum rank, determinant, and trace of the codes are also listed. It is shown that codes A and B have full rank and the same determinant of 4, whereas code C is not of full rank and, therefore, its determinant is 0. On the other hand, the minimum trace for codes B and C is 10, whereas code A has a smaller minimum trace of 4. The performance of the codes with various numbers of receive antennas on slow Rayleigh fading channels is evaluated by simulation. The frame length was 130 symbols. The frame error rate (FER) performance versus the SNR per receive antenna (e.g., $SNR = 2 \times E_s/N_0$) is shown in Figure 5.2.

From Figure 5.2, it can be observed that codes A and B outperform code C if one receive antenna is employed. This is explained as follows. When the number of independent subchannels $M_T M_R$ is small, the minimum rank of the code dominates the code performance. Since both code A and code B are of full rank ($r = 2$) and code C is not ($r = 1$), codes A and B achieve a better performance relative to code C. It can also be seen that the performance curves for codes A and B have an asymptotic slope of -2, whereas the slope for code C is -1, consistent with the diversity order of 2 for codes A and B and 1 for code C. At a FER of 10^{-2}, codes A and B outperform code C by about 5 dB due to a larger diversity order. It clearly indicates that the minimum rank is much more important in determining the code performance for systems with a small number of independent subchannels.

However, when the number of receive antennas is four, code C performs better than code A, as shown in Figure 5.2, which means the code with full rank is worse than the code with a smaller rank. This occurs as the diversity gain rM_R in this case is 8 and 4 for codes A and C, respectively. According to the trace criterion, code C is superior to code A due to a larger minimum trace value. At a FER of 10^{-2}, the advantage of code C relative to code A is about 1.3 dB.

Table 5.2 Parameters of the Codes With Two Transmit Antennas

	(a_0^1, a_0^2)	(a_1^1, a_1^2)	(b_0^1, b_0^2)	(b_1^1, b_1^2)	det	tr	Rank
Code A	(0, 2)	(2, 0)	(0, 1)	(1, 0)	4.0	4.0	2
Code B	(0, 2)	(1, 2)	(2, 3)	(2, 0)	4.0	10.0	2
Code C	(0, 2)	(2, 2)	(2, 3)	(0, 2)	0	10.0	1

From: [2]. © 2001 IEEE.

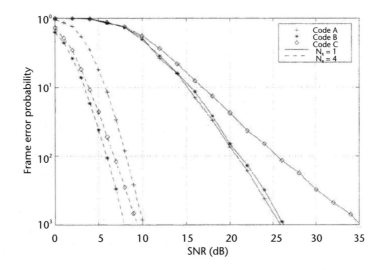

Figure 5.2   FER performance of the 4-state space-time coded QPSK with two transmit antennas. The solid line indicates one receive antenna. The dash indicates four receive antennas. (*From:* [10]. © 2003 John Wiley & Sons Ltd. Reproduced with permission.)

From Figure 5.2, we can also see that code B is about 0.8 dB better than code C at the FER of 10^{-2}, although they have the same minimum trace. This is due to the fact that code B has the same minimum trace and a larger rank than code C. Therefore, code B can achieve a larger diversity, which is manifested by a steeper error rate slope for code B than for code C.

This example has clearly verified the code design criteria for slow fading channels. We include a flow chart in Figure 5.3 as an *aide-memoire*. Readers are advised to memorize this.

From the preceding discussion, we know that the value of rM_R determines whether the TSC criteria or the trace criterion should be used. When $rM_R < 4$, the TSC criteria are applicable and when $rM_R > 3$, the trace criterion is applicable. However, in the code design, the number of receive antennas is not considered a design parameter. Considering the relationship between the maximum achievable rank, the number of transmit antennas, and the memory order of an STTC as shown in Table 5.1, we can visualize the cases in which each criteria set is applicable in the code design. The boundary between rank and determinant criteria and the trace criterion is illustrated in Figure 5.4. [2]. The points in the rectangular blocks are the cases where rank and determinant criteria are to be employed. The trace criterion can be used for all other cases. The figure suggests that the rank and determinant criteria only apply to the systems with one receive antenna.

5.4.3   Encoding/Decoding of STTCs for Quasi-Static Flat Fading Channels

The encoding for STTCs is similar to trellis coded modulation except that at the beginning and end of each frame, the encoder is required to be in the zero state. At each time t, depending on the state of the encoder and the input bits, a transition branch is selected. If the label of the transition branch is $s_t^1, s_t^2, \ldots,$ $s_t^{M_T}$, then transmit antenna i is used to send the constellation symbols

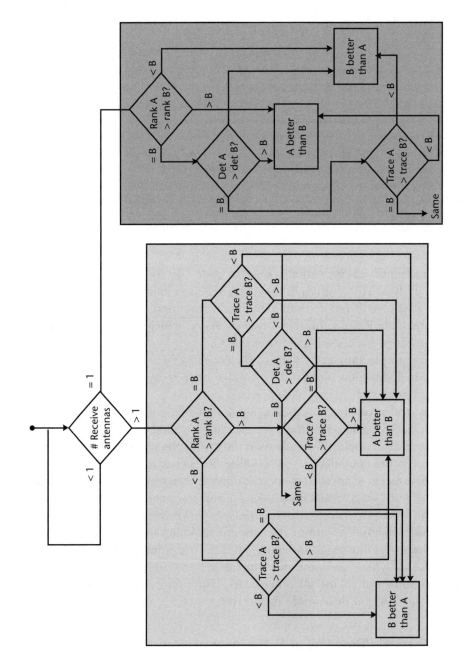

Figure 5.3   Flow chart: Is code A better than code B?

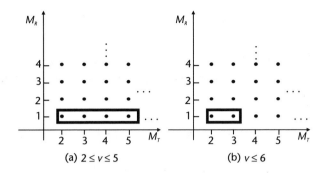

Figure 5.4   The boundary for applicability of the TSC and the trace criteria. (*From:* [2]. © 2001, IEEE.)

s_t^i, $i = 1, 2, \ldots, M_T$ and all these transmissions are in parallel. An example of an STTC encoder is shown in Figure 5.5 for 4-PSK. The encoder coefficient set, denoted by

$$g^i = \left[\left(g_{0,1}^i, g_{0,2}^i, \ldots, g_{0,M_T}^i\right), \left(g_{1,1}^i, g_{1,2}^i, \ldots, g_{1,M_T}^i\right), \ldots, \left(g_{v_i,1}^i, g_{v_i,2}^i, g_{v_i,M_T}^i\right)\right]$$

$$(5.27)$$

is usually found next to the trellis diagram of the trellis code. Each $g_{L,k}^i$ is an element of the 4-PSK constellation set $\{0, 1, 2, 3\}$ and v_i is the memory order of the ith shift register. Multiplier outputs are added modulo 4.

The encoder in Figure 5.5 can be described in generator polynomial format. The input binary sequence to the upper shift register can be represented as

$$u^1(D) = u_0^1 + u_1^1 D + u_2^1 D^2 + u_3^1 D^3 + \ldots \qquad (5.28)$$

Similarly, the binary input sequence to the lower shift register can be written as

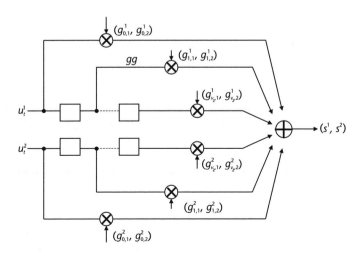

Figure 5.5   Space-time trellis code encoder for 4-PSK.

$$u^2(D) = u_0^2 + u_1^2 D + u_2^2 D^2 + u_3^2 D^3 + \ldots \qquad (5.29)$$

where u_j^k, $j = 0, 1, 2, 3, \ldots$, $k = 1, 2$ are binary symbols 0, 1. The feed-forward generator polynomial for the upper encoder and transmit antenna i, where $i = 1, 2$ can be written as

$$G_i^1(D) = g_{0,i}^1 + g_{1,i}^1 D + \ldots + g_{\nu_1,i}^1 D^{\nu_1} \qquad (5.30)$$

where $g_{j,i}^1$, $j = 0, 1, \ldots, \nu_1$ are nonbinary coefficients that can take values from a constellation such as 4-PSK as $1, -j, -1, j$ and ν_1 is the memory order of the upper encoder. Similarly, the feed-forward generator polynomial for the lower encoder and transmit antenna i, where $i = 1, 2$ can be written as

$$G_i^2(D) = g_{0,i}^2 + g_{1,i}^2 D + \ldots + g_{\nu_2,i}^2 D^{\nu_2} \qquad (5.31)$$

where $g_{j,i}^1$, $j = 0, 1, \ldots, \nu_2$ are nonbinary coefficients that can take values from a constellation such as 4-PSK as $1, -j, -1, j$ and ν_2 is the memory order of the lower encoder. The encoded symbol sequence transmitted from antenna i is given by

$$s^i(D) = u^1(D) G_i^1(D) + u^2 G_i^2(D) \bmod 4 \qquad (5.32)$$

We can also express this as

$$s^i(D) = [u^1(D) \quad u^2(D)] \begin{bmatrix} G_i^1(D) \\ G_i^2(D) \end{bmatrix} \bmod 4 \qquad (5.33)$$

We shall use this form later in this chapter.

Assuming that r_t^j is the received signal at antenna j at time t, the branch metric is given by

$$\sum_{j=1}^{M_R} \left| r_t^j - \sum_{i=1}^{M_T} h_{i,j} q_t^i \right|$$

The Viterbi algorithm is used to compute the path with the lowest branch metric. In the absence of ideal CSI, we estimate the CSI based on training symbols.

5.4.4 Code Construction for Quasi-Static Flat Fading Channels

Note that:

- If a space-time trellis code guarantees a diversity advantage of r for the quasi-static flat fading channel model (given one receive antenna), then it is called an r-STTC.

- The constraint length of an r-STTC is at least $r - 1$.
- If the diversity advantage is $M_T M_R$, then the transmission rate is at most b bit/s/Hz with a 2^b signal constellation. Thus 4-PSK, 8-PSK, and 16-QAM will be upper-bounded by 2, 3, 4 bit/s/Hz, respectively.
- If b is the transmission rate, the trellis complexity is at least $2^{b(r-1)}$.
- An STTC is shown to be geometrically uniform [11] and its performance is independent of the transmitted code word.

5.4.5 Example Using 4-PSK

STTCs are an extension of conventional trellis codes [12] to multiantenna systems. These codes are handcrafted to extract diversity gain and coding gain using the criteria described in Section 5.4.4. Each STTC can be described using a trellis. The number of nodes in a trellis diagram corresponds to the number of states in the trellis. Figure 5.6 shows the trellis diagram for a simple 4-PSK, four-state trellis code for $M_T = 2$ with rate 2 bit/s/Hz. The trellis has four nodes corresponding to four states. There are four groups of symbols at the left of every node since there are four possible inputs (4-PSK constellation). Each group has two entries corresponding to the symbols to be output through the two transmit antennas. At the top of the diagram we have the binary input bits that drive these symbols, which are output from the transmit antennas. These symbols come in pairs (for a two-antenna transmitter), wherein the first digit corresponds to the symbol transmitted from antenna 1 and the second from antenna 2. The encoder is required to be in the zero state at the beginning and at the end of each frame (block). Beginning at state 0, if the incoming two bits are 10, the encoder outputs a 0 on antenna 1 and a 2 on antenna 2 and changes to state 2. Thereafter, it waits at state 2 for the next number. If the next incoming two bits are 01, the encoder outputs a 2 on antenna 1 and a 1 on antenna 2 and changes to state 1 and so on.

Input bits	00	01	10	11		State #
State 0 Output for antenna 1, antenna 2	00	01	02	03		0
State 1 Output for antenna 1, antenna 2	10	11	12	13		1
State 2 Output for antenna 1, antenna 2	20	21	22	23		2
State 3 Output for antenna 1, antenna 2	30	31	32	33		3

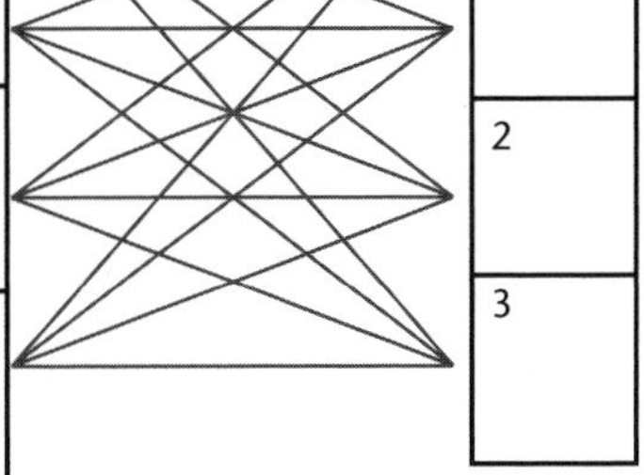

Figure 5.6 Example of 2 transmit space-time trellis code with 4 states (4-PSK constellation with spectral efficiency of 2 bit/s/Hz).

Mathematically, if (b_t, a_t) are the input binary sequence, the output signal pair $s_1^t s_2^t$ at time t is given by

$$\left(s_1^t, s_2^t\right) = b_{t-1}(2, 0) + a_{t-1}(1, 0) + b_t(0, 2) + a_t(0, 1) \bmod 4 \qquad (5.34)$$

$$= ((2b_{t-1} + a_{t-1}) \bmod 4, (2b_t + a_t) \bmod 4)$$

For the diversity advantage to be 2 (to qualify as a 2-space-time code), the rank of $B(s, e)$ has to be 2. This can be seen from (5.34), since if the paths corresponding to code words s and e diverge at time t_1 and remerge at time t_2, then the vectors $\left(e_1^{t_1} - s_1^{t_1}, e_2^{t_1} - s_2^{t_1}\right)$ and $\left(e_1^{t_2} - s_1^{t_2}, e_2^{t_2} - s_2^{t_2}\right)$ are linearly dependent on each other and with $e_1^{t_1} - s_1^{t_1} = e_2^{t_1} - s_2^{t_1} = 0$ and $e_1^{t_2} - s_1^{t_2} \neq 0$ and $e_2^{t_2} - s_2^{t_2} \neq 0$.

To compute the coding advantage, we need to find code words s and e such that the determinant

$$\det\left(\sum_{t=1}^{L} \left(e_t^1 - s_t^1, e_t^2 - s_t^2\right)^\dagger \left(e_t^1 - s_t^1, e_t^2 - s_t^2\right)\right) \qquad (5.35)$$

is minimized. Using the theorem that the code is geometrically uniform, we can assume that we start with the all-zeros code word.

We can express the edge labels $(s_1 s_2)$ by the complex equivalent $\left(j^{s_1}, j^{s_2}\right)$ where $j = \sqrt{-1}$. By substituting $\left(j^{s_1}, j^{s_2}\right)$ into (5.27), we obtain

$$\det\left(\sum_{t=1}^{L} \left(j^{s_1} - 1, j^{s_2} - 1\right)^* \left(j^{s_1} - 1, j^{s_2} - 1\right)\right) \qquad (5.36)$$

since a zero code word maps to $j^0 \leftrightarrow 1$. Then

$$\det\left(\sum_{t=1}^{L} \binom{j^{-s_1} - 1}{j^{-s_2} - 1}\left(j^{s_1} - 1, j^{s_2} - 1\right)\right) \qquad (5.37)$$

and, therefore, the inner product takes the form

$$\begin{pmatrix} \left(j^{-s_1} - 1\right)\left(j^{s_1} - 1\right) & \left(j^{-s_1} - 1\right)\left(j^{s_2} - 1\right) \\ \left(j^{-s_2} - 1\right)\left(j^{s_1} - 1\right) & \left(j^{-s_2} - 1\right)\left(j^{s_2} - 1\right) \end{pmatrix} \qquad (5.38)$$

If we transpose this matrix (it does not affect the determinant value), we obtain,

$$\begin{pmatrix} \left(j^{-s_1} - 1\right)\left(j^{s_1} - 1\right) & \left(j^{-s_2} - 1\right)\left(j^{s_1} - 1\right) \\ \left(j^{-s_1} - 1\right)\left(j^{s_2} - 1\right) & \left(j^{-s_2} - 1\right)\left(j^{s_2} - 1\right) \end{pmatrix} \qquad (5.39)$$

Figure 5.7 shows the corresponding labeling. The reader is advised to refer to Figure 5.6. In that example, we dealt with the case of a transition form state 0 to state 2, given that the input bits were 10. In such a case, $s_1 = 0$ and $s_2 = 2$. Substituting these values into (5.39), we obtain $\begin{bmatrix} 0 & 0 \\ 0 & 4 \end{bmatrix}$. This is clearly seen in the state diagram in Figure 5.7 when we follow the arrow from state 0 (00) to state 2 (10) at the diagonally opposite end of the figure. Hence, diverging from the zero state contributes a matrix of the form

$$\begin{bmatrix} 0 & 0 \\ 0 & t \end{bmatrix}$$

and remerging to the zero state contributes a matrix of the form

$$\begin{bmatrix} s & 0 \\ 0 & 0 \end{bmatrix}$$

with $s, t \geq 2$. This fact can be seen when we refer to Figure 5.6. Transitioning from state 2 to state 0 requires that $s_1 = 2$ and $s_2 = 0$, when the input bits are 00. This yields a matrix [after substituting in (5.39)], $\begin{bmatrix} 4 & 0 \\ 0 & 0 \end{bmatrix}$.

Similarly, suppose we are in state 1 and the binary input number is 11. In such a case, $s_1 = 1$ and $s_2 = 1$. This yields a matrix $\begin{bmatrix} 2 & 2 \\ 2 & 2 \end{bmatrix}$, and the state remains at state 1. This is the loop at the bottom left-hand corner of the diagram. Proceeding

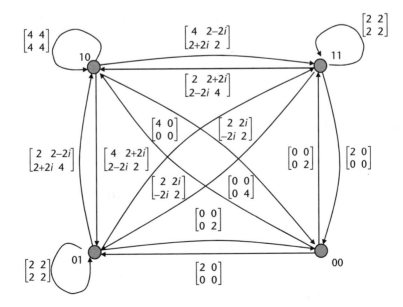

Figure 5.7   State diagram for 4-PSK example. (*From:* [1]. © 1998, IEEE.)

along similar lines, the reader can verify the remaining matrixes. Thus (5.35) can be written as

$$\det\left(\begin{pmatrix} s & 0 \\ 0 & t \end{pmatrix} + \begin{pmatrix} a & b \\ \bar{b} & d \end{pmatrix}\right)$$

with a, $d \geq 0$, $|b|^2 \leq ad$. So the minimum determinant is 4 (see Section 5.4.2). Given that the diversity is rM_R, we wish to maximize the minimum determinant. We have achieved this goal by ensuring that this minimum determinant is 4. The rank of the matrix B is full rank (i.e., $r = 2$). If we require full diversity of $M_T M_R$, then it is important that this rank criterion is satisfied. In such a case, for this example, the minimum determinant will be 4. If it were less, then the code is useless. Recall that the value of the minimum determinant defines the coding gain. The higher this minimum determinant, the more the coding gain. We should, therefore, strive to make this value as high as possible. It does not, however, provide an accurate estimate of the *true* coding gain (i.e., there is no direct relationship for us to actually be able to predict the realizable coding gain).

The *design rules* that guarantee diversity for the 4-PSK and 8-PSK code are:

- Transitions departing from the same state differ in the second symbol.
- Transitions arriving at the same state differ in the first symbol.

The reader can verify these rules with reference to our analysis, wherein transiting from state 0 to state 2 required a code of 02 and from state 2 to state 0 required a code of 20 (see Figure 5.7), yielding matrixes $\begin{bmatrix} 0 & 0 \\ 0 & 4 \end{bmatrix}$ and $\begin{bmatrix} 4 & 0 \\ 0 & 0 \end{bmatrix}$, respectively.

Using (5.33), (5.34) can also be expressed as

$$s^1 s^2 = [b_t \quad a_t] \begin{bmatrix} g^1 \\ g^2 \end{bmatrix}$$

$$= [b_t \quad a_t] \begin{bmatrix} (02), (20) \\ (01), (10) \end{bmatrix}$$

where g^1 and g^2 are generator polynomials.

The second column of this matrix is the $t - 1$ state. If we expand the matrix, we obtain (5.34) as

$$\left(s_1^t, s_2^t\right) = b_{t-1}(2, 0) + a_{t-1}(1, 0) + b_t(0, 2) + a_t(0, 1) \bmod 4$$

This method of representing codes as generator polynomials is useful and compact and is used extensively.

Before concluding this section, we show some useful generator polynomials using the rank and determinant criteria, as well as trace criterion as appropriate (see Tables 5.3 and 5.4). The reader is referred to [10] for a more comprehensive listing.

Table 5.3   Generator Sequences for Varying Number of Transmit Antennas Based on Rank and Determination Criteria

Modulation	v	Number of Transmit Antennas	Generator Sequences	Rank (r)	det	tr
QPSK	2	2	g^1 [(0, 2), (2, 0)] g^2 [(0, 1), (1, 0)]	2	4.0	
QPSK	4	2	g^1 [(0, 2), (2, 0), (0, 2)] g^2 [(0, 1), (1, 2), (2, 0)]	2	12.0	
QPSK	4	3	g^1 [(0, 0, 2), (0, 1, 2), (2, 3, 1)] g^2 [(2, 0, 0), (1, 2, 0), (2, 3, 3)]	3	32	16
8PSK	3	2	g^1 [(0, 4), (4, 0)] g^2 [(0, 2), (2, 0)] g^3 [(0, 1), (5, 0)]	2	2	4
8PSK	4	2	g^1 [(0, 4), (4, 4)] g^2 [(0, 2), (2, 2)] g^3 [(0, 1), (5, 1), (1, 5)]	2	3.515	6

From: [10]. © 2003 John Wiley & Sons, Ltd. Reproduced with permission.

Table 5.4   Generator Sequences for Varying Number of Transmit Antennas Based on Trace Criterion

Modulation	v	Number of Transmit Antennas	Generator Sequences	Rank (r)	det	tr
QPSK	2	2	g^1 [(0, 2), (1, 2)] g^2 [(2, 3), (2, 0)]	2	4.0	10.0
QPSK	4	2	g^1 [(1, 2), (1, 3), (3, 2)] g^2 [(2, 0), (2, 2), (2, 0)]	2	8.0	16.0
QPSK	2	4	g^1 [(0, 2, 2, 0), (1, 2, 3, 2)] g^2 [(2, 3, 3, 2), (2, 0, 2, 1)]	2	—	20.0
8PSK	4	2	g^1 [(2, 4), (3, 7)] g^2 [(4, 0), (6, 6)] g^3 [(7, 2), (0, 7), (4, 4)]	2	0.686	8.0
8PSK	4	4	g^1 [(2, 4, 2, 2), (3, 7, 2, 4)] g^2 [(4, 0, 4, 4), (6, 6, 4, 0)] g^3 [(7, 2, 2, 0), (0, 7, 6, 3), (4, 4, 0, 2)]	2	—	20.0

From: [10]. © 2003 John Wiley & Sons, Ltd. Reproduced with permission.

The reader is encouraged to try experimenting with these codes using the accompanying software.

Figure 5.8 shows a 4-PSK 8 state trellis code.

The analysis of this trellis diagram is similar.

5.5   Design of Space-Time Trellis Codes on Fast Fading Channels

5.5.1   Error Probability on Fast Fading Channels

The analysis for slow fading channels in the previous sections can be directly applied to fast fading channels. At each time t, we define a *space-time symbol difference vector*, $F(s, e)$ as

Space-time code, 4PSK, 8 states, 2 bit/s/Hz

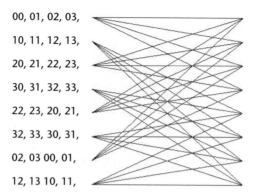

00, 01, 02, 03,

10, 11, 12, 13,

20, 21, 22, 23,

30, 31, 32, 33,

22, 23, 20, 21,

32, 33, 30, 31,

02, 03 00, 01,

12, 13 10, 11,

Figure 5.8   Example of 2 transmit space-time trellis code with 8 states (4-PSK constellation with spectral efficiency of 2 bit/s/Hz).

$$F(s, e) = \left[s_t^1 - e_t^1, s_t^2 - e_t^2, \ldots, s_t^{M_T} - e_t^{M_T} \right]^T \qquad (5.40)$$

Following the derivation in [1], we consider a $M_T \times M_T$ matrix $S = S(s, e)$, defined as $S = F(s, e)F^{\dagger}(s, e)$. It is clear that S is Hermitian, so there exists a unitary matrix V_t and a real diagonal matrix D_t such that

$$V_t S V_t^{\dagger} = D_t$$

The diagonal entries of D_t, $\{D_t^i, i = 1, \ldots, M_T\}$ are the eigenvalues of S; the rows of V_t, $\{v_t^i, i = 1, \ldots, M_T\}$ are the eigenvectors of S, which form a complete orthonormal basis of an M_T-dimensional vector space.

Note that S is a rank 1 matrix (since we are dealing on a symbol basis) if $s \neq e$ and is rank 0 otherwise. It follows that $M_T - 1$ elements in the list $\{D_t^i, i = 1, \ldots, M_T\}$ are zeros and, consequently, we can let the single nonzero element in this list be D_t^1, which is equal to the squared Euclidian distance between the two space-time symbols s_t and e_t.

$$D_t^1 = |s_t - e_t|^2 = \sum_{i=1}^{M_T} |s_t^i - e_t^i|^2 \qquad (5.41)$$

The eigenvector of $S(s_t, e_t)$ corresponding to the nonzero eigenvalue D_t^1 is denoted by v_t^i.

We define h_t^j as

$$h_t^j = \left(h_{j,1}^t, h_{j,2}^t, \ldots, h_{j,M_T}^t \right) \qquad (5.42)$$

Now,

$$d^2(s, e) = \left\| H \cdot (E - S) \right\|^2$$

$$= \sum_{t=1}^{L} \sum_{j=1}^{M_R} \left| \sum_{i=1}^{M_T} h_{j,i}^t \left(e_t^i - s_t^i \right) \right|^2$$

This can be rewritten as

$$d^2(s, e) = \sum_{t=1}^{L} \sum_{j=1}^{M_R} \sum_{i=1}^{M_T} \left| \beta_{j,i}^t \right|^2 \cdot D_t^i \tag{5.43}$$

where $\beta_{j,i}^t = h_t^j \cdot v_t^j$.

Since at each time t there is, at most, only one nonzero eigenvalue, D_t^1, the expression (5.43) can be represented by

$$d^2(s, e) = \sum_{t \in \rho(s,e)} \sum_{j=1}^{M_R} \left| \beta_{j,i}^t \right|^2 \cdot D_t^1 \tag{5.44}$$

$$= \sum_{t \in \rho(s,e)} \sum_{j=1}^{M_R} \left| \beta_{j,i}^t \right|^2 \cdot \left| s_t - e_t \right|^2$$

where $\rho(s, e)$ denotes the set of time instances $t = 1, 2, \ldots, L$ such that $|s_t - e_t| \neq 0$. Substituting (5.44) into (5.19), we obtain

$$P\left(s \to E | H\right) \leq \frac{1}{2} \exp\left(- \sum_{t \in \rho(s,e)} \sum_{j=1}^{M_R} \left| \beta_{j,1}^t \right|^2 \cdot \left| s_t - e_t \right|^2 \frac{E_s}{4N_0} \right) \tag{5.45}$$

Since $h_{i,j}$ are samples of a complex Gaussian random variable with mean $Eh_{i,j}$ let

$$K^j = (Eh_{1,j}, Eh_{2,j}, \ldots, Eh_{M_T,j})$$

Since V is unitary, this implies $\beta_{i,j}$ are independent complex Gaussian random variables with variance 0.5 per dimension and with mean $K^j \cdot v_i$.

If we define δ_H as the number of space-time symbols in which the two code words s and e differ, then at the right-hand side of inequality (5.45), there are $\delta_H M_R$ independent random variables.

Once again, like in the slow fading cases, we discuss two situations, such as when $\delta_H M_R < 4$ and $\delta_H M_R \geq 4$.

Case When $\delta_H M_R \geq 4$

According to the central limit theorem [7], the expression $d^2(s, e)$ in (5.43) can be approximated by a Gaussian random variable with the mean

$$\mu_d = \sum_{t \in \rho(s,e)} \sum_{j=1}^{M_R} |s_t - e_t|^2 \tag{5.46}$$

and the variance

$$\sigma_d^2 = \sum_{t \in \rho(s,e)} \sum_{j=1}^{M_R} |s_t - e_t|^4 \tag{5.47}$$

By averaging (5.45) over the Gaussian random variable and using

$$\int_{D=0}^{\infty} \exp(-\gamma D) p(D) \, dD = \exp\left(\frac{1}{2} \gamma^2 \sigma_D^2 - \gamma \mu_D\right) Q\left(\frac{\gamma \sigma_D^2 - \mu_D}{\sigma_D}\right), \; \gamma > 0$$

we obtain the PEP as

$$P(s \to e) \le \frac{1}{2} \exp\left(\frac{1}{2}\left(\frac{E_s}{4N_0}\right)^2 \sigma_D^2 - \frac{E_s}{4N_0}\mu_d\right) Q\left(\frac{E_s}{4N_0}\sqrt{M_R D^4} - \frac{\sqrt{M_R d_E^2}}{\sqrt{D^4}}\right) \tag{5.48}$$

where d_E^2 is the accumulated squared Euclidian distance between the two space-time symbol sequences, given by

$$d_E^2 = \sum_{t \in \rho(s,e)} |s_t - e_t|^2 \tag{5.49}$$

and D^4 defined as

$$D^4 = \sum_{t \in \rho(s,e)} |s_t - e_t|^4 \tag{5.50}$$

Case When $\delta_H M_R < 4$
When the value of $\delta_H M_R < 4$, the central limit theorem argument is no longer valid and the average error probability can be expressed as

$$P(s \to e) \le \int \cdots \int_{|\beta_{j,1}^t|=0}^{\infty} P(s \to E|H) p\left(\left|\beta_{1,1}^1\right|\right) p\left(\left|\beta_{2,1}^1\right|\right), \ldots, p\left(\left|\beta_{M_R,1}^L\right|\right) \tag{5.51}$$

$$\cdot \, d\left|\beta_{1,1}^1\right| d\left|\beta_{2,1}^1\right|, \ldots, d\left|\beta_{M_R,1}^L\right|$$

where $\left| \beta_{j,1}^{t} \right|$, $t = 1, 2, \ldots, L$ and $j = 1, 2, \ldots, M_R$ are independent Rayleigh distributed random variables. By integrating (5.51) term by term, the PEP becomes [1]

$$P(s \to e) \leq \prod_{t \in \rho(s,e)} \left(\frac{1}{1 + |s_t - e_t|^2 \frac{E_s}{4N_0}} \right)^{M_R} \tag{5.52}$$

$$\leq \left(d_p^2 \right)^{-M_R} \left(\frac{E_s}{4N_0} \right)^{-\delta_H M_R}$$

where d_p^2 is the product of the squared Euclidian distances between the two space-time symbol sequences, given by

$$d_p^2 = \prod_{t \in \rho(s,e)} |s_t - e_t|^2 \tag{5.53}$$

At high SNRs, the frame error probability is dominated by the PEP with the minimum product $\delta_H M_R$. The exponent of the SNR term, $\delta_H M_R$, is called the *diversity gain* for fast Rayleigh fading channels and

$$G_s = \frac{\left(d_p^2 \right)^{1/\delta_H}}{d_u^2} \tag{5.54}$$

is called the *coding gain* for fast Rayleigh fading channels, where d_u^2 is the squared Euclidian distance of the reference uncoded system. Note that both diversity and coding gains are obtained as the minimum $\delta_H M_R$ and $\left(d_p^2 \right)^{1/\delta_H}$ over all pairs of distinct code words because this becomes the worst case.

We now formally define the design criteria for fast Rayleigh fading channels [1].

- Case when $\delta_H M_R < 4$
  This is applicable if $\delta_H M_R$ is small. The design criteria can be summarized as:
  - Maximize the minimum space-time symbol-wise Hamming distance δ_H between all pairs of distinct code words.
  - Maximize the minimum product distance d_p^2, along the path with the minimum symbol-wise Hamming distance δ_H.
- Case when $\delta_H M_R \geq 4$
  This is applicable for large values of $\delta_H M_R$. In such cases the pairwise error probability is upper-bounded by (5.48). In the case of high SNRs,

$$\frac{E_s}{4N_0} \geq \frac{d_E^2}{D^4}$$

where d_E^2 and D^4 are given by (5.49) and (5.50), respectively. By using the inequality $Q(x) \leq \frac{1}{2} e^{-x^2/2}$, $x \geq 0$, (5.48) can be approximated by

$$P(s \to e) \leq \exp\left(-M_R \frac{E_s}{4N_0} \sum_{t=1}^{L} \sum_{i=1}^{M_T} \left|s_t^i - e_t^i\right|^2\right) \qquad (5.55)$$

$$= \exp\left(-M_R \frac{E_s}{4N_0} d_E^2\right)$$

On inspection of (5.55) it becomes apparent, that the frame error rate probability at high SNRs is dominated by the PEP with the minimum squared Euclidian distance d_E^2. To minimize the PEP on fading channels, the codes should satisfy the following criteria [5]:

- Make sure that the product of the minimum space-time symbol-wise Hamming distance and the number of receive antennas, $\delta_H M_R$, is large enough (larger than or equal to 4).
- Maximize the minimum Euclidian distance among all pairs of distinct code words.

This view is similar to the requirement for trace criterion in slow fading channels.

5.6   Performance Analysis in a Slow Fading Channel

The performance of the STTC on slow fading channels is evaluated through simulations.

These curves have been obtained using the accompanying software. In these simulations for both slow and fast fading channels, the rank and determinant criteria have been employed in formulating codes if the number of receive antennas is one and trace criteria for all other cases. Figure 5.9 shows that all the codes achieve the same diversity order of 2, demonstrated by the same slope of the FER performance curves. The code performance is improved by increasing the number of states. Figure 5.10 shows the relative performance of 4-PSK 4-state and 8-state codes with $M_T = M_R = 2$ and $M_T = 4$ and $M_R = 2$. Figure 5.10 shows that increasing the number of transmit antennas also increases the *margin* of the coding gain compared with the coding gains in the top- half graph. This is evident from (5.18), wherein the value of rM_R (ideally $M_T M_R$), defines the amount of coding gain. Similarly, if we keep M_T constant and increase M_R we achieve the same result with something more, in that part of this margin is also due to the array gain through multiple receive antennas. Furthermore, the diversity order realized with this scheme in the lower half is twice that in the upper half. Proceeding logically, as the number of receiver antennas increases, the diversity order increases proportionately and the channel tends to AWGN due to the increased diversity. A similar effect can be observed by keeping the number of receive antennas constant and increasing the number of transmit antennas. Finally, in the presence of a large

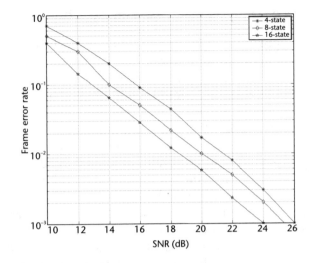

Figure 5.9 Performance comparison of 4-PSK codes based on the rank and determinant criteria on slow fading channels with two transmit and one receive antennas.

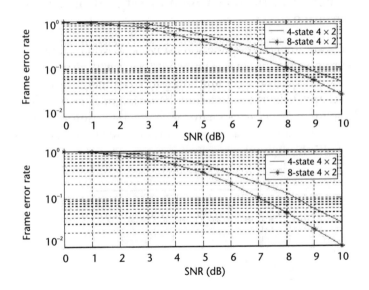

Figure 5.10 Performance comparison of 4-PSK codes based on trace criterion on slow fading channels with two transmit and two receive antennas and four transmit and two receive antennas.

number of receive antennas, increasing the number of transmit antennas does not produce that much of an increase in performance, as seen in the case when the number of receive antennas is limited and we increase the number of transmit antennas. The reader is encouraged to examine these issues using the software supplied with this book by adding coding pertaining to four transmit antenna systems. See [1] for details.

5.7 Performance Analysis in a Fast Fading Channel

The performance of the STTC on fast fading channels is evaluated through simulations. These simulations were carried out by Vucetic [5] and the curves are shown here for inspection. Figure 5.11 shows the FER performance of QPSK STTC with a bandwidth efficiency of 2 bit/s/Hz in Rayleigh channel. The number of receive antennas was one in the simulations. We can see that 16-state QPSK codes are better than 4-state codes by 5.9 dB at a FER of 10^{-2} for two transmit antennas. Once again, as the number of states increases, the coding gain increases and so does the performance. The error rate curves of the codes are parallel, as predicted by the same value of δ_H. Different values of d_p^2 yield different coding gains, which are represented by the horizontal shifts of the FER curves. In Figure 5.12 we examine the case of three transmit antennas. We note that 16-state QPSK codes are superior to 4-state codes by 6.8 dB at a FER of 10^{-2}. This means that the performance relative to two transmit antennas has improved. The conclusion here is that as the number of the transmit antennas gets larger, the performance gain achieved from increasing the number of states becomes larger.

5.8 The Effect of Imperfect Channel Estimation on Code Performance

We carry out imperfect estimation using the MMSE technique discussed in Chapter 4. The results are shown in Figure 5.13 for 4-PSK 4-state code with two transmit and two receive antennas and imperfect channel estimation in a slow Rayleigh fading channel. In the simulation 10 orthogonal signals in each data frame are used as pilot sequence to estimate the channel state information at the receiver. From the figure, we can see that the deterioration due to imperfect channel estimation is about 5 dB throughout.

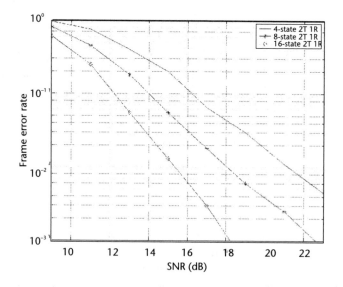

Figure 5.11 Performance of the QPSK STTC on fast fading channels with two transmit and one receive antennas. (*From:* [10]. © 2003 John Wiley & Sons, Ltd. Reproduced with permission.)

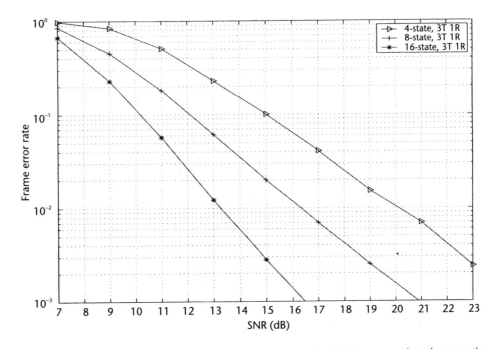

Figure 5.12   Performance of QPSK STTC on fast fading channels with three transmit and one receive antennas. (*From:* [10]. © 2003 John Wiley & Sons, Ltd. Reproduced with permission.)

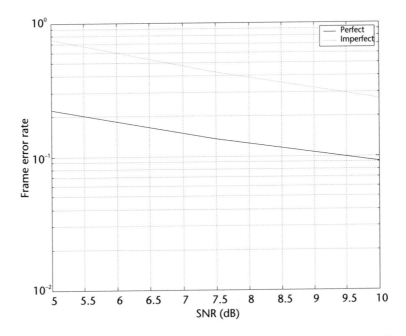

Figure 5.13   Performance of the 4-state 4-PSK code on slow Rayleigh fading channels with two transmit and two receive antennas and imperfect channel estimation.

5.9 Effect of Antenna Correlation on Performance

Figure 5.14 shows the performance of 4-PSK 4-state code. This example has been implemented using the code based on trace criterion. The performance gap is 0.5 dB throughout for both a correlation factor of 0.75 as well as unity. The reader is encouraged to try this with the accompanying software using the code provided in the simulation exercises and with reference to the generator sequences listed in Section 5.4.5.

5.10 Delay Diversity as an STTC

The delay diversity scheme discussed in Chapter 4 can be recast as an STTC. Assume a system with two transmit antennas and one receive antenna. In the delay diversity scheme, it will be recalled, we transmit one symbol from one antenna and then transmit the *same* symbol from the second antenna, but after a short delay of one symbol. Figure 5.15 shows the trellis diagram for such a delay diversity code for 8-PSK transmission over a system with two transmit antennas.

If we assume the input sequence as

$$s = (010, 101, 111, 000, 001, \dots)$$

The output sequence generated by the space-time trellis encoder is given by

$$s = (02, 25, 57, 70, 01, \dots)$$

using the technique discussed earlier. The transmitted signal sequences from the two transmit antennas are

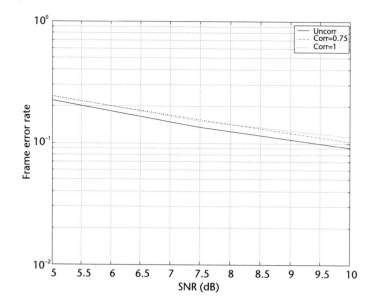

Figure 5.14 Performance of the 4-PSK 4-state code on correlated slow Rayleigh fading channels with two transmit and two receive antennas.

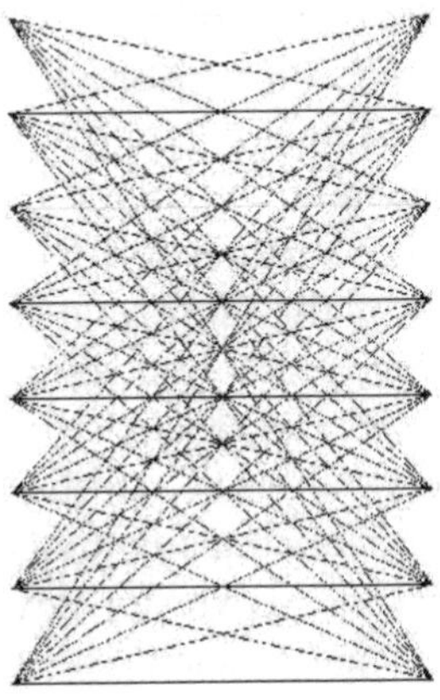

Space-time code, 8PSK, 8 states, 3 bit/s Hz

00, 01, 02, 03, 04, 05, 06, 07,

10, 11, 12, 13, 14, 15, 16, 17,

20, 21, 22, 23, 24, 25, 26, 27,

30, 31, 32, 33, 34, 35, 36, 37,

40, 41, 42, 43, 44, 45, 46, 47,

50, 51, 52, 53, 54, 55, 56, 57,

60, 61, 62, 63, 64, 65, 66, 67,

70, 71, 72, 73, 74, 75, 76, 77,

Figure 5.15 Trellis diagram for delay diversity code with 8-PSK transmission and $M_T = 2$.

$$s^1 = (0, 2, 5, 7, 0, \dots)$$

$$s^2 = (2, 5, 7, 0, 1, \dots)$$

Very clearly, this is delay diversity, since the signal sequence transmitted from the first antenna is a delayed version of the signal sequence from the second antenna. If we express s^1 and s^2 as a matrix,

$$S = \begin{bmatrix} 0 & 2 & 5 & 7 & 0 \\ 2 & 5 & 7 & 0 & 1 \end{bmatrix}$$

it is easily verified that the rank of this matrix is 2. Hence, applying the rank criterion for the space-time code word design discussed earlier, the delay diversity transmission extracts the full diversity order of $2M_R$.

5.11 Comparison of STBC and STTC

We have learned that space-time block codes and space-time trellis codes are two very different transmit diversity schemes. Space-time block codes are constructed from known orthogonal designs, achieve full diversity, and are easily decodable by maximum likelihood decoding via linear processing at the receiver, but they suffer from a lack of coding gain. On the other hand, space-time trellis coders possess both diversity and coding gain, yet are complex to decode (since we use *joint* maximum likelihood sequence estimation) and arduous to design. In both cases, however, the code design for a large number of transmit antennas remains an open question.

A pressing question is the performance comparison between STBC and STTC. In such a case, it is only fair to use concatenated STBC since STBC inherently lacks coding gain. Concatenated codes that have been used so far include AWGN Trellis codes [13] or turbo codes [14]. An attempt to provide such a comparison is found in [15], but the comparison was unfair in terms of the data rate and computational complexity. A fair comparison was given in [16], but only for the fast fading channel. Bauch concatenated turbo codes with STBC in [14] but did not provide a performance comparison with STTC. Sumeet et al. [17] later attempted a fair comparison of the performance of STBC with STTC over the flat fading quasi-static channel, presenting results in terms of the FER while keeping the transmit power and spectral efficiency constant. We shall now discuss her results.

In general, any STC can be analyzed in the same way as STTC using diversity advantage and coding advantage. Both of these advantages affect the performance curve differently. Diversity advantage causes the slope of the FER versus SNR graph to change in such a way that the larger the diversity, the more negative the slope. Coding advantage shifts the graph horizontally: the greater the coding advantage, the larger is the left shift. In [18], full diversity codes were used and, hence, the slopes of their FER graphs were identical.

To further examine the coding advantage aspect, consider a high SNR regime (typically 4 dB to 18 dB). We first take the logarithm of the PEP expression in (5.18) for the kth code. This yields $P_k = \log(PEP) \approx -M_T M_R s_k - M_T M_R c_k$, where $M_T M_R$ is the full diversity advantage, $s_k = \log\left(\dfrac{E_s}{4N_0}\right)$ is the SNR term and $c_k = \log\left(\displaystyle\prod_{t=1}^{M_T} \lambda_i^{1/M_T}\right)$ is the coding advantage term. If we let $\delta_P = P_k - P_L$, $\delta_c = c_k - c_L$ and $\delta_s = s_k - s_L$ for the kth and Lth code, then

$$\delta_P = -M_T M_R \delta_s - M_T M_R \delta_c$$

If k is a better code, then $\delta_c > 0$. At a given SNR, $\delta_s = 0$ and the PEP for k is less than that for L by $\delta_P \approx M_T M_R \delta_c$. Clearly, this difference increases with M_R, the number of receive antennas. Thus, the effect of coding advantage improves when more receive antennas were used. This aspect was earlier discussed in Section 5.4. In the simulations conducted by Sumeet [17], the performance comparison between STBC and STTC was carried out for a system with two transmit and one, two, and three receive antennas with 4-PSK modulation using STTC-Grimm [19] and STTC-Yan [20]. These STTC codes were used instead of STTC-Tarokh because they have the best possible coding advantage in the class of feed-forward convolutional (FFC) codes [19]. The block code used was Alamouti, which is a full-rate code. This was concatenated with outer AWGN trellis codes. The spectral efficiency was maintained at 2 bit/s/Hz throughout the simulation. The FER is given for 130 symbols/frame. The channel is a quasi-static Rayleigh fading channel. The results from this paper are shown in Figures 5.16 to 5.18.

In Figure 5.16 (4-state code) we note that STBC by itself performs as well or better than all the STTCs, even though it provides no coding gain. This can be explained by the multidimensional structure of STBCs, since each code word spans two time symbols and averages out the noise over time. Other interesting observations are:

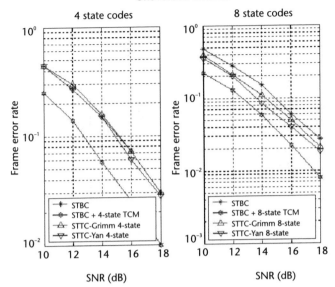

Figure 5.16   Performance of STBC, STBC + TCM, and STTC using 4 and 8 state codes with two transmit and one receive antenna. (© IEEE, 2001.)

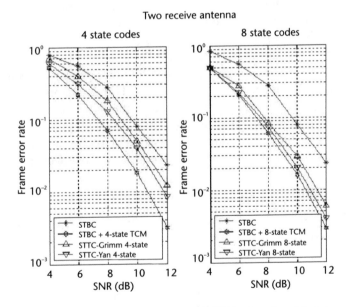

Figure 5.17   Performance of STBC, STBC + TCM, and STTC using 4 and 8 state codes with two transmit and two receive antennas. (© IEEE, 2001.)

- With the same number of trellis states, concatenated STBCs outperform STTCs at SNRs of interest (i.e., 4 to 12 dB for the case of two receivers and 10 to 18 dB for the case of one receiver).
- With increasing number of antennas and trellis states, STTC begins to outperform concatenated STBC.

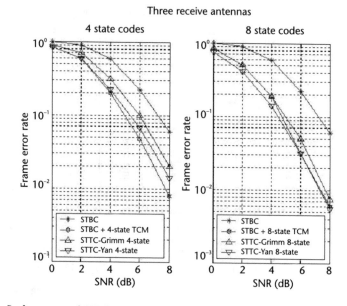

Figure 5.18 Performance of STBC, STBC + TCM, and STTC using 4 and 8 state codes with two transmit and three receive antennas. (© IEEE, 2001.)

If the number of receive antennas are one or, at most, two, a simple concatenation of STBC with traditional AWGN trellis codes can significantly outperform STTC with the same number of states. This has very important implications for design and implementation of MIMO systems.

STBC + TCM curves lose in performance with increasing receive antennas, because STBCs incur a loss in capacity [21] over channels with rank greater than one. Since TCM codes are *outer* codes, they are unable to recover performance after the signal has been encoded and decoded using space-time block codes. Examples of block codes that preserve capacity are discussed in [7].

Table 5.5 summarizes the performance of STBC versus STTC.

5.12   Simulation Exercises

1. If the number of transmit antennas is fixed and we increase the number of receive antennas, the margin of coding gain increases. Prove this statement

Table 5.5 Comparison of STBC Versus STTC

STBC	*STTC*
1. No coding gain.	Has coding gain.
2. Easily decodable by maximum likelihood decoding via linear processing.	Preserves capacity irrespective of the number of antennas.
3. Simple to design based on orthogonal sequences.	Difficult to design.
4. For one receive antenna and 4-state code, performance is similar to STTC.	STTC outperforms with increasing antennas and trellis states.
5. Easily lends itself to industrial applications because of its simplicity.	Complex to deploy.
6. Loses capacity with two or more receive antennas.	Preserves capacity irrespective of the number of antennas.

using the accompanying software and the generator codes given in Section 5.4.5. Remember the rule, that the higher the trace value with multiple receive antennas, the better the performance, regardless of the rank and determinant of other codes.

2. Figure 5.13 shows the performance of a 4-state 4-PSK code on slow Rayleigh fading channels with two transmit and two receive antennas and imperfect channel estimation using MMSE. Repeat this exercise using a 16-state 4-PSK code.

3. Figure 5.14 shows the performance of 4-PSK codes in the presence of correlation between receive antennas. Check out the performance using higher trace codes from the tables given in Section 5.4.5. How do STTC codes compare with STBC codes from the point of view of robustness in the presence of correlation?

References

[1]   Tarokh, V., N. Seshadri, and A. R. Calderbank, "Space-Time Codes for High Data Rate Wireless Communication: Performance Criterion and Code Construction," *IEEE Trans. Inform. Theory,* Vol. 44, No. 2, March 1998, pp. 744–765.

[2]   Chen, Z., J. Yuan, and B. Vucetic, "Improved Space-Time Trellis Coded Modulation Scheme on Slow Rayleigh Fading Channels," *Proceedings of ICC 2001,* Vol. 4, pp. 1110–1116.

[3]   Chen, Z., et al., "Space-Time Trellis Coded Modulation With Three and Four Transmit Antennas on Slow Fading Channels," *IEEE Commun. Letters,* Vol. 6, No. 2, February 2002, pp. 67–69.

[4]   Yuan, J., et al., "Performance Analysis and Design Space-Time Coding on Fading Channels," submitted to *IEEE Trans. Commun.,* 2000.

[5]   Firmanto, W., B. Vucetic, and J. Yuan, "Space-Time TCM With Improved Performance on Fast Fading Channels," *IEEE Commun. Letters,* Vol. 5, No. 4, April 2001, pp. 154–156.

[6]   Vucetic, B., and J. Nicolas, "Performance of M-PSK Trellis Codes Over Nonlinear Mobile Satellite Channels," *IEE Proceedings I,* Vol. 139, August 1992, pp. 462–471.

[7]   Hassibi, B., and B. M. Hochwald, "High-Rate Codes That are Linear in Space and Time," *Proc. 38th Annual Allerton Conference on Communication, Control and Computing,* 2000.

[8]   Ventura-Traveset, J., et al., "Impact of Diversity Reception on Fading Channels With Coded Modulation-Part I: Coherent Detection," *IEEE Trans. Commun.,* Vol. 45, No. 5, May 1997, pp. 563–572.

[9]   Vucetic, B., and J. Nicolas, "Performance of M-PSK Trellis Codes Over Nonlinear Mobile Satellite Channels," *IEE Proc. I,* Vol. 139, August 1992, pp. 462–471.

[10] Vucetic, B., and J. Yuan, *Space-Time Coding,* Chichester, UK: John Wiley & Sons, 2003.

[11] Forney, Jr., G. D. "Geometrically Uniform Codes," *IEEE Trans. Inform. Theory,* Vol. 37, No. 5, September 1991, pp. 1241–1260.

[12] Biglieri, E., et al., *Introduction to Trellis-Coded Modulation With Applications,* New York: Macmillan, 1991.

[13] Yan, Q., and R. S. Blum, "Optimum Space-Time Convolutional Codes," in *Proc. IEEE WCNC'00,* Chicago, IL, September 2000, pp. 1351–1355.

[14] Naguib, A., et al., "A Space-Time Coding Modem for High-Data-Rate Wireless Communications," *IEEE J. Select. Areas Commun.,* Vol. 16, No. 10, October 1998, pp. 1459–1478.

[15] Guey, J. C., "Concatenated Coding for Transmit Diversity Systems," *Proceedings VTC 1999*, Vol. 5, pp. 2500–2504.

[16] Tarokh, V., et al. "Space-Time Codes for Data Rate Wireless Communication: Performance Criteria in the Presence of Channel Estimation Errors, Mobility, and Multiple Paths," *IEEE Trans. Commun.*, Vol. 47, No. 2, February 1999, pp. 199–207.

[17] Sandhu, S., and A. J. Paulraj, "Space-Time Block Codes Versus Space-Time Trellis Codes," *Proceedings of ICC*, 2001.

[18] Yongacoglu, A., and M. Siala, "Space-Time Codes for Fading Channels," *Proceedings of VTC*, Vol. 5, pp. 2495–2499.

[19] Grimm, J., *Transmitter Diversity Code Design for Achieving Full Diversity on Rayleigh Fading Channels*, Ph.D. thesis, Purdue University, 1998.

[20] Yan, Q., and R. S. Blum, "Optimum Space-Time Convolutional Codes," *Proceedings WCNC*, 2000.

[21] Sandhu, S., and A. Paulraj, "Space-Time Block Codes: A Capacity Perspective," *IEEE Commun. Lett.*, Vol. 4, No. 12, December 2000, pp. 384–86.

CHAPTER 6

Layered Space-Time Codes

6.1 Introduction

In Chapters 4 and 5 we considered codes that had a spatial rate of unity or less than unity. These codes did not have any multiplexing gain with respect to a SISO channel, but, on the other hand, unlike a SISO channel, they possessed a diversity order of ideally $M_T M_R$. Hence, these codes were excellent for improving the link quality by combating deep fades. In this chapter, we shall examine codes that are expressly meant for improving the multiplexing gain by transmitting M_T independent data streams. This yields a spatial rate and multiplexing gain of M_T. Foschini [1] proposed a layered space-time (LST) architecture. In this landmark paper, he proposed to exploit the delay spreads existing in a wideband channel. Basically, the reader should imagine delay spreads as independent highways existing through the channel between the transmitter and receiver antennas. Depending on the frequency and bandwidth of the transmitted signals, they take a particular path through the channel. The time taken to traverse that path is delay spread for that path. These paths by their very nature are independent and each path does not "see" the other. Foschini decided to exploit this property existing in nature by transmitting signals through these paths. Hence, we do not need to deliberately make the data streams orthogonal like we do in space-time block coding. The orthogonal property just exists in nature and we simply exploit it. In view of the narrowband nature of the transmission, each data stream follows only one route to the receiver and, consequently, there are no multipaths experienced by the individual data streams. This is a very important criterion in that the signals are narrowband. Indeed it is applicable to all MIMO transmission systems. This technology is called spatial multiplexing (SM). However, the reality is far from this ideal. The data streams are not truly independent and they do possess a certain amount of interaction with each other, giving rise to a phenomenon called multistream interference (MSI) and the usual channel fading and additive noise.

6.2 LST Transmitters: Types of Encoding

There are two major types of classification of SM technology—horizontal encoding (HE) and vertical encoding (VE). We shall now examine these.

6.2.1 Horizontal Encoding

The schematic for this method is shown in Figure 6.1. This is also called horizontal layered space-time code (HLST).

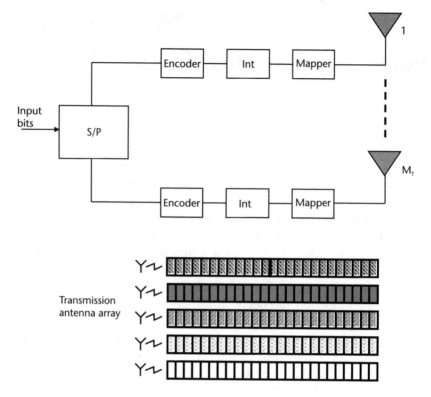

Figure 6.1 Horizontal encoding.

In this method the bit stream is first demultiplexed into M_T data streams. Each data stream is thereafter block encoded and interleaved. This is followed by mapping into the chosen modulation scheme from a constellation. The temporal encoding is independent and each data stream is transmitted from its individual antenna. The spatial rate is, therefore, obviously M_T. If r_m is the modulation rate depending on the type of constellation chosen, and r_c the convolution rate, then the signaling rate becomes $r_m r_c M_T$ bits/transmission. In this scheme each transmitted stream is received by M_R antennas. Hence, the maximum diversity attainable is M_R order diversity. This makes this system suboptimal, since we would have preferred $M_T M_R$ order diversity. We have assumed no channel knowledge at the transmitter and perfect channel knowledge at the receiver, yielding an array gain of M_R at the receiver. There is also a coding gain accruing from the encoder. There are two variants of this scheme made with a view to increasing the diversity order as close to $M_T M_R$ as possible. These are diagonal encoding (DE) and threaded encoding (TE).

6.2.1.1 Diagonal Encoding

This type of encoding, also called diagonal layered space-time code (DLST), is shown in Figure 6.2.

The initial signal processing is just like that for HE. However, before going to the antenna, the signal is stream rotated in a round-robin fashion so that the

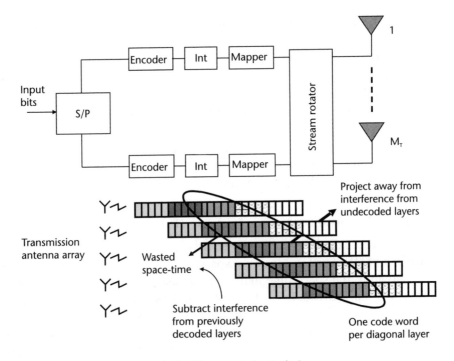

Figure 6.2 Diagonal encoding or D-BLAST transmission technique.

bit-stream and antenna association is periodically recycled. If the code word is large enough we can ensure that it is transmitted over all M_T antennas. This imparts a diversity of M_T at the transmitter. This is the Diagonal Bell Laboratories Layered Space-Time (D-BLAST) scheme [1]. Note the wasted space-time area in Figure 6.2, where no transmission takes place. This initial wastage is required to enable optimal decoding. This is explained as follows. D-BLAST architecture arranges the layers diagonally in space and time. Each block represents a transmitted symbol. Each layer is represented with a different shade and runs diagonally through the antenna elements as time progresses. Instead of committing each data stream to a single antenna like in Figure 6.1, the diagonal approach ensures that none of the layers miss out because of a poor transmission path. The BLAST receiver uses a multiuser detection strategy based on a combination of interference, cancellation, and suppression. In D-BLAST, each diagonal layer constitutes a complete code word, so decoding is performed layer-by-layer. Consider the code word matrix in Figure 6.2. The entries below the first diagonal layer are zeros. To decode the first diagonal, the receiver generates a soft-decision statistic for each entry in that diagonal. In doing so, the interference from the upper diagonals is suppressed by projecting the received signal onto the null space of the upper interfaces. The soft statistics are then used by the corresponding channel decoder to decode this diagonal. The decoder output is then fed back to cancel the first diagonal contribution in the interface while decoding the next diagonal. The receiver then proceeds to decode the next diagonal in the same manner. In spite of all this, D-BLAST can achieve $M_T M_R$ order diversity if we use Gaussian code blocks with infinite block size. As usual, coding gain will depend on the encoder and array gain of M_R is realizable at the receiver, since it has perfect knowledge of the channel. We have stated that D-BLAST can achieve $M_T M_R$ order diversity if we use Gaussian code blocks with infinite block size. In reality this is not possible and so a new type of SM coding called threaded layered space-time code (TLST) was proposed [2]. This method mixes the signal more thoroughly across the antennas than does the D-BLAST diagonal system. The only requirement here is that during each symbol period any given layer is transmitting over, at the most, one antenna. As a result all spatial interference will come from other layers. However, unlike D-BLAST we cannot deal with the signal processing at the receiver one layer at a time but we need to carry out joint decoding of multiple threads. Hence, it is more complex to implement.

6.2.1.2   Threaded Space-Time Encoding (TSTE)

This is also called TLSLT coding. This is shown in Figure 6.3. This is again a variant of HE. The last block is a spatial interleaver, which interleaves the symbols as shown in the space-time matrix. Each shade represents a thread. We have one code word per thread. In the first column, the symbols of each layer are not shifted. In the second column they are shifted once in a cyclic manner. In the third column they are shifted twice and so on. We define the space-time code matrix $\mathbf{A}$. The $M_T \times \ell$ matrix $\mathbf{A}$ contains the symbols transmitted over the M_T transmit antennas for ℓ symbol periods. We can describe each layer in general by specifying a set of elements from $\mathbf{A}$. Let $\mathcal{L} = \{L_1, L_2, \ldots, L_{M_T}\}$ be a set of indices specifying the elements of $\mathbf{A}$. Mathematically L_i is defined as [2]

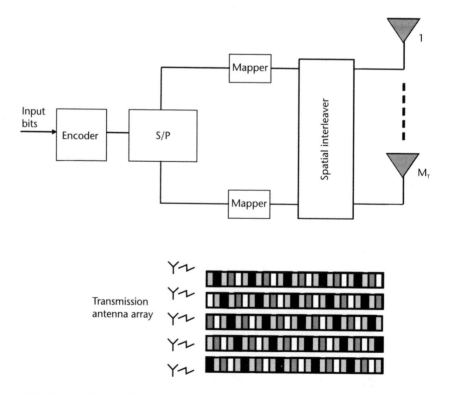

Figure 6.3   Threaded layered space-time (TLST).

$$L_i = \left\{ ([t + i - 1]_{M_T} + 1, \ell): 0 \le t \le \ell \right\} \qquad (6.1a)$$

To avoid spatial interference between layers, if we specify element (a, t) as belonging to L_i, then (b, t) cannot belong to L_i since $a \ne b$. We define the spatial and temporal spans of a layer so as to enable the layer to take advantage of the spatial and temporal diversity in the system. The spatial span of layer i is the difference between the $\max(a)$ and $\min(a)$ for all (a, t) in L_i. In other words, the spatial span denotes the range of antenna elements used for transmission by a layer. Similarly, the temporal span of layer i is the difference between $\max(t)$ and $\min(t)$ for all (a, t) in L_i. Therefore, to take full advantage of the diversity available, each layer should use the full range of antenna elements over the full range of symbol periods. Putting it in another way, we require that each layer has a spatial span of M_T and a temporal span of ℓ.

Now consider the M_T data streams that are demultiplexed and encoded. If the length of the data stream before demultiplexing was k, then after demultiplexing its length would be k/M_T. After block encoding, let each of the M_T independent data streams have a length of N/M_T where $N > k$. Finally, we modulate the encoded vectors using M_T spatial modulators (mapping), which map the encoded data into $M_T \times \ell$ matrix **A**. The relationship between ℓ and N/M_T is defined by the type of constellation being mapped. We are now in a position to analyze the new TLST architecture. This approach maximizes the diversity available through the encoding and interleaving of layers. To take full advantage of the resources, the TLST architecture requires that each layer transmit one symbol during each symbol interval and use on an average each transmission antenna equally often. Hence, all layers have equal use of the system resources. In the space-time matrix **A** in Figure 6.3, the spatial resources are shared by rotating the layers through the antennas symbol by symbol. Note that a single layer or thread contains multiple rotations through the array of antennas. This approach helps impart full transmit diversity M_T to the transmitted signal.

The TLST architecture is a combination of the coding process and the layering process. The transmission matrix **A** is of size $M_T \times \ell$. Since each layer has equal use of the transmission resources, $(1/M_T)$th or ℓ of the symbols in **A** will belong to L_i Furthermore the symbols from each layer will be spread evenly over all antennas so that each antenna will transmit ℓ/M_T symbols from L_i. The reader should bear in mind that each symbol is a signal constellation point corresponding to b alphabet symbols (bits) that will be modulated into the l constellation points in L_i. The design of this encoding process can be represented using a matrix construction [2]. Let g be a mapping such that $g(\underline{x})$ is the encoded vector of length bl. We can specify $g(\underline{x})$ by

$$g(\underline{x}) = \underline{x}M_1 \big| \underline{x}M_2 \big| \cdots \big| \underline{x}M_{M_T} \big| \qquad (6.1b)$$

where $M_1, \ldots, M_{M_T}$ are binary matrixes of dimension $k \times bl/M_T$. Thus, for any arbitrary input $\underline{x}$, $g(\underline{x})$ is the associated code word for the block code specified by $M_1, \ldots, M_{M_T}$. This code word can be split up into M_T segments corresponding to the encoded vectors to be modulated over each of the M_T antennas. In other words, $\underline{x}M_j$ is mapped by the spatial modulator f_i into the l/M_T symbols of L_i to be transmitted over antenna j. The complete transmission matrix corresponding to layer i is given by $c_i = f_i(g(\underline{x}))$.

If we bear in mind the above formulation for the block code used in encoding the data of layer i, the matrixes $M_1, \ldots, M_{M_T}$ can be designed to achieve the desired level of diversity. It is shown in [2] that a space-time code specified by the encoding matrixes $M_1, \ldots, M_{M_T}$ and the spatial modulator f_i achieves spatial diversity equal to dM_R in a quasi-static fading channel if and only if d is the largest integer such that the encoding matrixes M have the property that:

$$\forall a_1, \ldots a_{M_T} \in \{0, 1\} \text{ such that } a_1 + a_2 + \ldots + a_{M_T} = M_T - d + 1:$$

$$M = \left[a_1 M_1 a_2 M_2 \ldots a_{M_T} M_{M_T} \right] \text{ has rank } k \text{ over the binary field}$$

The implication here is that if for some reason the matrix M is singular, then the code difference matrix between two information streams will have an all-zero row for every nonzero coefficient a_i. Since there are $n - d + 1$ nonzero coefficients, this difference matrix will have a rank less than d. Hence, dM_R diversity is not achieved. We conclude that full spatial diversity $M_T M_R$ is achieved *iff* $M_1, M_2, \ldots, M_{M_T}$ are of rank k over the binary field. Thus, by designing the encoding matrixes as discussed, dM_R level diversity can be obtained for a given layer within a system with M_T transmit and M_R receive antennas.

Given a space-time code with code words given by (6.1b), if the code achieves a diversity of level dM_R, then any set of linear transformations on the segments of $g(\underline{x})$ will result in a new code of the same diversity level [2]. Thus, once a code has been designed, new codes of equal diversity gain can be found quite easily.

Generally, in the interests of coding gain, we prefer to use trellis codes instead of block codes within the space-time architecture. Trellis codes provide higher coding gain but come at the cost of increased decoding complexity. For this case, suppose we code with a binary convolutional code of rate k/M_T. Here, the encoder takes k input sequences and produces M_T output sequences. Let $x_i(t)$ and $y_i(t)$ denote the ith input and output sequences, respectively. The convolutional code is defined by an "impulse response" $g_{ij}(t)$ that relates the ith input $(x_i(t))$ to the jth output $(y_j(t))$. From [2], if we define the D-transform of the sequence $\{x(t)\}$ as

$$X(D) = x(0) + x(1)D + x(2)D^2 + \ldots$$

then the encoder action in the D-domain is given by the matrix equation:

$$\underline{Y}(D) = \underline{X}(D)G(D) \tag{6.1c}$$

where $\underline{Y}(D)$ is an n-dimensional vector of the output sequences $Y_1(D) \ldots Y_n(D)$, $\underline{X}(D)$ is a k-dimensional vector of the input sequences $X_1(D) \ldots X_k(D)$, and $G(D)$ is given by:

$$G(D) = \begin{bmatrix} G_{1,1}(D) & G_{1,2}(D) & \cdots & G_{1,n}(D) \\ G_{2,1}(D) & G_{2,2}(D) & \cdots & G_{2,n}(D) \\ \vdots & \vdots & \ddots & \vdots \\ G_{k,1}(D) & G_{k,2}(D) & \cdots & G_{k,n}(D) \end{bmatrix}$$

We now define the jth column of $G(D)$ as $\underline{F}_j(D)$ and write the jth output of the encoder as:

$$Y_j(D) = \underline{X}(D)\underline{F}_j(D)$$

Bearing in mind the above code, the M_T outputs of the encoder, $Y_1(D) \ldots Y_n(D)$, can be assigned directly to the M_T transmission layers in a space-time layered architecture. In this case, it is shown in [2] that the resulting space-time code, defined by the convolutional code and a spatial modulator f, will have a diversity level of dM_R with $d = M_T - v + 1$ if v is the smallest integer such that whenever $a_1 + a_2 + \ldots + a_{M_T} = v$ for all $a_1, \ldots, a_{M_T}$ {0, 1} then the matrix

$$\begin{bmatrix} a_1\underline{F}_1(D) & a_2\underline{F}_2(D) & \cdots & a_{M_T}\underline{F}_{M_T}(D) \end{bmatrix}$$

has full rank. By properly choosing the columns of the convolutional code "impulse response," one can design a layered space-time code with suitable diversity level.

Until now, we have designed the space-time code for a given layer and its performance was considered independently of other layers. But we need to take into account the interference from other layers. The overall performance of TST architecture will depend largely on the ability of the receiver to separate layers [2]. Thus, the signal processing scheme at the receiver is an integral part of system design.

The signal processing for the space-time modulated signal at the receiver can be regarded as a multiuser detection problem. Each layer must be independently detected and decoded. The detection and decoding tasks can be combined into an efficient iterative process. For example, an MMSE receiver can be used to generate soft decisions estimating the symbols received from the M_T transmission antennas. The M_T streams of estimates can be unwoven into the M_T layers, which can then be decoded. Feeding the results back to the MMSE receiver, the *a priori* probabilities can be updated and interference to other layers can be canceled. The estimate of the symbol sent by the ith antenna at time t can be written as:

$$y^{(i)} = \underline{w}_f^{(i)^T} \underline{r} + w_b^{(i)} \tag{6.1d}$$

where $\underline{w}_f$ is a vector of weighting coefficients suppressing interference from other layers, and w_b is the feedback coefficient to cancel the effects from previously decoded symbols. The MMSE solutions for these coefficients are given in [2].

The authors in [2] go on to analyze the performance of a TST architecture employing a space-time code of diversity level dM_R and the iterative MMSE receiver already discussed. This system is compared with several other popular architectures, including D-BLAST. The proposed architecture is similar to D-BLAST in that there is equal antenna usage between layers, but the general approach of the TST allows for greater temporal diversity. When compared with a horizontal layering approach, known as H-BLAST, we see that the TST provides greater spatial diversity through the efficient distribution of layers across antenna elements.

Simulation results were also provided in [2] to compare the performance of TST with D-BLAST and a multilayered approach proposed by Tarokh et al. in [3]. In each case, convolutional codes with rate 1/2 were used in the coding of input signals, with decoding algorithms based on the Viterbi algorithm. The MMSE receiver was used for the TST architecture. The simulation results (frame error rate versus signal-to-noise ratio) demonstrated a gain in performance for the TST architecture of 3–8 dB over the D-BLAST and multilayering architectures. These improvements are due primarily to the TST architecture's greater ability to take advantage of the diversity available.

6.2.2 Vertical Encoding

The configuration for vertical encoding is shown in Figure 6.4. In this method the bit stream is encoded, interleaved, and mapped before being fed to a demultiplexer. It is then split into M_T streams. The implication in the design is that each information bit can be spread across all antennas. This kind of transmission, however, requires joint decoding at the receiver, making it very complex. The spatial rate is M_T and the signaling rate is $r_m r_c M_T$ bits/transmission. Since the transmission bits are spread over M_T antennas and each stream is received by M_R antennas, the diversity order is $M_T M_R$. Furthermore, as usual, we also have coding gain depending on the temporal code and array gain at the receiver, which has perfect knowledge of the channel.

6.2.2.1 V-BLAST

A variant of the virtual encoding (VE) is the vertical BLAST (V-BLAST) algorithm, shown in Figure 6.5.

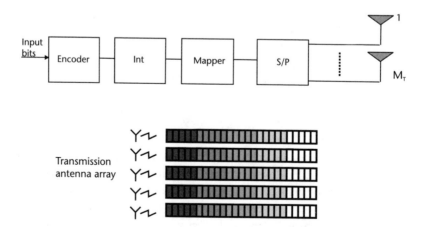

Figure 6.4 Vertical encoding.

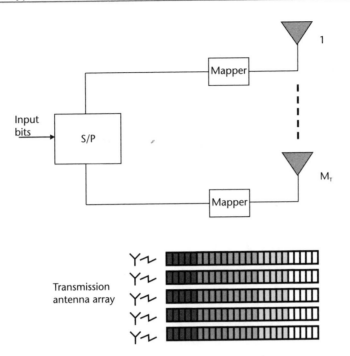

Figure 6.5   V-BLAST configuration.

The basic idea here is to usefully exploit the multipath, rather than mitigate it, by considering the multipath itself as a source of diversity that allows the parallel transmission of substreams from the same user. This approach is based on a MIMO $(M_T M_R)$ system (M_T transmitting and M_R receiving antennas) that supports transmit space diversity obtaining higher data rate while using the same total power and bandwidth adopted for the 1-D (SISO) system. The indoor environment is the ideal rich scattering environment necessary to get the best performance promised by this approach. The Bell Labs Layered Space-Time (BLAST) [1, 4] architecture uses multielement antenna arrays at both transmitter and receiver to provide high-capacity wireless communications in a rich scattering environment. It has been shown that the theoretical capacity approximately increases linearly as the number of antennas is increased [4]. Two types of BLAST realizations have been widely publicized: V-BLAST [4] and D-BLAST [1]. The V-BLAST is a practical algorithm shown to achieve a large fraction of the MIMO channel capacity in the case of narrowband point-to-point communication scenarios. The V-BLAST algorithm implements a nonlinear detection technique based on zero forcing (ZF) combined with symbol cancellation to improve the performance. The idea is to look at the signals from all the receive antennas simultaneously, first extracting the strongest substream from the received signals, then proceeding with the remaining weaker signals, which are easier to recover once the strongest signals have been removed as a source of interference. This is called successive interference cancellation (SIC) and is somewhat analogous to decision feedback equalization. When symbol cancellation is used, the order in which the substreams are detected becomes important for the overall performance of the system. In fact, the transmitted symbol with the smallest postdetection SNR will dominate the error performance of the system.

Postdetection SNR is determined by ordering. The optimal ordering is based on the result that simply choosing the best postdetection SNR at each stage of the detection process leads to the maximization of the worst SNR over all possible orderings [4].

For simplicity, we base our explanation on Figure 6.6. Suppose the number of transmitters is M_T and the number of receivers is M_R. QAM transmitters 1 to M_T operate cochannel at symbol rate $1/T$ symbols, with synchronized symbol timing. This collection of transmitters constitutes a vector drawn from a QAM constellation. Receivers 1 to M_R are individually conventional QAM receivers. The receivers also operate cochannel, each receiving the signals radiated from all M_T transmit antennas. Flat fading is assumed and the matrix channel transfer function is $\mathbf{H}^{M_R \times M_T}$, where $h_{i,j}$ is the complex transfer function from transmitter j to receiver i and $M_T \le M_R$. We assume that the transmission is organized in bursts of L symbols and that the channel time variation is negligible over the L symbol periods, comprising a burst, and that the channel is estimated accurately using training symbols embedded in each burst.

Let $\mathbf{a} = (a_1 \quad a_2 \quad \ldots \quad a_M)^T$ denote the vector of transmit symbols. Then the corresponding received M_R vector $\mathbf{i}$

$$r_1 = \mathbf{Ha} + \mathbf{v} \tag{6.2}$$

where $\mathbf{v}$ is a wide sense stationary (WSS) noise vector with i.i.d. components.

Step 1: Using nulling vector w_{k_1}, form a linear combination of the components of r_1 to yield y_{k_1}:

$$y_{k_1} = w_{k_1}^T r_1 \tag{6.3}$$

Step 2: Slice y_{k_1} to obtain $\hat{a}_{k_1}$:

$$\hat{a}_{k_1} = Q(y_{k_1}) \tag{6.4}$$

where $Q(\cdot)$ denotes the quantization (slicing) operation appropriate to the constellation in use.

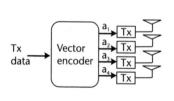
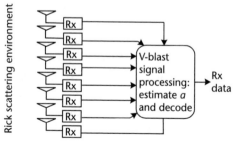

Figure 6.6 V-BLAST system.

Step 3: Assuming that $\hat{a}_{k_1} = a_{k_1}$, cancel a_{k_1} from the received vector r_1; resulting in modified received vector r_2:

$$r_2 = r_1 - \hat{a}_{k_1}(H)_{k_1} \tag{6.5}$$

where $(H)_{k_1}$ denotes the k_1th column of **H**. Steps 1–3 are then performed for components $k_2 \ldots k_{M_T}$ by operating in turn on the progression of modified received vectors $r_2, r_3, \ldots, r_{M_T}$.

The specifics of the detection process depend on the criterion chosen to compute the nulling vectors w_{k_i}, the most common choices being minimum MMSE and ZF. The detection process used in this section is the latter and is simpler. The k_ith ZF nulling vector is defined as the unique minimum norm vector satisfying

$$w_{k_i}^T(H)_{k_j} = \begin{cases} 0 & j > i \\ 1 & j = i \end{cases} \tag{6.6}$$

Thus, the k_ith ZF nulling vector is orthogonal to the subspace spanned by the contributions to r_i, due to those symbols not yet estimated and canceled. It can be easily shown that the unique vector satisfying (6.6) is just the k_ith row of $H_{k_{j-1}}^{\pm}$, where the notation $H_{k_j}^{-}$ denotes the matrix obtained by zeroing columns $k_1, k_2, \ldots, k_j$ of **H** and + denotes the Moore-Penrose pseudoinverse [5].

The full ZF detection algorithm is a recursive procedure, including determination of optimal ordering:

$$\text{initialization: } \mathbf{G}_1 = \mathbf{H}^+ \tag{6.7a}$$

$$i = 1 \tag{6.7b}$$

$$\text{recursion: } k_j = \arg\min \left\| (\mathbf{G}_i)_j \right\|^2 \tag{6.7c}$$

where $j \notin \{k_1 \quad k_2 \quad \ldots \quad k_{i-1}\}$

$$w_{k_i} = (\mathbf{G}_i)_{k_i} \tag{6.7d}$$

$$y_{k_i} = w_{k_i}^T r_i \tag{6.7e}$$

$$\hat{a}_{k_i} = Q(y_{k_i}) \tag{6.7f}$$

$$r_{i+1} = r_i - \hat{a}_{k_i}(H)_{k_i} \tag{6.7g}$$

$$\mathbf{G}_{i+1} = \mathbf{H}_{k_i}^{\pm} \tag{6.7h}$$

$$i = i + 1 \tag{6.7i}$$

where $(G_i)_j$ is the jth row of $\mathbf{G}_i$. Thus (6.7c) determines the elements the optimal row sequence that maximizes the performance of the BER curve. The proof for this is given in [4]. Equation (6.7d–f) compute the ZF nulling vector, the decision statistic, and the estimated component of $\mathbf{a}$, respectively. Equation (6.7g) performs the cancellation of the detected component from the received vector and (6.7h) computes the new pseudoinverse for the next iteration. This new pseudoinverse is based on the "deflated" version of $\mathbf{H}$ in which the columns $k_1, k_2, \ldots, k_j$ have been zeroed. This is because these columns correspond to components of $\mathbf{a}$, which have already been estimated and canceled.

We shall return to this topic later in this chapter when we discuss LST receivers.

6.3   Layered Space-Time Coding: Design Criteria

Consider the HLST encoder in Figure 6.1. We will now derive the design criteria for LST systems using a procedure similar to the one discussed in Chapter 5 for STTC systems. This section is based on the work done by Vucetic et al. (*From:* [6]. © 2001, IEEE. Reproduced with permission.)

Assume that at any given time t a symbol s_t^i, $i \in \{1, 2, \ldots, M_T\}$ is transmitted out of transmit antenna i. If we assume a Rayleigh fading channel, receive antenna j receives symbol r_t^j, which can be expressed as [6],

$$r_t^j = \sum_{i=1}^{M_T} h_t^{i,j} s_t^i + n_t^j \tag{6.8}$$

where $h_t^{i,j}$ is the channel coefficient between the ith transmit and the jth receive antenna at time t. n_t^j is an additive white Gaussian noise sample at receive antenna j at time t, which is an independent sample of a zero mean complex Gaussian random variable with variance σ^2 per dimension.

The modulated signal estimate $\tilde{s}_t^k$ of the kth layer at time t can be written as [6]

$$\tilde{s}_t^k = \mathbf{w} \left(\mathbf{r}_t - \sum_{l=k+1}^{M_T} \mathbf{h}_t^l s_t^k \right) \tag{6.9}$$

for some linear combination coefficients $\mathbf{w}$ an M_R dimensional row vector whose values are determined by using a ZF function or MMSE criterion. The received column vector $\mathbf{r}_t = \begin{bmatrix} r_t^1 & r_t^2 & \ldots & r_t^{M_R} \end{bmatrix}^T$ contains the received symbols at the M_R receive antennas. The M_R-dimensional column vector $\mathbf{h}_t^l = \begin{bmatrix} h_t^{l,1} & h_t^{l,2} & \ldots & h_t^{l,M_R} \end{bmatrix}^T$ contains the channel coefficients between the lth layer and M_R receive antennas at time t. We carry out QR decomposition [5] on the $M_R \times M_T$ channel matrix $\mathbf{H}_t = \begin{bmatrix} \mathbf{h}_t^1 & \mathbf{h}_t^2 & \ldots & \mathbf{h}_t^{M_T} \end{bmatrix}$ and decompose it into an unitary matrix $\mathbf{U}_t$ and an upper triangular matrix $\mathbf{R}_t$ such that

$$\mathbf{H}_t = \mathbf{U}_t \mathbf{R}_t \tag{6.10}$$

where

$$
\mathbf{R}_t = \begin{bmatrix}
\left(R_1^1\right)_t & \left(R_1^2\right)_t & \left(R_1^3\right)_t & \cdots & \left(R_1^{M_T}\right)_t \\
0 & \left(R_2^2\right)_t & \left(R_2^3\right)_t & \cdots & \left(R_2^{M_T}\right)_t \\
0 & 0 & \left(R_3^3\right)_t & \cdots & \left(R_3^{M_T}\right)_t \\
\vdots & & \ddots & & \vdots \\
0 & 0 & 0 & \cdots & \left(R_{M_R}^{M_T}\right)_t
\end{bmatrix}
\tag{6.11}
$$

is an $M_R \times M_T$ upper triangular matrix and $\mathbf{U}_t$ is of size $M_R \times M_R$.
We define $\mathbf{y}_t = \begin{bmatrix} y_t^1 & y_t^2 & \cdots & y_t^{M_R} \end{bmatrix}$ as

$$
\mathbf{y}_t = \mathbf{U}_t \mathbf{r}_t
\tag{6.12}
$$

$$
= \mathbf{R}_t \mathbf{s}_t + \hat{\mathbf{n}}_t
$$

where $\hat{\mathbf{n}}_t = \mathbf{U}_t \mathbf{n}_t$ is an M_R-dimensional column vector of AWGN noise samples and $\mathbf{s}_t$ is an M_T-dimensional column vector of the transmitted modulated symbols at time t. Since $\mathbf{R}_t$ is upper triangular,

$$
y_t^k = \left(R_k^k\right)_t s_t^k + \hat{n}_t^k
\tag{6.13}
$$

$$
+ \text{ interference from } s_t^{k+1}, s_t^{k+2}, \ldots, s_t^{M_T}
$$

Assuming that hard decisions from the constituent decoders are correct, they can be used to completely suppress the interference term in (6.13). Therefore, the decision variable $\tilde{s}_t^k$ can be expressed as

$$
\tilde{s}_t^k = \left(R_k^k\right)_t s_t^k + \hat{n}_t^k
\tag{6.14}
$$

for $k \in \{1, 2, \ldots, M_T\}$.

6.3.1 Performance Analysis of an HLST System

If we assume code words of length L, the pairwise error probability of the kth layer $P(s^k, e^k)$ is the probability that the decoder selects as its output the sequence $e^k = \left(e_1^k, e_2^k, \ldots, e_t^k, \ldots, e_L^k\right)$, when the transmitted sequence on the kth layer was in fact $s^k = \left(s_1^k, s_2^k, \ldots, s_t^k, \ldots, s_L^k\right)$. This occurs if

$$
\sum_{t=1}^{L} \left\| \tilde{s}_t^k - \left(R_k^k\right)_t s_t^k \right\|^2 \geq \sum_{t=1}^{L} \left\| \tilde{s}_t^k - \left(R_k^k\right)_t e_t^k \right\|^2
\tag{6.15}
$$

or equivalently

$$\sum_{t=1}^{L} 2\Re\left\{\hat{n}_t^k \left(R_k^k\right)_t \left(s_t^k - e_t^k\right)\right\} \geq \sum_{t=1}^{L} \left\| \left(R_k^k\right)_t \left(s_t^k - e_t^k\right) \right\|^2 \tag{6.16}$$

where $\Re\{\cdot\}$ indicates the real part of a complex number.

For a given realization of the channel matrix $\mathbf{H}$ or equivalently the matrix $\mathbf{R}$, the right-hand side of (6.16) is a constant. We call this constant $d^2(s^k, e^k)$. The left-hand side of the expression is zero mean Gaussian random variable with variance $4\sigma^2 d^2(s^k, e^k)$. Therefore, for a given $\mathbf{H}$, the conditional pairwise error probability can be expressed as

$$P(s^k \rightarrow e^k | \mathbf{H}) = Q\left(\sqrt{\frac{d^2(s^k, e^k)}{8\sigma^2}} \right) \tag{6.17}$$

where $Q(x)$ is the complementary error function, approximated as,

$$Q(x) = \frac{1}{\sqrt{2\pi}} \int_0^\infty e^{-t^2/2} \, dt \tag{6.18}$$

By using the inequality

$$Q(x) \leq \frac{1}{2} e^{-x^2/2}, \; x \geq 0 \tag{6.19}$$

we obtain

$$P(s^k \rightarrow e^k | \mathbf{H}) \leq \frac{1}{2} \exp\left\{ \frac{-d^2(s^k, e^k)}{8\sigma^2} \right\} \tag{6.20}$$

where $d^2(s^k, e^k)$ is defined by

$$d^2(s^k, e^k) = \sum_{t=1}^{L} \left\| \left(R_k^k\right)_t \left(s_t^k - e_t^k\right) \right\|^2 \tag{6.21}$$

6.3.1.1 Behavior in Slow Fading Channels

From (6.20) and given the channel coefficient matrix $\mathbf{H}$, the probability that the decoder decides erroneously in favor of a modulated sequence e^k can be expressed as

$$P(s^k \rightarrow e^k | \mathbf{H}) \leq \exp\left(\frac{-\sum_{t=1}^{L} \left\| \left(R_k^k\right)_t \right\|^2 \left\| s_t^k - e_t^k \right\|^2}{8\sigma^2} \right) \tag{6.22}$$

where the matrix $\mathbf{R}$ comes from the QR decomposition of $\mathbf{H}$ and σ^2 is the noise variance per dimension.

If the channel is a slow fading one, the fading coefficients remain constant over a frame. This makes $\left\| (R_k^k)_t \right\|^2 = \left\| (R_k^k) \right\|^2$ and

$$P(s^k \rightarrow e^k \,|\, \mathbf{H}) \leq \exp\left(-\frac{1}{8\sigma^2} \left\| R_k^k \right\|^2 \sum_{t=1}^{L} \left\| s_t^k - e_t^k \right\|^2 \right) \qquad (6.23)$$

where $\left\| R_k^k \right\|^2$ is a sum of $2(M_R - k)$ zero mean Gaussian distributed random variables, each with variance $1/2$.

Let $D_k = d_{E_k}^2 \left\| R_k^k \right\|^2$ where $d_{E_k}^2 = \sum_t \left\| s_t^k - e_t^k \right\|^2$. If $2(M_R - k) \geq 4$, then based on the Central Limit Theorem D_k approaches a Gaussian distribution with mean $\mu_{D_k} = (M_R - k) d_{E_k}$ and variance $\sigma_{D_k}^2 = (M_R - k) d_{E_k}^2$.

Therefore, the conditional error probability given in (6.23) can be rewritten as

$$P(s^k \rightarrow e^k \,|\, \mathbf{H}) \leq \exp\left(-\frac{1}{8\sigma^2} D_k \right) \qquad (6.24)$$

The PEP is obtained by averaging (6.24) over the probability density function of D_k, $p(D_k)$ resulting in

$$P(s^k \rightarrow e^k) \leq \exp\left\{ \frac{(M_R - k) d_{E_k}^2}{2(8\sigma^2)^2} - \frac{(M_R - k) d_{E_k}^2}{8\sigma^2} \right\} \qquad (6.25)$$

$$\cdot Q\left(\frac{\sqrt{M_R - k}\left(d_{E_k}^2 - 8\sigma^2 \right)}{8\sigma^2} \right)$$

If $2(M_R - k) < 4$, the Gaussian approximation on the probability density functions of D_k no longer holds. In such a case D_k is a chi-square distributed with $2(M_R - k)$ degrees of freedom. Therefore, the PEP is obtained by averaging (6.23) over a chi-square distribution resulting in [6]

$$P(s^k \rightarrow e^k) \leq \left(1 + \frac{\left\| s^k - e^k \right\|^2}{8\sigma^2} \right)^{-(M_R - k)} \qquad (6.26)$$

6.3.1.2 Behavior in Fast Fading Channels

For a channel coefficient matrix $\mathbf{H}$, the PEP of the kth layer $P(s^k, e^k)$ can be expressed as

$$P(s^k \rightarrow e^k \,|\, \mathbf{H}) \leq \exp\left(\frac{-\sum_{t=1}^{L} \left\| (R_k^k)_t \right\|^2 \left\| s_t^k - e_t^k \right\|^2}{8\sigma^2} \right) \qquad (6.27)$$

where the matrix $\mathbf{R}$ comes from the QR decomposition of $\mathbf{H}$.

We introduce a variable $b_{t,k} = \left\|(R_k^k)_t\right\|^2 \left\|s_t^k - e_t^k\right\|^2$ and $D_k = \Sigma_t b_{t,k}$. This makes $b_{t,k}$ a chi-square distributed random variable with $2(M_R - k)$ degrees of freedom. Let d_H be the Hamming distance between the two code words and $d_H \geq 4$. Then, by virtue of the Central Limit Theorem, D_k is a Gaussian random variable with mean $\mu_{D_k} = (M_R - k)\Sigma_t\left\|s_t^k - e_t^k\right\|^2$ and variance $\sigma_{D_k}^2 = (M_R - k)\Sigma_t\left\|s_t^k - e_t^k\right\|^4$.

Since we know the probability density function of D_k, the PEP can be obtained by averaging the conditional error probability over the PDF $p(D_k)$ as

$$P(s^k \to e^k) \leq \int_0^\infty \exp\left\{-\frac{1}{8\sigma^2}D_k\right\}p(D_k)d(D_k)$$

$$\leq \exp\left\{\frac{(M_R - k)D^4}{2(8\sigma^2)^2} - \frac{(M_R - k)d^2}{8\sigma^2}\right\} \tag{6.28}$$

$$\cdot Q\left(\frac{\sqrt{(M_R - k)D^4}}{8\sigma^2} - d^2\sqrt{\frac{(M_R - k)}{D^4}}\right)$$

where $d^2 = \Sigma_t\left\|s_t^k - e_t^k\right\|^2$ and $D^4 = \Sigma_t\left\|s_t^k - e_t^k\right\|^4$.

If the Hamming distance, $d_H < 4$, the Gaussian assumption on the PDF of D_k is no longer valid. In such a case, D_k is chi-square distributed with $2(M_R - k)$ degrees of freedom.

Hence, the PEP becomes

$$P(s^k \to e^k) \leq \prod_{t \in \eta(s^k, e^k)} E\left[\exp\left\{\frac{-\left\|(R_k^k)_t\right\|^2 \left\|s_t^k - e_t^k\right\|^2}{8\sigma^2}\right\}\right] \tag{6.29}$$

$$\leq \prod_{t \in (s^k, e^k)} \left(1 + \frac{\left\|s_t^k - e_t^k\right\|^2}{8\sigma^2}\right)^{k - M_R}$$

where $E[\cdot]$ denotes the expectation operator and $\eta(s^k, e^k) = \left\{t \mid s_t^k \neq e_t^k\right\}$.

6.3.2 Performance Analysis of a DLST System

Given a channel coefficient matrix $\mathbf{H}$, the probability that the receiver decides in favor of a distinct diagonal $e = \left(e_1^1, e_2^2, \ldots, e_t^k, \ldots, e_{M_T}^{M_T}\right)$, while the transmitted code word is $s = \left(s_1^1, s_2^2, \ldots, s_t^k, \ldots, s_{M_T}^{M_T}\right)$, can be expressed as,

$$P(s \to e \mid \mathbf{H}) \leq \exp\left\{\frac{-\sum_t^{M_T}\left\|(R_t^t)_t\right\|^2 \left\|(s_t^k - e_t^k)\right\|^2}{8\sigma^2}\right\} \tag{6.30}$$

where the matrix $\mathbf{R}$ comes from the QR decomposition of $\mathbf{H}$.

Remember that because each code word occupies a diagonal in the transmission resource array, $k = t$, $s_t^k = s_t^t$ and $e_t^k = e_t^t$. Similarly, $(R_k^t)_t = (R_t^t)_t$. Equation (6.30) is valid on both fast and slow fading channels because each subchannel undergoes *independent fading* (they are transmitted one diagonal at a time and orthogonal in time). Equation (6.30) is valid for both fast and slow fading channels because each subchannel undergoes independent fading.

Let $D = \sum_{t+1}^{L} \left\| (R_t^t)_t \right\|^2 \left\| (s_t^t - e_t^t) \right\|^2$, where $\left\| (R_t^t)_t \right\|^2$ is a chi-square distributed random variable with $2(M_R - t)$ degrees of freedom. If d_H is the Hamming distance between the two code words and $d_H \geq 4$, then by virtue of the central limit theorem, D is a Gaussian random variable with mean μ_D and variance σ_D^2 given by

$$\mu_D = \sum_t (M_R - t) \left\| s_t^t - e_t^t \right\|^2 \tag{6.31}$$

$$\sigma_D^2 = \sum_t (M_R - t) \left\| (s_t^t - e_t^t) \right\|^4 \tag{6.32}$$

If we know the PDF of D, the PEP can be obtained by averaging the conditional error probability over the PDF $p(D)$ as

$$P(s \rightarrow e) \leq \int_0^\infty \exp\left\{ -\frac{1}{8\sigma^2} D \right\} p(D)\, dD$$

$$\leq \exp\left\{ \frac{\sum_t (M_R - t) \left\| s_t^t - e_t^t \right\|^2 \left(\frac{1}{8\sigma^2} - 1 \right)}{2 \times 8\sigma^2} \right\} \tag{6.33}$$

$$\cdot Q\left\{ \frac{\sum_t (M_R - t) \left\| s_t^t - e_t^t \right\|^2 \left(\left\| s_t^t - e_t^t \right\| - 8\sigma^2 \right)}{8\sigma^2 \sqrt{\sum_t (M_R - t) \left\| s_t^t - e_t^t \right\|^4}} \right\}$$

If $d_H < 4$ the PDF of D is no longer approaching Gaussian distribution and instead is chi-square distributed. The error probability is obtained by taking the expectation of conditional PEP over a chi-square distribution as [6]

$$P(s \rightarrow e) \leq E\left[\exp\left\{ -\frac{1}{8\sigma^2} \sum_t \left\| (R_t^t)_t \right\|^2 \left\| s_t^t - e_t^t \right\|^2 \right\} \right] \tag{6.34}$$

$$\approx \prod_{t \in \eta(s,e)} \left(\left\| s_t^t - e_t^t \right\|^2 \right)^{-(M_R - t)} \left(\frac{1}{8\sigma^2} \right)^{-\sum_{t \in \eta(s,e)} (M_R - t)}$$

where $E[\cdot]$ denotes the expectation operator and $\eta(s, e) = \{ t \mid s_t^t \neq e_t^t \}$.

6.3.3 Code Design Criteria

6.3.3.1 HLST on Fast Fading Channels

Using (6.19) and assuming a high SNR, the PEP when $d_H \geq 4$ as given in (6.28) can be further approximated as

$$P(s^k - e^k) \leq \frac{1}{2}\left\{-\frac{1}{2}\frac{(M_R - k)d^2}{8\sigma^2}\right\} \qquad (6.35)$$

Hence, to minimize the PEP when $d_H \geq 4$, the minimum value of $d^2 = \Sigma_t \left\|s_t^k - e_t^k\right\|^2$ has to be maximized.

Again, at high SNRs, the PEP when $d_H < 4$ as given in (6.29) can be further approximated as

$$P(s^k \to e^k) \leq \prod_{t \in \eta(s^k, e^k)} \left(\frac{\left\|s_t^k - e_t^k\right\|^2}{8\sigma^2}\right)^{(-M_R - k)} \qquad (6.36)$$

To minimize the PEP, the minimum of η has to be maximized over all code word pairs. This is equivalent to maximizing the Hamming distance, d_H, of the code. Second, the minimum product distance $\prod_{t \in \eta(s^k, e^k)} \left\|s_t^k - e_t^k\right\|^2$ over all code word pairs that are d_H Hamming distance apart has to be maximized.

We now formulate the design criteria for HLST fast fading channels as follows:

- If the Hamming distance of the code is greater than or equal to four, the minimum value of $d^2 = \Sigma_t \left\|s_t^k - e_t^k\right\|^2$ has to be maximized.
- If the Hamming distance of the code is less than four, first the Hamming distance of the code has to be maximized. Second, the minimum product distance $\prod_{t \in \eta(s^k, e^k)} \left\|s_t^k - e_t^k\right\|^2$ has to be maximized.

6.3.3.2 HLST on Slow Fading Channels

At high SNRs the PEP on the kth layer given in (6.25) can be further approximated as

$$P(s^k \to e^k) \leq \frac{8\sigma^2}{\sqrt{2\pi(M_R - k)}d_{E_K}^2} \cdot \exp\left\{\frac{-(M_R - k)}{2}\right\} \qquad (6.37)$$

which suggests that the performance of HLST on a slow fading channel when $2(M_R - k) \geq 4$ is dominated by the minimum value of $d_{E_k}^2$. Hence, to minimize the PEP requires that the minimum value of $d_{E_k}^2$ be maximized.

When $2(M_R - k) < 4$, however, the code performance is bounded by (6.26). It therefore suggests that to minimize the PEP the minimum squared Euclidian distance $\left\|s^k - e^k\right\|^2$ has to be maximized. Hence, in slow fading channels, regardless of the value of $2(M_R - k)$ the design criterion of an HLST is guided by maximizing the minimum value of $d_{E_k}^2$.

6.3.3.3   DLST on Slow and Fast Fading Channels

The equation in (6.33) upper-bounds the error performance of a diagonally layered code when the Hamming distance $d_H \geq 4$. Using (6.19) this can be further approximated as

$$P(s \rightarrow e) \leq \frac{1}{2} \exp \left\{ -\frac{1}{8\sigma^2} \sum_t (M_R - t) \left\| s_t^t - e_t^t \right\|^2 \right\} \qquad (6.38)$$

$$\leq \frac{1}{2} \prod_{t \in \eta(s,e)} \exp \left\{ \frac{-(M_R - t) \left\| s_t^t - e_t^t \right\|^2}{8\sigma^2} \right\}$$

where $\eta(s, e) = \left(t \mid s_t^t \neq e_t^t \right)$, which suggests that the PEP of DLST in slow and fast fading channels is minimized when the value of $\sum_t (M_R - t) \left\| s_t^t - e_t^t \right\|^2$ is maximized.

In the case when $d_H < 4$ the code performance in fast and slow fading channels is bounded by (6.34). We now define the *truncated multidimensional effective length* (TMEL) and the *truncated multidimensional product distance* (TMPD) between two distinct code words s and e as $\text{TMEL} = \sum_{t \in \eta(s,e)} M_R - t$ and $\text{TMPD} = \prod_{t \in \eta(s,e)} \left\| s_t^t - e_t^t \right\|^{2(M_R - t)}$ [6]. To minimize the PEP, the minimum value of TMEL over all code word pairs has to be maximized. Second, of all the code word pairs with the maximum minimum TMEL, the minimum value of TMPD ought to be maximized.

We now formulate the design criteria for DLST codes in fast and slow fading channels as:

- If the Hamming distance of the code is greater than or equal to four, the minimum value of $\sum_t (M_R - t) \left\| s_t^t - e_t^t \right\|^2$ has to be maximized.
- If the Hamming distance of the code is less than four, first the minimum value of $\text{TMEL} = \sum_{t \in \eta(s,e)} M_R - t$ has to be maximized. Second, the minimum value of $\text{TMPD} = \prod_{t \in \eta(s,e)} \left\| s_t^t - e_t^t \right\|^{2(M_R - t)}$ has to be maximized.

6.3.3.4   Performance Evaluation

Vucetic [6] assumed that the receiver has perfect knowledge of the channel. The modulation adopted is QPSK. Ring codes were used in the encoder [7]. Her results are shown here for inspection. The channel is a slow fading Rayleigh channel. The performance of HLST with three different (8, 4) ring codes are examined. Each code has a Hamming distance of 3 and minimum squared Euclidian distance of 6.0, 8.0, and 10.0. Six transmit and 10 receive antennas are assumed. This yielded a bandwidth efficiency of 6 bit/s/Hz. A frame is assumed to consist of a code word of eight 4-PSK symbols. In Figure 6.7 we note that the best performance is achieved when the minimum squared Euclidian distance is the largest. In this case, it is the ring code $d^2_{E_k} = 10.0$ which offers the 2-dB gain at a frame error rate of 10^{-4}.

The codes in Figure 6.7 have a Hamming distance of 3 (which is less than four). Figure 6.8 shows the case of a fast fading channel using (9, 3) ring codes with 8-PSK modulation. The codes have a Euclidian distance of 6.515 and 4.343, respectively. Their product distances are 0.081 and 0.235, respectively. Once again,

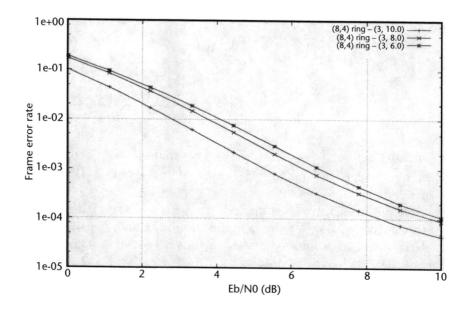

Figure 6.7 FER comparison for HLST codes in a slow fading channel. (© 2001, IEEE.)

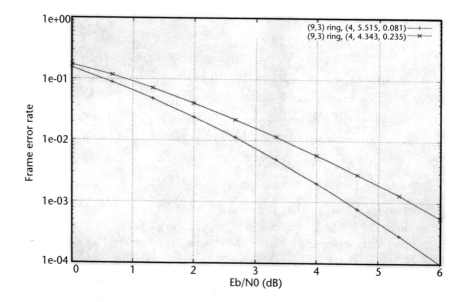

Figure 6.8 FER comparison for HLST codes in a fast fading channel. (© 2001, IEEE.)

the code with the highest Euclidian distance performs best despite its much lower value of product distance. This verifies the design criteria.

6.4 LST Receivers

The reality of MIMO receivers is that we need to contend with MSI, since the transmitted streams interfere with each other. In addition to this, we have the usual problem of channel fading and additive noise. Initially we assume uncoded SM (i.e., the data stream comprises uncoded data, where no temporal coding has been employed, but only mapping).

The three most common receivers for uncoded SM are ML ZF, and MMSE. We shall now briefly examine these. We assume a flat fading environment following the law for the received signal

$$\mathbf{r} = \mathbf{Hs} + \mathbf{n} \tag{6.39}$$

where $\mathbf{r}$ is the received $M_R \times 1$ vector, $\mathbf{H}$ is the channel matrix of size $M_R \times M_T$, $\mathbf{s}$ is the $M_T \times 1$ transmitted signal and $\mathbf{n}$ is the $M_R \times 1$ ZMCSCG noise vector with covariance matrix $N_0 \mathbf{I}_{M_R}$. In this we are following the notation used in earlier chapters.

6.4.1 ML Receiver

This is an optimum receiver. If the data stream is temporally uncoded, the ML receiver solves

$$\hat{s} = \arg \min_s \left\| \mathbf{r} - \mathbf{Hs} \right\|^2 \tag{6.40}$$

where $\hat{s}$ is the estimated symbol vector. The ML receiver searches through all the vector constellation for the most probable transmitted signal vector. This implies investigating S^{M_T} combinations, a very difficult task. Hence, these receivers are difficult to implement, but provide full M_R diversity and zero power losses as a consequence of the detection process. In this sense it is optimal. There have been developments based on fast algorithms employing sphere decoding [8, 9]. The interested reader is referred to these for further details.

6.4.2 Zero-Forcing Receiver

The ZF receiver is a linear receiver. It behaves like a linear filter and separates the data streams and thereafter independently decodes each stream. We assume that the channel matrix $\mathbf{H}$ is invertible and estimate the transmitted data symbol vector as

$$\hat{\mathbf{s}} = (\mathbf{H}^H \mathbf{H})^{-1} \mathbf{Hs} = \mathbf{H}^\dagger \mathbf{s} \tag{6.41}$$

where † represents pseudoinverse [5]. Since an inverse of $\mathbf{H}$ can only exist if the columns of $\mathbf{H}$ are independent, it is assumed that $\mathbf{H} = \mathbf{H}_\omega$ (i.e., the entries are i.i.d). The noise in the separated streams is correlated and consequently the SNRs are not independent. The SER on any one channel averaged over all channel instances is upper-bounded by [10]

$$\overline{P}_e \leq \overline{N}_e \left(\frac{\rho d_{\min}^2}{2M_T} \right)^{-(M_R - M_T + 1)} \tag{6.42}$$

Equation (6.42) shows that the diversity order of each stream is $M_R - M_T + 1$. The ZF receiver decomposes the link into M_T parallel streams, each with diversity gain and array gain proportional to $M_R - M_T + 1$. Hence, it is suboptimum.

6.4.3 MMSE Receiver

We examine another linear detection algorithm to the problem of estimating a random vector s on the basis of observations y is to choose a matrix $\mathbf{B}$ that minimizes the mean square error [11]

$$\epsilon^2 = E\left[(\mathbf{s} - \hat{\mathbf{s}})^T (\mathbf{s} - \hat{\mathbf{s}}) \right] = \left[(\mathbf{s} - \mathbf{By})^T (\mathbf{s} - \mathbf{By}) \right] \tag{6.43}$$

The solution of the linear MMSE is given by

$$\hat{\mathbf{s}} = \mathbf{B} \times \mathbf{r} = \left(\frac{1}{SNR} \mathbf{I}_{M_R} + \mathbf{H}^H \mathbf{H} \right)^{-1} \mathbf{H}^H \times \mathbf{r} \tag{6.44}$$

where the superscript H denotes the complex conjugate transpose. The ZF receiver perfectly separates the cochannels' signals at the cost of noise enhancement. The MMSE receiver, on the other hand, can minimize the overall error caused by noise and mutual interference between the cochannel signals, but this is at the cost of separation quality of the signals [11].

6.4.4 Successive Cancellation Receiver

The successive cancellation receiver (SUC) algorithm is usually combined with V-BLAST receivers. This provides improved performance at the cost of increased computational complexity. Rather than jointly decoding the transmitted signals, this nonlinear detection scheme first detects the first row of the signal and then cancels its effect from the overall received signal vector. It then proceeds to the next row. The reduced channel matrix now has dimension $M_R \times (M_T - 1)$ and the signal vector has dimension $(M_T - 1) \times 1$. It then does the same operation on the next row. The channel matrix now reduces to $M_R \times (M_T - 2)$ and the signal vector reduces to $(M_T - 2) \times 1$ and so on. If we assume that all the decisions at each layer are correct, then there is no error propagation. Otherwise, the error rate performance is dominated by the weakest stream, which is the first stream decoded by the receiver. Hence, the improved diversity performance of the succeeding layers does not help. To get around this problem the ordered successive cancellation (OSUC) receiver was introduced. In this case, the signal with the strongest signal-to-interference-noise (SINR) ratio is selected for processing. This improves the quality of the decision and reduces the chances of error propagation. This is like an inherent form of selection diversity wherein the signal with the strongest SNR is selected.

The OSUC algorithm is as follows:

- *Ordering:* Determine the optimal detection order by choosing the row with minimum Euclidian norm (strongest SINR).
- *Nulling:* Estimate the strongest transmit signal by nulling out all the weaker transmit signals.
- *Slicing:* Detect the value of the strongest transmit signal by slicing to the nearest signal constellation value.
- *Cancellation:* Cancel the effect of the detected signal from the received signal vector to reduce the detection complexity for the remaining signals.

The OSUC algorithm is usually combined with the ZF or MMSE receiver, as in the V-BLAST algorithm discussed in Section 6.2.2.1.

6.4.5 Zero Forcing V-BLAST Receiver

This has already been discussed in Section 6.2.2.1.

6.4.6 MMSE V-BLAST Receiver

The OSUC is combined with the MMSE algorithm to yield a performance superior to the ZF with OSUC combination. The MMSE receiver suppresses both the interference and noise components, whereas the ZF receiver removes only the interference components. This implies that the mean square error between the transmitted symbols and the estimate of the receiver is minimized. Hence, MMSE is superior to ZF in the presence of noise. The algorithm is as follows:

Initialization

$$i \leftarrow 1$$

$$\mathbf{r}_1 = \mathbf{r}$$

$$\mathbf{G}_1 = \left(\mathbf{H}^H\mathbf{H} + \sigma^2\mathbf{I}_{M_T}\right)^{-1}\mathbf{H}^H$$

$$k_1 = \arg\min_j \left\|(G_i)_j\right\|^2$$

Recursion

$$\mathbf{w}_{k_1} = (\mathbf{G}_i)_{k_i}$$

$$y_{k_i} = \mathbf{w}_{k_i}^T \mathbf{r}_i$$

$$\hat{a}_{k_i} = Q(y_{k_i})$$

$$\mathbf{r}_{i+1} = \mathbf{r}_i - \hat{a}_{k_i}(\mathbf{H})_{k_i}$$

$$\mathbf{G}_{i+1} = \left(\mathbf{H}_i^H \mathbf{H}_i + \sigma^2 \mathbf{I}_{\mathrm{M_T}}\right)^{-1} \mathbf{H}_i^{\mathrm{H}}$$

$$k_{i+1} = \operatorname*{arg\,min}_{j \notin \{k_1,\dots,k_i\}} \left\|(G_{i+1})_j\right\|^2$$

$$i = i + 1$$

where σ^2 is the variance of i.i.d. complex Gaussian random noise with zero mean. It can be seen that the detection ordering is based on the SINR.

6.4.7   Simulation Results

Figure 6.9 shows the comparative simulation results for ML, MMSE (OSUC), and ZF (OSUC) using V-BLAST systems. This was a 2×2 system with QPSK modulation. The results show that ML is the best in performance followed by MMSE (OSUC) and ZF (OSUC). The power is normalized across the transmit antennas and the channel is a slow Rayleigh fading channel.

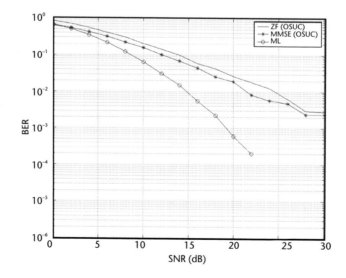

Figure 6.9 Comparison of ML, MMSE (OSUC), and ZF (OSUC) for a 2×2 system using QPSK modulation.

6.4.8   Receivers for HLST and DLST Systems

Receivers for such systems are the same as previously discussed. The main difference is that they should decode on block basis instead of on symbol basis. In a block, the information and parity bits are included in the same block. Therefore, the diversity order will remain as $M_R - M_T + 1$ for ZF and MMSE receivers. Coding improves the quality of reception and will mitigate the error propagation phenomenon. The ML receiver achieves M_R order diversity. In addition, temporal coding provides coding gain.

6.5   Iterative Receivers

Ideally, in order to achieve high capacities, we would like to use very large code words followed by ML decoders at the receiver. Reality precludes us from these goals, because the block sizes are finite in length and ML receivers are too complex. In encoded systems the complexity of ML receivers is further enhanced because it has to perform joint detection and decoding on an overall trellis obtained by combining the trellises of the layered space-time coded and the channel code. The complexity of the receiver is an exponential function of the product of the number of transmit antennas and the code memory order. Iterative receivers can approach optimal performance with a tolerable receiver complexity [12–14].

Figure 6.10 is a block diagram of an iterative receiver.

This receiver can be applied only to coded LST systems. The decoders are soft input/soft output decoders. The outer coded bits are then subtracted from the input and interleaved. The interleaved output is canceled *a posteriori* from the preceding received signal. Interleaving helps receiver convergence. This is called soft iterative interference cancellation. The decoder can apply the maximum *a posteriori* algorithm (MAP) among many possible. This algorithm is optimum in the sense that it minimizes the bit error probability at the decoder output. This ends our discussion

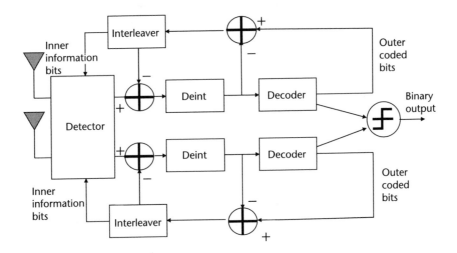

Figure 6.10   Generic block diagram of an iterative receiver.

on iterative receivers. Further analysis is beyond the scope of this book. The interested reader is referred to [15] and the references listed therein.

6.6 The Effect of Imperfect Channel Estimation on Code Performance

We carry out imperfect estimation using the MMSE technique discussed in the previous chapters. The results are shown in Figure 6.11 for QPSK code with one transmit and two receive antennas (SIMO channel) and imperfect channel estimation in a slow Rayleigh fading channel. In the simulation, 10 orthogonal signals in each data frame of 130 symbols are used as pilot sequence (preamble) to estimate the channel state information at the receiver. From the figure, we can see that the deterioration due to imperfect channel estimation is about 0.7 dB throughout.

6.7 Effect of Antenna Correlation on Performance

Figure 6.12 shows the performance of QPSK code.

We note that there is degradation in bit error rates with rising correlation across the receiver antennas. The channel is a slow Rayleigh fading channel and it is assumed that there is no correlation at the transmitter. The modulation is QPSK.

6.8 Diversity Performance of SM Receivers

Before concluding this chapter, we summarize the diversity performance of spatial multiplexing receivers in Table 6.1.

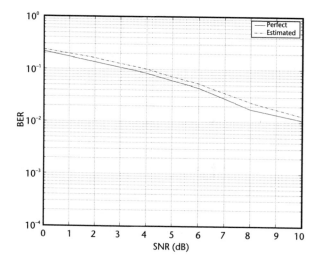

Figure 6.11 Imperfect channel estimation for QPSK SIMO channel.

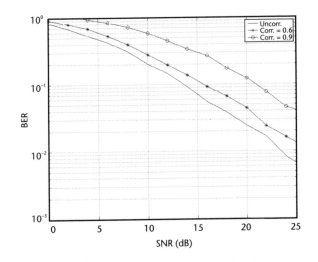

Figure 6.12 Receiver correlation effect on a 2 × 2 system with QPSK modulation.

Table 6.1 Diversity Order and SNR Performance of SM Receivers

Receiver	Diversity Order	SNR Loss
ZF (OSUC)	$\geq M_R - M_T + 1 \leq M_R$	Low
MMSE (OSUC)	$\geq M_R - M_T + 1 \leq M_R$	Low
ML	M_R	Zero

Both the ZF and MMSE receivers in the OSUC mode show a diversity order of more than $M_R - M_T + 1$, but less than M_R. The ML receiver is the optimum receiver and shows the full diversity order of M_R.

6.9 Summary

In this chapter we examined the spatial multiplexing problem. We noted that essentially there are two broad categories of LST transmitters, those with horizontal encoding and those with vertical encoding. The former is subdivided into a diagonal encoding scheme called D-BLAST and a threaded encoding scheme called TLST coding. We determined that TLST codes yielded the maximum transmit diversity. The vertical encoded scheme also incorporated an uncoded scheme called V-BLAST, which has gained a lot of popularity because of its simplicity. We then examined the design criteria for HLST and DLST codes in both slow and fast fading channels. In the analysis, we considered component codes with large as well as small Hamming distances. We determined that the best performance is determined by the code with the highest Euclidian distance in both slow and fast fading channels. We then examined various types of LST receivers. It was found that ML is the best in performance, followed by MMSE (OSUC) and ZF (OSUC). We also briefly introduced the reader to the generic block diagram of iterative receivers as a means of achieving good performance with tolerable complexity. Finally, we examined the behavior of SM systems in the presence of channel imperfections and antenna correlation.

6.10   Simulation Exercises

1. Using the simulator provided with this book, recover the curves given in the chapter for V-BLAST systems. Study the coding.
2. Plot the MMSE curves for V-BLAST with a correlation of 0.6. How does this curve compare with a MMSE curve with no correlation?

References

[1]   Foschini, G., "Layered Space-Time Architecture for Wireless Communication in a Fading Environment When Using Multielement Antennas," *Bell Labs Technical Journal*, Autumn 1996, pp. 41–59.

[2]   El Gamal, H., and A. R. Hammons, "A New Approach to Layered Space-Time Coding and Signal Processing," *IEEE Trans. Inf. Theory*, Vol. 47, No. 6, September 2001, pp. 2321–2334.

[3]   Tarokh, V., et al., "Combined Array Processing and Space-Time Coding," *IEEE Trans. on Inform. Theory*, Vol. 45, May 1999, pp. 1121–1128.

[4]   Golden, G. D., et al., "Detection Algorithm and Initial Laboratory Results Using the V-BLAST Space-Time Communication," *Electronic Letters*, Vol. 35, No. 1, January 7, 1999, pp. 14–16.

[5]   Strang, G., *Linear Algebra and Its Applications*, Fort Worth, TX: Saunders College Publishing, Brace Jovanovich College Publishers, 3rd edition, 1988.

[6]   Firmanto, W., et al., "Layered Space-Time Coding: Performance Analysis and Design Criteria," *Globecom 2001*, Vol. 2, November 2001, pp. 1083–1087.

[7]   Massey, J., and T. Mitterholzer, "Convolutional Codes Over Rings," *Proc. of the Fourth Joint Sweden-USSR Int. Workshop on Inform. Theory*, Gotland, Sweden, 1989.

[8]   Viterbo, E., and J. Boutros, "A Universal Lattice Code Decoder for Fading Channels," *IEEE Trans. Inf. Theory*, Vol. 45, July 1999, pp. 1639–1642.

[9]   Damen, G., A. Chkeif, and J. Belfiore, "Lattice Code Decoder for Space-Time Codes," *IEEE Commn. Letters*, Vol. 4, No. 5, May 2000, pp. 161–163.

[10] Paulraj, A., R. Nabar, and D. Gore, *An Introduction to Space-Time Wireless Communications*, Cambridge, UK: Cambridge University Press, 2003.

[11] Bolcskei, H., and A. Paulraj, "Multiple-Input Multiple-Output (MIMO) Wireless Systems," *Communications Handbook*, CRC Press, 2001.

[12] Berrou, C., A. Glavieux, and P. Thitimajshima, "Near Shannon Limit Error Correcting Coding and Decoding: Turbo-codes," *Proc. IEEE ICC*, Switzerland, May 1993, pp. 1064–1070.

[13] Wang, X., and V. Poor, "Iterative (Turbo) Soft Interference Cancellation and Decoding for Coded CDMA," *IEEE Trans. Commn.*, Vol. 47, No. 7, July 1999, pp. 1046–1067.

[14] Ten Brink, S., J. Speidel, and R. H. Yan, "Iterative Demapping for QPSK Modulation, *Electronic Letters*, Vol. 34, No. 15, July 1998, pp. 1459–1460.

[15] Vucetic, B., and J. Yuan, *Space-Time Coding*, Chichester, UK: John Wiley & Sons Ltd., 2003.

CHAPTER 7

Orthogonal Frequency Division Multiplexing

7.1 Introduction

This book is not about OFDM. The reader will find excellent sources listed in the references section on this topic. However, in the interests of continuity, it is necessary to cover certain basic aspects of OFDM to enable the reader to understand and correctly appreciate subsequent chapters. Toward this end, we shall cover the basics of OFDM in this chapter. First, we will discuss the need for a multicarrier scheme and introduce the OFDM technique as a very efficient multicarrier scheme. We will then discuss OFDM modulation and demodulation with respect to a typical OFDM transmitter/receiver chain. The chapter will then examine popular synchronization schemes and channel estimation techniques common to OFDM systems. Finally, we will examine the impact of OFDM on packet transmission systems.

7.2 Basic Principles

7.2.1 Data Transmission over Multipath Channels

In a classical wireless transmission scenario, the transmitted signal arrives at the receiver using various paths of different lengths. This is shown in Figure 7.1. Since the multiple versions of the signal interfere with each other [intersymbol interference (ISI)], it becomes extremely difficult to extract the original information.

The common representation of the multipath channel is the channel impulse response of the channel, which is the signal at the receiver if a single pulse is transmitted, as shown in Figure 7.2.

If the system transmits information at discrete time intervals T, then the critical measure concerning the multipath channel is the delay τ_{max} of the longest path with respect to the earliest path. A received symbol can theoretically be influenced by τ_{max}/T previous symbols. This influence has to be estimated and compensated for in the receiver.

7.2.2 Single Carrier Approach

In Figure 7.3, the general structure of a single carrier transmission is shown. The transmitted symbols are pulse-formed by a transmitter filter. After passing the

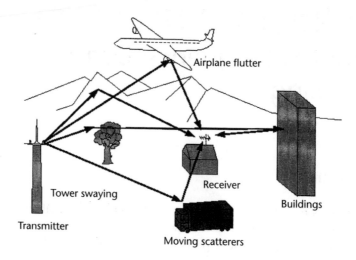

Figure 7.1   Multipath transmission in a broadcasting application.

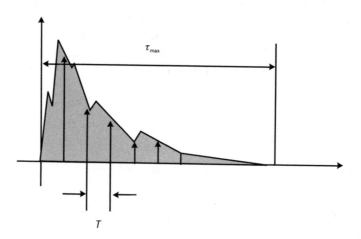

Figure 7.2   Effective length of channel impulse response.

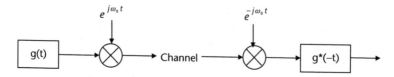

Figure 7.3   Basic structure of a single carrier system.

multipath channel in the receiver, a filter matched to the channel is used to maximize the SNR.

Suppose the scenario is characterized by the following conditions:

- Transmission rate: $R = (1/T) = 8$ Msym/sec
- Maximum channel delay: $\tau_{max} = 220$ μsec

For the single carrier system, this results in an ISI of $\frac{\tau_{\max}}{T} \approx 1,800$.

The complexity involved in removing this interference is tremendous. The problem gets drastically reduced in a multicarrier approach.

7.2.3 Multicarrier Approach

Figure 7.4 shows the general structure of a multicarrier system.

The original data stream of rate R is multiplexed into N parallel data streams of rate $R_{mc} = 1/T_{mc} = R/N$. Each of the data streams is modulated with a different frequency and the resulting signals are transmitted together in the same band. Correspondingly, the receiver consists of N parallel receiver paths. Due to the increased distance between transmitted symbols, the ISI for each subsystem reduces to

$$\frac{\tau_{\max}}{T_{mc}} = \frac{\tau_{\max}}{N \cdot T}$$

For the case of $N = 8,192$, we have an ISI of

$$\frac{\tau_{\max}}{T_{mc}} = 0.2$$

Such a little ISI can often be tolerated and no extra measure such as an equalizer is needed. However, implementing an 8,192 multichannel receiver is not feasible. This has given rise to the concept of OFDM.

7.3   OFDM

OFDM [1, 2] is a multicarrier transmission technique, which divides the available spectrum into many carriers, each one being modulated by a low-rate data stream. Unlike FDMA, OFDM uses the spectrum much more efficiently by spacing the

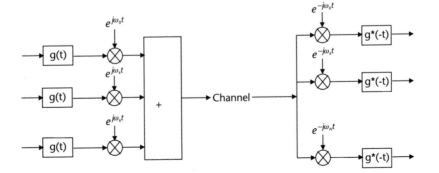

Figure 7.4   Basic structure of a multicarrier system.

channels closer together. This is achieved by making all the carriers orthogonal to one another, preventing interference between closely spaced carriers.

FDMA wastes the spectrum owing to the requirement of inserting guard bands between channels for channel isolation and filtering. In a typical system, up to 50% of the total spectrum is wasted in this manner. This problem becomes worse as the channel bandwidth becomes narrower and the frequency band increases.

OFDM overcomes this problem by splitting the available bandwidth into many narrow band channels. The carriers for each channel are made orthogonal to one another, allowing them to be spaced very close together with no overhead, as in FDMA. This is shown in Figure 7.5 for four carriers.

The orthogonality of the carriers means that each carrier has an integer number of cycles over a symbol period. Due to this, the spectrum of each carrier has a null at the center frequency of each of the other carriers in the system. This results in no interference between the carriers. Each carrier in an OFDM system signal has a very narrow bandwidth (e.g., 1 KHz), thus resulting in low symbol rate. This results in the signal having a high tolerance to multipath delay spread, as the delay spread must be very long to cause significant ISI (e.g., >500 μsecs).

7.4 OFDM Generation

To generate OFDM successfully, the relationship between all the carriers must be carefully controlled to maintain the orthogonality between the carriers. Because of this, OFDM is generated by first choosing the spectrum required, based on the input data and modulation scheme used. Each carrier to be produced is assigned some data to transmit. The required amplitude and phase of the carrier is then calculated, based on the modulation scheme (typically, BPSK, QPSK, or QAM). The required spectrum is then converted back to its time domain signal using an

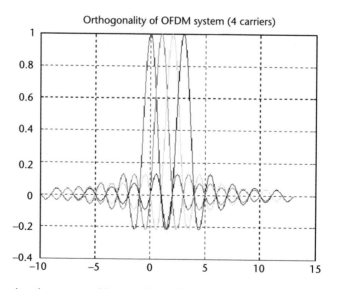

Figure 7.5 Showing the spectra of four carriers orthogonal to each other.

IFFT. The IFFT performs the transformation very efficiently and provides a simple way of ensuring the orthogonality of the carrier signals.

The FFT transforms a cyclic time domain signal into its equivalent frequency spectrum. This is done by finding the equivalent waveform generated by a sum of orthogonal sinusoidal components. The amplitude and phase of the sinusoidal components represent the frequency spectrum of the time domain signal. The IFFT performs the reverse process, transforming a spectrum (amplitude and phase of each component) into a time domain signal. The orthogonal carriers required for the OFDM signal can be easily generated by setting the amplitude and phase of each bin, then performing the IFFT. Since each bin of an IFFT corresponds to the amplitude and phase of a set of orthogonal sinusoids, the reverse process guarantees that the carriers generated are orthogonal.

The system block diagram of an OFDM system is shown in Figure 7.6.

The input is a binary serial data stream. This is encoded using any suitable modulation technique. In this case we have used M-QAM modulation to convert the data stream into a multilevel data stream. The binary input will be transformed into a multilevel signal reducing the symbol rate to, $D = \dfrac{R}{\log_2 M}$ symbols/sec, where R is the bit rate of the data stream in bits/sec.

If we convert this serial data to parallel, the data rate gets further reduced by N, where N is the number of parallel channels. Hence, these parallel channels are essentially low data rate channels and since they are narrowband, they experience flat fading. This is the greatest advantage of the OFDM technique.

This parallel data stream is then subjected to an IFFT and then summed. This constitutes the OFDM modulation. Mathematically it can be expressed as,

$$X(t) = \sum_{n=1}^{N} (a_n + jb_n)(\cos \omega_n t + j \sin \omega_n t) \tag{7.1}$$

After OFDM modulation, a guard interval is inserted to suppress ISI caused by multipath distortion. This guard interval is also called cyclic prefix. A cyclic prefix is a copy of the last part of the OFDM symbol, which is "prepended" to the transmitted symbol. This makes the transmitted symbol periodic, which plays a decisive role in identifying frames correctly, so as to avoid ISI and intercarrier interference (ICI). The guard interval allows time for the multipath signals from the previous symbol to die away before the information from the current symbol is gathered. Using this cyclic extended symbol, the samples required for performing the FFT (to decode the symbol) can be taken anywhere over the length of the cyclic prefix. This provides multipath immunity as well as symbol time synchronization tolerance. As long as the multipath delay echoes stay within the guard period duration, there is strictly no limitation regarding the signal level of the echoes: they may even exceed the signal level of the shorter path. The signal energy from all the paths just add at the input to the receiver and since the FFT is energy conservative, the whole available power feeds the decoder. If the delay spread is longer than the guard interval, then ISI is caused. The price we pay for this advantage is the loss of signal-to-noise ratio, as the cyclic prefix constitutes "noise" so far as the signal is concerned and provides less power to the signal frame by occupying

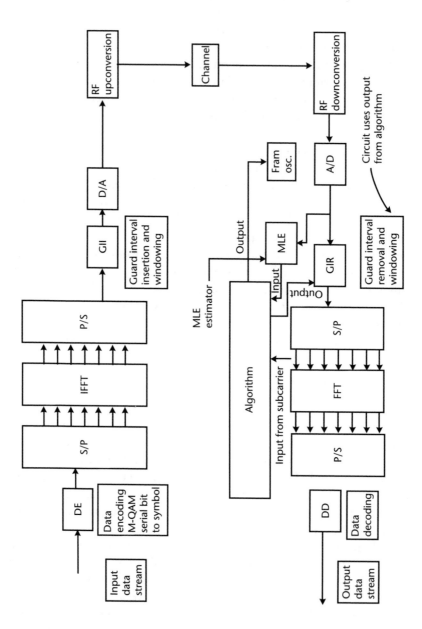

Figure 7.6 OFDM system configuration.

signal space. The windowing applied after guard time insertion is to reduce the out-of-band radiation when the OFDM symbol would have a "discontinuous" end or beginning. Normally, it is a raised cosine window. The signal is then D/A converted to produce the analog baseband signal, upconverted to RF, and then transmitted. The transmission is usually amplitude modulated (AM). The reception is the converse and is self-explanatory. A major disadvantage of OFDM is that the envelope is not constant. Due to the summation of sine waves, we have a large peak to average ratio (PAR). This poses problems regarding the linear bandwidth of the RF amplifiers. Various techniques have been investigated with a view to reducing the problems caused by PAR. Throughout this chapter, the following assumptions are made regarding the OFDM system:

- A cyclic prefix is used.
- The impulse response of the channel is shorter than the cyclic prefix. Otherwise there will be ISI.
- Channel noise is additive, white, and complex Gaussian.
- The fading is slow enough for the channel to be considered constant *during one OFDM symbol* interval unless otherwise mentioned.

It was previously stated that OFDM has a synchronization problem. This aspect will be examined further in this chapter. However, we shall briefly examine the synchronization process in Figure 7.6. In this approach, we have an algorithm in the receiver, which examines the cyclic prefixes emerging from the A/D converter. It then makes maximum likelihood estimates (MLE) [3] based on these cyclic prefixes. However, MLE is just one method of synchronizing the channel and is the one shown in this figure. There are others, and we shall briefly review these later. These MLE estimates manifest themselves as large peaks, which we can then use to determine the start of the OFDM frames. The algorithm then examines these MLE estimates and simultaneously compares the timing obtained from these MLE estimates with the timing obtained from the pilot sequences in the subcarriers. It then adjusts the frame oscillator to the corrected sampling frequency. Simultaneously, it sends a pulse to the guard interval removal (GIR) circuit to remove the cyclic prefixes.

7.5 Synchronization Issues

We carry out a brief survey of existing synchronization techniques. We begin the survey by deriving an expression, which defines the symbol timing and carrier frequency offset between the OFDM transmitter and OFDM receiver. From this expression we determine the salient points to be borne in mind during synchronization of OFDM systems. We then study and analyze the behavior of popular synchronization schemes.

7.5.1 Symbol Time and Frequency Carrier Offset Derivation

The OFDM transmitter shown in Figure 7.7 [4] first converts the high-rate data streams into parallel data of N subchannels. The transmission data is normally

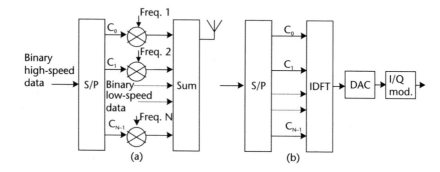

Figure 7.7 (a) Basic OFDM modulator, and (b) its equivalent IDFT implementation configuration.

modulated at this stage using QPSK, DQPSK, QAM or any other modulation scheme. In our approach this modulation is carried out before S/P conversion. Considering QPSK, the quadrature modulated data sequence of N subchannels is denoted $(x_{-N/2}, \ldots, x_{-1}, x_1, x_2, \ldots, x_{N/2})$, where each x_i is a complex number $x_i = a_i + jb_i$ with a_i, $b_i = \pm 1$, with $-N/2 \leq i \leq N/2$ and $i \neq 0$. This process is also called symbol mapping in contrast to OFDM modulation being the next step.

The OFDM modulation can be performed using the IDFT, which is represented by the IFFT instead of generating the carriers separately.

These mapped data are fed into the IDFT circuit and the OFDM signal is generated. One transmitted OFDM symbol is given by,

$$s(t) = \begin{cases} \displaystyle\sum_{i=-N/2, i \neq 0}^{N/2} x_i \cdot e^{(j2\pi f_i)} & \text{for } kT - T_{\text{win}} - T_g \leq t \leq kT + T_{FFT} + T_{\text{win}} \\ 0 & \text{elsewhere} \end{cases}$$

$$(7.2)$$

where N is the number of used subcarriers (excluding dc-carrier, which can be distorted by the electronics) in the FFT points for the OFDM modulation, x_i the complex modulated data, f_i the frequency of the ith subcarrier, T_{FFT} the duration of the total number of FFT-points, T_g the duration of the guard interval, T_{win} the duration of the windowing, and i the index on the subcarrier.

The frequency of the ith subcarrier can be replaced using $f_i = f_0 + i/T_{FFT}$ with f_0 as the center frequency of the occupied frequency spectrum, with $-N/2 \leq i \leq N/2$.

The transmitter pulse prototype $(w(t))$ of each OFDM symbol is smoothed with a window around the edges of the OFDM frame to prevent out-of-band radiation resulting from sinc-function sidelobes. This is defined by,

$$w(t) = \begin{cases} \frac{1}{2}[1 - \cos \pi(t - T_{\text{win}} - T_g)/T_{\text{win}}] & \text{for } -T_{\text{win}} - T_g \leq t \leq -T_{\text{guard}} \\ 1 & \text{for } -T_{\text{guard}} \leq t \leq T_{FFT} \\ \frac{1}{2}[1 - \cos \pi(t - T_{FFT})/T_{\text{win}}] & \text{for } T_{FFT} \leq t \leq T_{FFT} + T_{\text{win}} \end{cases}$$

$$(7.3)$$

For a certain number of transmitted OFDM symbols, (7.2) shifts in time and becomes,

$$s_k(t - kT) = \begin{cases} w(t - kT) \displaystyle\sum_{i=-N/2, i \neq 0}^{N/2} x_{i,k} \cdot e^{\left(j2\pi\left(f_0 + \frac{i}{T_{FFT}}\right)(t - kT)\right)} & \begin{array}{l} \text{for } kT - T_{win} \leq \\ t \leq kT + T_{FFT} + \\ T_{win} \end{array} \\ 0 & \text{elsewhere} \end{cases}$$

(7.4)

where T is the duration of a complete OFDM symbol and $x_{i,k}$ is the complex modulated data for the ith subcarrier and kth symbol.

The continuous sequence of transmitted OFDM symbols (see Figure 7.8) is expressed as

$$s(t) = \sum_{k=-\infty}^{\infty} s_k(t - kT)$$

(7.5)

Furthermore, the complex equivalent lowpass signal will be used, which is represented by

$$s_{lp,k}(t - kT) = \begin{cases} w(t - kT) \displaystyle\sum_{i=-N_{FFT}/2}^{N_{FFT}/2 - 1} x_{i,k} \cdot e^{\left(j\frac{2\pi.i}{T_{FFT}}(t - kT)\right)} & \begin{array}{l} \text{for } kT - T_{win} - T_g \\ \leq t \leq kT + T_{FFT} \\ + T_{win} \end{array} \\ 0 & \text{elsewhere} \end{cases}$$

(7.6)

In this equation, N_{FFT} is used instead of N, assuming that $x_i = 0$ at the band edges.

The transmitted OFDM signal experiences influences of the time-varying multipath fading propagation channel. The channel is expressed by its equivalent lowpass impulse response $h(\tau, t)$ plus AWGN $n(t)$,

$$r_{LP}(t) = h(\tau, t) * s_{LP}(t) + n(t) = \int_0^{\tau_{max}} h(\tau, t) \cdot s_{LP}(t - \tau) \, d\tau + n(t)$$

(7.7)

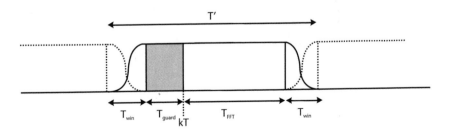

Figure 7.8 Continuous sequence of OFDM symbols.

where $h(\tau, t)$ denotes the impulse response at time $t = \tau$ (τ denotes the excess delay of the channel impulse response), which is zero outside the range of this convolutional integral $\tau = [0, \tau_{max}]$. The square value of the impulse response can be considered a power delay intensity function and yields useful information about the multipath characteristics of the channel. This power delay profile may be treated as a delay density function, where the delay is weighted by the signal level at that delay-instant [2].

The power delay profile for a typical channel model is shown in Figure 7.9(a) and a realistic power delay profile is shown in Figure 7.9(b).

The excess delay of the channel at $\tau = 0$ is defined as the delay time at which the first wave arrives at the receiver. Thus, over an ideal channel, the transmit and receive time instants are defined to be equal.

Two assumptions are made to simplify the derivation of the received OFDM signal. First, the channel will be considered time-invariant during the transmission of one OFDM symbol (i.e., $h(\tau, t)$ simplifies to $h(\tau)$). Second, the maximum excess delay is defined as $\tau_{max} < T_{guard}$. This results in no interference of the previous OFDM symbol into the effective period of the current one. Thus, ISI is suppressed in case of perfect time synchronization.

The demodulation of the OFDM signal is performed by the inverse operation to (7.2) (i.e., the extraction of the constellation points uses DFT, which is represented by the FFT). This is shown by,

$$\hat{x}_i = \frac{1}{T_{FFT}} \int_{t=0}^{T_{FFT}} s(t) \cdot e^{-j2\pi f_i t} \, dt \tag{7.8}$$

where T_{FFT} denotes the duration of one OFDM symbol and f_i the frequency of the ith subcarrier.

The frequency spacing of $1/T_{FFT}$ preserves the orthogonality of the complex signals between the carriers, which is seen from,

$$\frac{1}{T_{FFT}} \int_{t=0}^{T_{FFT}} e^{-j\frac{2\pi}{T_{FFT}}(i-i')t} \, dt = \begin{cases} 1 & \text{for } i = i' \\ 0 & \text{for } i \neq i' \end{cases} \tag{7.9}$$

an expression that occurs in the demodulation of (7.2).

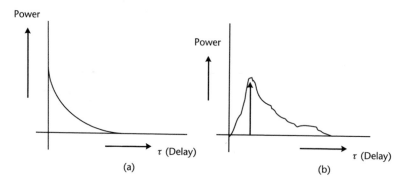

Figure 7.9 Power delay profile of (a) channel model, and (b) realistic channel.

From the received OFDM signal ($r_{LP}(t)$), the transmitted signal constellation points x_i can be extracted. Knowing the exact time instant kT at which the kth OFDM symbol starts, the received signal constellations $y_{i,k}$ can be written as [3],

$$y_{i,k} = \frac{1}{T_{FFT}} \int_{t=kT}^{kT+T_{FFT}} r_{LP}(t) \cdot e^{-j2\pi i \frac{(t-kT)}{T_{FFT}}} \, dt \tag{7.10}$$

$$= \frac{1}{T_{FFT}} \int_{t=kT}^{kT+T_{FFT}} \left[\int_0^{\tau_{max}} h(\tau) \cdot s_{LP}(t-\tau) \, d\tau + n(t) \right] \cdot e^{-j2\pi i \frac{(t-kT)}{T_{FFT}}}$$

Note that $w(t-kT) = 1$ in the range of integration and thus is left out of the equation.

The received OFDM signal suffers from synchronization errors in time and frequency that must be considered in the derivation. Due to a frequency mismatch between the oscillators of the transmitter and receiver, a frequency offset δf and a carrier phase offset θ_0 must be accounted for as a frequency shift in the lowpass equivalent received signal

$$r'_{LP}(t) = r_{LP}(t) \cdot e^{j(2\pi\delta ft + \theta_0)} \tag{7.11}$$

Furthermore, a symbol time offset δt changes the ideal demodulation interval of $t \in [kT, kT + T_{FFT}]$ to $t \in [kT + \delta t, kT + T_{FFT} + \delta t]$, which can be accounted for by shifting the integration interval in (7.10), written as

$$y_{i,k} = \frac{1}{T_{FFT}} \int_{t=kT+\delta t}^{kT+T_{FFT}+\delta t} r'_{LP}(t) \cdot e^{-j2\pi i \frac{(t-kT-\delta t)}{T_{FFT}}} \, dt \tag{7.12}$$

Combining (7.11) and (7.12) results in,

$$y_{i,k} = \frac{1}{T_{FFT}}$$

$$\cdot \int_{t=kT+\delta t}^{kT+T_{FFT}+\delta t} \left(\left(\int_0^{\tau_{max}} h(\tau) \cdot \left(\int_{i'=-N_{FFT}/2}^{N_{FFT}/2-1} x_{i',k} \cdot e^{j2\pi \left(\frac{i'}{T_{FFT}}\right)(t-kT-\tau)} \right) d\tau \right) \cdot e^{j(2\pi\delta ft + \theta_0)} \right)$$

$$\cdot e^{-j2\pi i \frac{(t-kT-\delta t)}{T_{FFT}}} \, dt$$

$$+ \frac{1}{T_{FFT}} \int_{t=kT+\delta t}^{kT+T_{FFT}+\delta t} n(t) \cdot e^{-j2\pi i \frac{(t-kT-\delta t)}{T_{FFT}}} \, dt \tag{7.13}$$

The second term in (7.13) denotes independent noise samples $n_{i,k}$ since the exponential terms represent orthogonal functions.

Substituting for $t' = t - KT - \delta t$ in (7.13) and changing the order of summation, the following equation is obtained,

$$y_{i,k} = e^{j\theta} \cdot x_{i',k} \cdot h_{i',k}$$

$$\cdot \frac{1}{T_{FFT}} \left[\int_{t'=0}^{T_{FFT}} e^{j2\pi\delta f t'} \cdot e^{j2\pi i' \frac{t'}{T_{FFT}}} \cdot e^{-j2\pi i' \frac{t'}{T_{FFT}}} \cdot e^{j2\pi\delta t \left(\frac{i'}{T_{FFT}}\right)} dt' \right]$$

If $i = i'$ we obtain

$$y_{i,k} = e^{j\theta} \cdot x_{i,k} \cdot h_{i,k} \cdot \frac{1}{T_{FFT}} \left[\int_{t'=0}^{T_{FFT}} e^{j2\pi\delta f t'} \cdot e^{j2\pi\delta t \left(\frac{i'}{T_{FFT}}\right)} dt' \right]$$

$$y_{i,k} = e^{j\theta} \cdot \sum_{i'=-N_{FFT}/2}^{N_{FFT}/2-1} x_{i',k} \cdot \frac{1}{T_{FFT}} \int_{t'=0}^{T_{FFT}} \left(\int_{0}^{\tau_{max}} h(\tau) \cdot e^{-j2\pi i' \left(\frac{\tau}{T_{FFT}}\right)} d\tau \right)$$

$$\cdot e^{j2\pi i' \frac{t'}{T_{FFT}}} \cdot e^{j2\pi i' \frac{\delta t}{T_{FFT}}} \cdot e^{j2\pi\delta f t'} \cdot e^{-j2\pi i \frac{t'}{T_{FFT}}} dt' + n_{i,k} \qquad (7.14)$$

where $\theta = \theta_0 + 2\pi\delta f(kT + \delta t)$. The inner integral in the last equation represents the Fourier transform of $h(\tau)$ at the frequency instants $i'/T_{FFT} = i'\Delta f$, (i.e., the channel transfer function at the subcarrier frequencies for the kth OFDM symbol, denoted as $h_{i',k} = H(i'\Delta f)$ where $\Delta f = 1/T_{FFT}$ is the subcarrier spacing). Using this notation (7.14) can be written as the following,

$$y_{i,k} = e^{j\theta} \cdot x_{i',k} \cdot h_{i',k} \qquad (7.15)$$

$$\cdot \frac{1}{T_{FFT}} \left[\int_{t'=0}^{T_{FFT}} e^{j2\pi\delta f t'} \cdot e^{j2\pi i' \frac{t'}{T_{FFT}}} \cdot e^{-j2\pi i \frac{t'}{T_{FFT}}} \cdot e^{j2\pi\delta t \left(\frac{i'}{T_{FFT}}\right)} dt' \right]$$

If $i = i'$ we obtain

$$y_{i,k} = e^{j\theta} \cdot x_{i,k} \cdot h_{i,k} \cdot \frac{1}{T_{FFT}} \left[\int_{t'=0}^{T_{FFT}} e^{j2\pi\delta f t'} \cdot e^{j2\pi\delta t \left(\frac{i'}{T_{FFT}}\right)} dt' \right]$$

In (7.15), if $i \neq i'$, then the first expression is the case in the presence of ICI. The ICI term can be seen as an additional noise term and thus can be represented as a degradation of the SNR.

The final expression for the received signal without the ICI term (i.e., $i = i'$), using

$$\frac{1}{T_{FFT}} \int_{t'=0}^{T_{FFT}} e^{j2\pi\delta f t'} \, dt' = e^{j\pi\delta f T_{FFT}}, \text{ sinc} (\delta f \cdot T_{FFT}) \tag{7.16}$$

yields,

$$y_{i,k} = x_{i,k} \cdot h_{i,k} \cdot \text{sinc} (\delta f \cdot T_{FFT}) \cdot e^{j\Psi_{i,k}} + n_{i,k} \tag{7.17}$$

$$\text{with } \Psi_{i,k} = \theta_0 + 2\pi\delta f \left(kT_s + \frac{T_{FFT}}{2} + \delta t \right) + 2\pi\delta t \left(\frac{i}{T_{FFT}} \right)$$

The evaluation of this expression shows a common phase rotation and attenuation of all subcarriers due to the carrier frequency offset (δf) and carrier phase offset (θ_0) and a progressive phase rotation (proportional to i) due to the symbol timing offset (δt).

Based on the preceding analysis, the following salient points should be noted during synchronization of OFDM systems:

- Synchronization errors in time and frequency in the receiver cause distortions of the received signal constellations.
- An error in the symbol time estimation (δt) (i.e., the starting position of the *effective* part of the OFDM frame is not known and will cause a phase rotation of the signal constellation points). This phase rotation is progressive [i.e., zero at the center frequency and then increasing linearly until it reaches a maximum at the edges of the frequency band (last term of Ψ)].
- The propagation channel, due to multipath fading and AWGN, will cause a phase rotation of the individual subcarriers that is expressed by the $h_{i,k}$ term in (7.17). As a consequence, it is difficult to estimate this channel phase rotation and timing offset separately. The effect of the propagation channel can be minimized by using channel estimation techniques and assuming that the timing offset is not too large. Another option is to use differential modulation to cancel out the effects of the channel by subtracting the phase distortions at consecutive received frames, which are assumed to be highly correlated.
- A small frequency mismatch due to a frequency offset (δf) between the oscillators of the transmitter and receiver or a carrier phase offset (θ_0) results in a common phase rotation of the signal constellation points at FFT outputs (see first two terms of ψ). Rotating each of the FFT outputs over an angle proportional to the carrier frequency error can compensate this. Note that the common phase rotation is advancing for consecutive OFDM symbols, seen in the term $2\pi\delta f k T_s$.
- The impact of a carrier frequency error is twofold. First, the amplitude of the desired signal is reduced and, second, the ICI arises from the adjacent subcarriers.

- Accurate timing information must be known to the receiver to produce reliable estimates of the transmitted data sequence. A timing error, which is considered to be very small ($\delta t < T_g$), implying that the channel impulse response still lies within the guard interval, will not introduce ISI (i.e., no energy is collected from the adjacent OFDM symbols and the quality of the received signal is not degraded). The timing error (δt) can be expressed in a multiple of time-samples, which is written as

$$\delta t' = \frac{\delta t}{T_s} \text{ where } T_s = \frac{T_{FFT}}{N_{FFT}} \text{ is the sampling period} \qquad (7.18)$$

The number of samples delay that corresponds to a phase shift of 90° is (neglecting the phase rotation due to the channel),

$$\arg\left(e^{-j\frac{2\pi}{N_{FFT}}i \cdot \delta i'}\right) = \frac{\pi}{2} \Rightarrow \delta t' = \frac{N_{FFT}}{4 \cdot i} \qquad (7.19)$$

Thus, a 90° phase shift corresponds to a timing error of 32 samples [at the first subcarrier ($i = 1$)] when $N_{FFT} = 128$.

The fractional timing error cannot be detected because there is always an ambiguity of ±0.5 sample time because the correlation maximum is always precisely on the sampling instance.

7.6 Survey of Synchronization Techniques

There are essentially three types of synchronization problems:

- Symbol synchronization.
- Carrier synchronization.
- Sampling frequency synchronization.

7.6.1 Symbol Synchronization

The objective here is to know when the symbol starts. A timing offset gives rise to a phase rotation of the subcarriers. This phase rotation is largest on the edges of the frequency band. If a timing error is small enough to keep the channel impulse response within the cyclic prefix, the orthogonality is maintained. In this case, the symbol timing delay can be viewed as a phase shift introduced by the channel and the phase rotations can be estimated by a channel estimator. If a time shift is larger than the cyclic prefix, ISI will occur.

There are two main methods for timing synchronization, based on pilots [5, 6] or on the cyclic prefix [7–9].

7.6.1.1 Pilots

Scheme Suggested by Schmidl
In this scheme [6], two symbols with identical data are used to estimate the frequency offset. Moose's work [10] is the basis for the Schmidl method. He derived a

maximum likelihood estimator to detect the carrier frequency offset that is calculated after the FFT in the frequency domain. The estimation technique involves repetition of data symbols and comparison of the phases of each of the carriers between the successive symbols. Since the modulation phase values are not changed, the phase shift of each of the carriers between the successive repeated symbols is due to the frequency offset and noise. The acquisition range in the Schmidl method is limited to $\pm 1/2$ of the carrier spacing and it is assumed that the symbol timing was estimated perfectly. In this sense, this method also falls under the carrier synchronization category. However, it is included in the symbol synchronization category because, it also provides a method of symbol synchronization not using a cyclic prefix.

In the paper by Schmidl, a method is presented to perform rapid synchronization with a relatively simplified computation in the time domain and an extended range for the acquisition of the carrier frequency offset. The algorithm presented here is suitable for continuous transmission (as in broadcasting) and for burst transmission (as in wireless LAN applications). The synchronization is performed on a training sequence of two OFDM symbols. The frame timing is performed using one unique OFDM symbol as the first symbol, which has a repetition within half a symbol period. In the burst mode this is very effective in determining the start of a burst of data.

The symbol timing recovery relies on searching for a training symbol with two identical halves in time domain, with the crucial assumption that the channel effects are identical, except that there will be a phase difference between them caused by the carrier frequency offset.

The training symbol configuration in time and frequency domain is shown in Figure 7.10.

The two halves of the training symbol are made identical by transmitting a PN sequence on the even subcarriers while zeros are inserted on the odd ones. In this way, the receiver can distinguish between symbols meant for synchronization and symbols that contain data. The transmitted data will not be mistaken as the start of the frame since the data symbol must contain data on the odd frequencies.

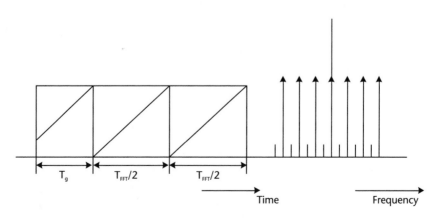

Figure 7.10   Training symbol with two identical halves in time domain resulting in nulls in odd frequencies in frequency domain.

The first half of the training symbol is considered to be identical to the second half, after passing through the channel, except for the progressive phase shift caused by the frequency offset. By multiplying the conjugate of the sample from the first half with the corresponding sample from the second half ($T_{FFT}/2$ seconds later), the arbitrary phases of the OFDM signal and the effect of the channel should cancel out. As a result, only the phase difference of $\phi = \pi T \delta f$ will remain, which can be used as an estimate for the frequency offset δf.

The phase difference is constant because the length between the samples (L) is constant. Figure 7.11 shows the computation scheme of the symbol timing estimation.

At the start of the frame, the products of each pair of samples will have approximately the same phase, so the magnitude of the sum will have a large value (like constructive interference). The sum of products can be expressed in the following equation:

$$P(d) = \sum_{m=0}^{L-1} \left(r_{d+m}^* \cdot r_{d+m+L} \right) \tag{7.20}$$

where $L = N_{FFT}/2$.

The sum of the correlation value is normalized with the received energy from the second half of the first training sequence. This energy is calculated as,

$$R(d) = \sum_{m=0}^{L-1} \left| r_{d+m+L} \right|^2 \tag{7.21}$$

Normalizing the sum of products over the received energy can then be expressed as a timing metric, that is,

$$M(d) = \frac{|P(d)|^2}{(R(d))^2} \tag{7.22}$$

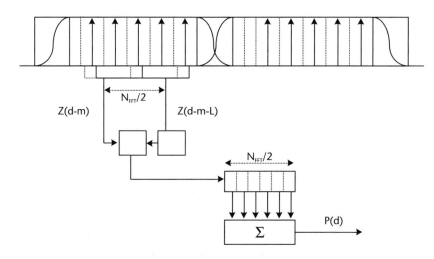

Figure 7.11 Computation of symbol timing estimation.

The maximum value will be reached as soon as the samples are pairs with distances of half a symbol period. This will result in a figure with a plateau of length equal to the guard interval, as shown in Figure 7.12. When the signal is propagating over a realistic channel, with multipath fading, the length of this plateau will be shortened by the length of the channel delay time.

The start of the frame can be taken anywhere on this plateau, as it will always be a "rough" estimation of the symbol timing error. The symbol timing error will have little effect as long as that part of the guard interval (GI) is discarded, which is corrupted by ISI.

The carrier frequency offset is estimated in two steps. First, the fractional part is detected and compensated for. Then the integer part is estimated and corrected. The fractional frequency offset is estimated calculating the angle of the sum of products (7.20) at the estimated starting position ($d_{\max}$):

$$\delta f = \frac{\Delta f}{\pi} \cdot angle\left(P(d_{\max})\right) \tag{7.23}$$

The reason for detecting only the fractional part of the frequency error is because there is an ambiguity in the angle of $P(d_{\max})$ expression of 2π, [i.e., an integer valued offset cannot be detected using (7.23)].

As a second step, the integer part of the frequency offset is estimated. The second training symbol, which contains a PN sequence modulated differentially with respect to the first training symbol, is used. This PN sequence can be retrieved and compared with a reference sequence.

Since the data transmitted on the training symbol is known, they will be used for channel estimation as well.

7.6.1.2 Cyclic Prefix

There are also synchronization algorithms based on the cyclic prefix. One method [5] exploits the difference between received samples spaced N samples apart [i.e., $r(k) - r(k + N)$]. When one of the received samples belongs to the cyclic prefix and the other one to the OFDM symbol from which it is copied, the difference

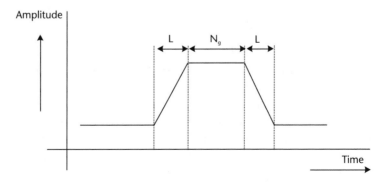

Figure 7.12 The timing metric. The maximum plateau indicates the start of the effective part of an OFDM symbol.

should be small. Otherwise the difference (between two uncorrelated random variables) will have twice the power and, hence, on average will be larger. By windowing this difference with a rectangular window of the same length as the cyclic prefix, the output signal has a minimum when a new symbol starts. There are three variations of this basic approach.

Correlation of a Pseudorandom Sequence

This method is attributed to Harada [7]. The principle is based on inserting a pilot PN sequence between symbols. The idea is to correlate between received data and delayed received data with the delay time of the period of the M sequence. In this case, during guard interval, the correlation is higher than during other intervals because the guard bits are extended by repeating information unit. The principle is shown in Figures 7.13(a) and 7.13(b).

The authors suggest the schematic as shown in Figure 7.14 to implement their technique.

One information unit is defined as the bits comprising the symbol to be transmitted plus one complete PN sequence (the pilot channel). The guard bits are inserted in the PN sequence of the pilot channel.

The initial frame separation is carried out in the same principle as in Figure 7.13(a) but employs the cyclic prefix in the pilot channel of the information unit. This determines the DFT start point and, thereafter, the algorithm separates the pilot sequence from the information sequence, as shown in Figure 7.14. This pilot sequence is then once again correlated [matched filter digital pulse compression, Figure 7.13(b)] to determine the exact start point and the delay profile. The correlation function will be sharper since this is an autocorrelation of a *complete* PN sequence and not the guard bits of a symbol. However, the important thing to note here is that the basic DFT start point is still determined based on the cyclic prefix of the information sequence and not by pulse compressing the entire PN sequence, as in Figure 7.13(b). In view of this, the authors find it necessary to integrate more than five symbols to get a reliable discrimination of the DFT start sequence during the implementation of the scheme in Figure 7.13(a). This implies that the authors require that the time and frequency offsets remain constant over at least five symbols. The PN sequence is only used to get an accurate delay profile since the PN sequence is known. The *entire* PN sequence is not used to determine the start of a new symbol sequence. This procedure is left for later in Figure 7.13(b), wherein the symbol start point indicates the start of the delay profile.

ML Estimation of Timing and Frequency Offset

This method is attributed to Sandell et al. [8]. Assume that we observe $2N + L$ consecutive samples of $r(k)$ and that these samples contain one complete $(N + L)$ sample OFDM symbol.

The position of this symbol within the observed block of samples, however, is unknown as the channel delay θ is unknown to the receiver. Define the index sets $I \triangleq [\theta, \ \theta + L - 1]$ and $I' \triangleq [\theta + N, \ \theta + N + L - 1]$, as shown in Figure 7.15. The set I' thus contains the indices of the data samples that are copied into the cyclic extension and the set I contains the indices of this extension. Collect the observed

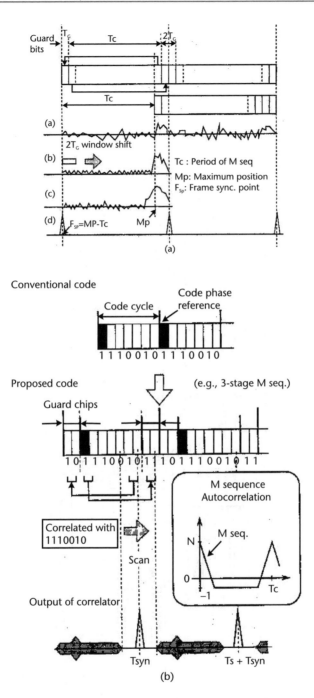

Figure 7.13 (a) Frame and DFT timing estimation method. (b) Correlation. (*From:* [9]. © 1997, IEEE.)

samples in the $(2N + L) \times 1 - vector \equiv [r(1) \ldots r(2N + L)]^T$. Notice that the samples in the cyclic extension and their replicas $k \in I \cup I'$ are pairwise correlated, (i.e., $\forall k \in I$):

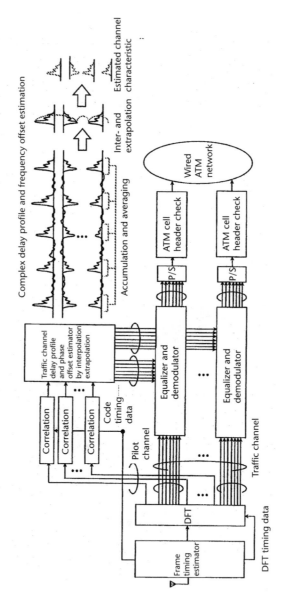

Figure 7.14   Configuration of the receiver. (*From:* [9]. © 1997, IEEE.)

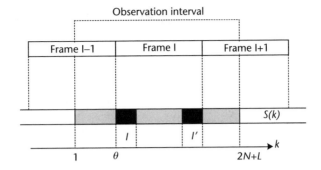

Figure 7.15   Structure of OFDM signal with cyclically extended frames, $s(k)$. (*From*: [8]. © 1997, IEEE.)

$$E\{r(k)r^*(k+m)\} = \begin{cases} \sigma_s^2 + \sigma_n^2 & m = 0 \\ \sigma_s^2\, e^{j2\pi\epsilon} & m = N \\ 0 & \text{otherwise} \end{cases} \tag{7.24}$$

where $\sigma_s^2 \equiv E\{|s(k)|^2\}$ and $\sigma_n^2 \equiv E\{|n(k)|^2\}$, whereas the remaining samples, $k \notin I \cup I'$, are mutually uncorrelated. We now explicitly exploit this correlation property and give the simultaneous ML estimates of θ and ϵ.

The log-likelihood function for θ and ϵ is the logarithm of the probability density function of the $2N + L$ observed samples in **r** given the arrival time θ and the carrier frequency offset ϵ. The ML estimation of θ and ϵ is the argument maximizing this function. Under the assumption that **r** is a jointly Gaussian vector, the log-likelihood function can be shown to be

$$\Lambda(\theta, \epsilon) = 2|\gamma(\theta)|\cos\{2\pi\epsilon + \angle\gamma(\theta)\} - \rho\epsilon(\theta) \tag{7.25}$$

where $\angle$ denotes the argument of a complex number,

$$\gamma(m) \equiv \sum_{k=m}^{m+L-1} r(k)r*(k+N), \qquad \epsilon(m) \equiv \sum_{k=m}^{m+L-1} |r(k)|^2 + |r(k+N)|^2 \tag{7.26}$$

are a correlation term and an energy term and $\rho \equiv \dfrac{\sigma_s^2}{\sigma_s^2 + \sigma_n^2}$ is the magnitude of the correlation coefficient between $r(k)$ and $r(k + N)$. The simultaneous ML-estimation of θ and ϵ becomes

$$\hat{\theta}_{ML} = \arg\max\{2|\gamma(\theta)| - \rho\epsilon(\theta)\} \tag{7.27}$$

$$\hat{\epsilon}_{ML} = -\frac{1}{2\pi}\angle\,\gamma(\hat{\theta}_{ML}) \tag{7.28}$$

A block scheme of the timing and frequency estimates is shown in Figure 7.16.

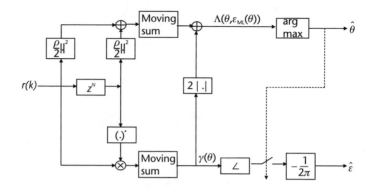

Figure 7.16 Overview of the ML estimates. (*From*: [8]. © 1997, IEEE.)

The close relation between the estimates can be exploited in an implementation. The processing is done continuously and two signals are generated. First, $\Lambda(\theta, \hat{\varepsilon}_{ML}(\theta))$, whose maximizing arguments are the estimated time instants $\hat{\theta}_{ML}$. Second, the signal $\gamma(\theta)$ defined in (7.26), whose complex arguments at time instants $\hat{\theta}_{ML}$ are proportional to the frequency estimate $\hat{\epsilon}_{ML}$. These two signals are shown in Figure 7.17.

The estimation can be improved if the parameters θ and ϵ can be considered constant over several OFDM symbols. Assume that the observation consists of $M(N+L)+N$ samples that contain M complete OFDM symbols. It can be shown

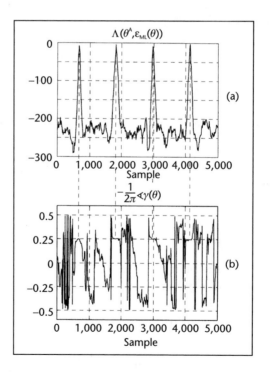

Figure 7.17 The signals that generate the ML estimates ($N = 1024$, $L = 64$, $\epsilon = 0.25$, and SNR = 15 dB): (a) The maximums of $\Lambda(\theta, \hat{e}(\theta))$, give the timing estimate $\hat{\theta}$; (b) At these time instants, the argument of $\gamma(\theta)$, give $\hat{e}$. (*From*: [8]. © 1997, IEEE.)

that the log-likelihood function of this observation is $\Lambda(\theta, \epsilon) = \sum_{m=0}^{M-1} \Lambda_m(\theta, \epsilon)$ where $\Lambda_m(\theta, \epsilon)$ is the log-likelihood function for symbol m in (7.26). Thus, to obtain the log-likelihood function we may sum the log-likelihood functions of the individual symbols. The choice of M, the number of symbols to average over, depends on the allowed complexity and how long the arrival time θ and frequency offset ϵ can be considered to be constant.

The Correlation Algorithm Using the Guard Interval

An OFDM symbol (see Figure 7.18) consists of an effective part (which contains the modulated information), the cyclic extension (the guard interval), and the windowing part, a smoothed prefix and postfix of the symbol. This technique was developed by Hanzo [7] and presents a simple and efficient algorithm for both symbol timing estimation and frequency tracking based on the received signals samples' cyclic nature, detected by means of autocorrelation functions. The algorithm uses the correlation function $G(d)$,

$$G(d) = \sum_{m=0}^{N_g-1} z(d - m) \cdot z(d - m - N_{FFT})^* \qquad (7.29)$$

which calculates the correlation over a sliding window of the received signal. The d denotes the index of the most recent input sample, $z(d)$ the received complex signal sample, N_{FFT} the number of FFT points used for the OFDM modulation, N_g is the length of the guard interval, and $*$ denotes the conjugate of a complex value.

$G(d)$ expresses the sum of correlation between a pair of samples in the two sequences of N_g samples length spaced by N_{FFT} samples. Figure 7.19 shows a schematic plot of the computation of $G(d)$.

The magnitude of $G(d_{max})$ is at its maximum, if the $z(d_{max})$ is the last sample of the current OFDM symbol (d_{max} denotes the index of the last sample of the effective part), since then the guard samples constituting the cyclic extension and their copies in the current OFDM symbol are perfectly aligned in the summation window.

The outcome of $G(d)$, under perfect channel conditions, shows correlation peaks (see Figure 7.20). The maximum value of the observed correlation peaks can be identified as the start of the effective part of the OFDM symbol. The

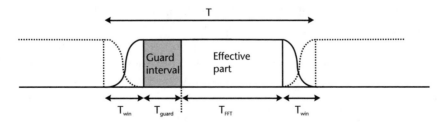

Figure 7.18   Transmitted OFDM frame; T_s is the sampling time $(1/f_s)$, N_{FFT} are the number of subcarriers in the effective part, T_{guard} is the guard interval (cyclic prefix), and T_{win} is the smoothed prefix and postfix.

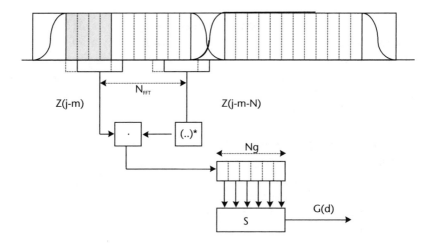

Figure 7.19 Schematic plot of the computation of the correlation function $G(d)$. The gray area represents the memory of the shift register.

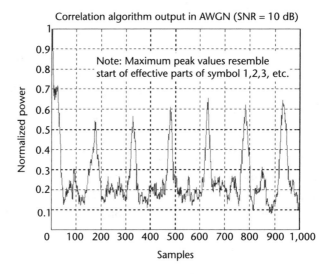

Figure 7.20 Outcome of correlation algorithm $G(d)$ [see (7.29)].

ensemble average of this correlation peak has a triangular shape. When windowing is applied to the OFDM signal, it will have a rounded top, whereas without windowing it should have a sharp edge.

Due to the effect of data that comprises the OFDM symbol and AWGN, the correlation shapes are not true triangles, but they have a rather irregular shape. Averaging can be used in channels with AWGN and multipath to avoid this distortion and give more truly triangular peaks. For signals in excess of about 7 dB SNR the result of the OFDM symbol timing synchronization in AWGN channel estimation is tightly concentrated around the perfect estimate with errors below $\pm 20T$ (T is the sampling rate) [6].

7.7   Frequency Offset Estimation

The carrier frequency offset (δf) results in a progressive phase rotation of the received time domain signal as given by (7.17), reproduced here for convenience as (7.30),

$$y_{i,k} = x_{i,k} \cdot h_{i,k} \cdot \text{sinc}(\delta f \cdot T_{FFT}) \cdot e^{j\psi_{i,k}} + n_{i,k}$$

where

  $y_{i,k}$ is the received sequence

  $x_{i,k}$ is the transmitted sequence

  $h_{i,k}$ is the channel impulse response

  $n_{i,k}$ is the independent noise

$$\psi_{i,k} = \theta_0 + 2\pi\delta f\left(kT_s + \frac{T_{FFT}}{2} + \delta t\right) + 2\pi\delta t\left(\frac{i}{T_{FFT}}\right) \qquad (7.30)$$

where

  θ_0 is the carrier phase offset

  T_s is the symbol time

  i is the subcarrier number

The evaluation of the expression in (7.30) shows that there is a common phase rotation and attenuation of all subcarriers due to the carrier frequency offset (δf) and carrier phase offset (θ_0) and a progressive phase rotation (proportional to i) due to the symbol timing offset (δt).

Let us assume that a certain sequence of the OFDM signal is repeated after a short delay (the channel should be quasi-static during that delay time). For instance, the GI is such a sequence, expressed as guard interval, $s(n) = s(n - N_{FFT})$, where the sample index n points at samples from the GI (i.e., $n \in [d_{\max} - N_g + 1, d_{\max}]$). Due to the frequency offset, these samples are received as (neglecting channel effects)

$$z(n) = s(n) \cdot e^{j(2\pi\delta fnT_s + \theta)} \qquad (7.31)$$

$$z(n - N_{FFT}) = s(n - N_{FFT}) \cdot e^{j(2\pi\delta f(n - N_{FFT})T_s + \theta)}$$

Plugging these equations into (7.29) at $d = d_{\max}$, yields

$$G(d_{\max}) = \sum_{m=0}^{N_g-1} z(d_{\max} - m) \cdot z(d_{\max} - m - N_{FFT})^*$$

$$= \sum_{m=0}^{N_g-1} s(d_{\max} - m) \cdot s(d_{\max} - m - N_{FFT})^* \cdot e^{j(2\pi\delta fnT_s + \theta)} \quad (7.32)$$

$$\cdot e^{-j(2\pi\delta f(n - N_{FFT})T_s + \theta)}$$

Assuming perfect knowledge of $d_{\max}$, the two transmitted sequences are equal and we finally obtain

$$G(d_{\max}) = \sum_{m=0}^{N_g-1} |s(d_{\max} - m)|^2 \cdot e^{j2\pi\delta fN_{FFT}T_s} \quad (7.33)$$

It is seen that except for the constant parameters N_{FFT} and T_s, the angle of this expression is determined by the frequency error. Thus, an estimate of the frequency error is obtained from

$$\delta f = \frac{\Delta f}{2\pi} \cdot \arg\{G(d_{\max})\} \quad (7.34)$$

where

$\Delta f = \dfrac{1}{T_{FFT}}$ is the subcarrier spacing

The frequency error must be smaller than $\Delta f/2$ due to the 2π ambiguity of $\arg\{G(d_{\max})\}$ [7].

Note that in case of multipath channels, the GI will be corrupted by the preceding OFDM symbol due to ISI. Thus there will be some differences between the received GI and its equivalent during the effective FFT period. Another critical assumption made is that perfect symbol timing estimation is already achieved, which is not always the case and quite difficult to achieve.

7.8 Carrier Synchronization

Frequency offsets are created by differences in oscillators in transmitter and receiver, Doppler shifts, or phase noise introduced by nonlinear channels. There are two destructive effects caused by a carrier frequency offset in OFDM systems. One is the reduction of signal amplitude (the sinc functions are shifted and no longer sampled at the peak) and the other is the introduction of ICI from other carriers (see Figure 7.21). The latter is caused by the loss of orthogonality between the subchannels.

Similar to symbol synchronization, carrier synchronization can also be divided into two categories: based on pilots or on the cyclic prefix.

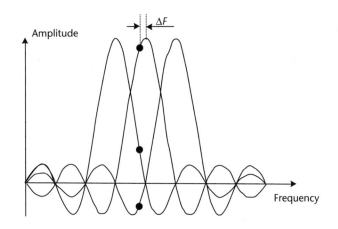

Figure 7.21   The carrier synchronization problem.

7.8.1   Pilots

This approach has been addressed in [11]. Here, the subcarriers are used for the transmission of pilots (usually a PN sequence). Using these known symbols, the phase rotations caused by the frequency offset can be estimated. Under the assumption that the frequency offset is less than half the subcarrier spacing, there is a one-to-one correspondence between the phase rotations and the frequency offset. To assure this, an algorithm is constructed by forming a function, which is sinc-shaped and has a peak for $f - \hat{f} = 0$. It was found that by evaluating this function in points $0.1/T$ apart, an acquisition can be obtained by maximizing that function. This was found to work well both for an AWGN channel and a fading channel.

7.8.2   Cyclic Prefix

The redundancy of the cyclic prefix can be used in several ways (e.g., by creating a function that peaks at zero offset and finding its maximizing value [12] or by doing maximum likelihood estimation, as previously discussed). If the frequency error is slowly varying compared with the OFDM symbol rate, a *phase-locked loop* (PLL) can be used to further reduce the error [13].

7.9   Sampling-Frequency Synchronization

The received continuous-time signal is sampled at instants determined by the receiver clock. There are two methods to deal with the mismatch in sampling frequency. In synchronized-sampling systems, a timing algorithm controls a voltage-controlled crystal oscillator to align the receiver clock with the transmitter clock. The other method is nonsynchronized sampling, where the sampling rate remains fixed, which requires postprocessing in the digital domain. The effect of clock frequency offset is twofold: the useful signal component is rotated and attenuated and, in addition, ICI is introduced. In [14] the BER performance of a nonsynchronized OFDM system has been investigated. It is shown that nonsynchronized sampling is much more sensitive to a frequency offset, compared with a synchro-nized-sampling system. For nonsynchronized sampling systems, it is shown that the degradation (in dB) due to a frequency sampling offset depends on the square of the carrier index and on the square of the relative frequency offset.

7.10 Performance Analysis of Synchronization Techniques

We now analyze the performance of the various synchronization schemes.

7.10.1 Symbol Synchronization

7.10.1.1 Pilots

Scheme Suggested by Schmidl
The timing metric obtained using this method exhibits a plateau, as shown in Figure 7.12. This leads to an uncertainty regarding the start of the effective part of the OFDM symbol. Moreover, the falling edge of this pulse is located $N_{FFT}/2$ away from the start of the frame. Schmidl concentrates on the procedure of finding the center of this plateau using the two techniques. One is finding the maximum of the timing metric. The other is to find the maximum and the points to the left and right in the time domain, which are 90% of the maximum, and then average these two 90% times to find the symbol timing estimate. The results obtained by using this latter method showed that there is no degradation in 10,000 runs for the AWGN channel and a degradation of just under 0.06 dB for the exponential channel at an SNR of 40 dB.

The samples that are used for estimation are protected from ISI by the cyclic prefix, which adversely also flattens the top of the correlation function. The array of samples used for FFT must start within this plateau to avoid loss of SNR. The fractional carrier frequency offset is then determined at the same point. The frequency acquisition method is robust and reliable but needs two symbols, and the symbol timing estimation is not accurate for reasons already discussed. How-ever, the plus point here is that this method does not rely on correlation using cyclic prefixes and, hence, the length of the correlation sum is much larger ($T_{FFT}/2$), which makes it resilient to multipath and low SNR.

7.10.1.2 Cyclic Prefix

Correlation of a Pseudorandom Sequence
The authors have simulated [9] QPSK-OFDM based wireless ATM transmission under Rayleigh fading environment and random distributed phase. This Rayleigh fading is evaluated by the parameter $f_d T$, in which f_d and T are Doppler frequency and the reciprocal of symbol rate per one OFDM parallel channel, respectively. For example, in the case of $f_d T = 6.25e - 4$ and $f_d = 80$ Hz, symbol rate at the output of guard bit insertion circuit is 128k symbol/s and if length of DFT is 128 and QPSK transmission, total transmission rate is 32.768 mbit/s. No channel coding is used.

Figure 7.22 shows the relationship between the number of accumulated symbols at the frame timing estimator and the BER performance. We note that if more than five symbols are accumulated to estimate the DFT start point, the BER is very close to the case of perfect synchronization. The authors discuss various other aspects in their paper, but these are with respect to their suggested scheme and are not of relevance here.

The important thing to note here is that the correlation by using the information cyclic prefix in the presence of accumulated symbols of five or more yields results close to the perfect synchronization case. Hence, the channel parameters need to be constant over at least five symbols.

7.11   ML Estimation of Timing and Frequency Offset

The authors [8, 15–18] used a microcellular scenario, an indoor wireless system operating at 2 GHz with a bandwidth of 10 MHz. An OFDM symbol consists of 1024 subcarriers with an additional 64-sample guardspace. A fading environment with additive white Gaussian noise is chosen; the channel has an exponentially decaying power delay spread with *rms* value equal to 2 μsec, corresponding to 20 samples. It is modeled to consist of 64 independent Rayleigh-fading taps. The mobile is assumed to be moving at a maximum speed of 5 km/h resulting in a maximum Doppler frequency of 10 Hz. This Doppler frequency is low enough to motivate the averaging of eight symbols in the synchronization algorithm, since the channel impulse response does not change significantly during this time interval. In the simulations, 20,000 symbols are used for each SNR value.

We note from Figure 7.23 that the dispersive fading channel has introduced an error floor. The error floor is reached at approximately 6 dB.

The timing offset estimator shows an error standard deviation of about three samples. This means that in a wireless application where the fading channel is

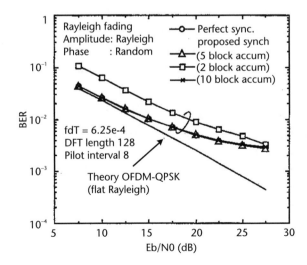

Figure 7.22   Relationship between accumulated blocks and BER performances. (*From:* [9]. © 1997, IEEE.)

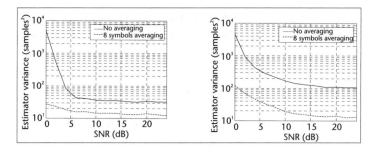

Figure 7.23   Performance of the timing (left) and frequency (right) estimator for the fading channel. The upper curves are without averaging and the lower are with eight frames averaging. (*From:* [8]. © 1997, IEEE.)

estimated continuously, the timing error must be adjusted in the equalizer or by a pilot-aided tracking algorithm. In slowly fading channels, the error variance (after averaging) is significantly small. It can be used to feed a PLL, generating the system frame clock for sampling-frequency synchronization.

The frequency offset estimator shows an error standard deviation of less than 1% of the intertone spacing. This satisfies the requirements in [10] for multiuser OFDM systems. The reason for the superior performance of the frequency estimator over the timing estimator is an implicit averaging [see (7.26) and (7.28)]. Moreover, if the frequency error is slowly varying compared with the OFDM symbol rate, a PLL can be used to further reduce the error. In the final analysis of this method:

- It enables us to simultaneously determine the timing and frequency offsets. If the fading is assumed constant over at least eight symbols, we can use a PLL to generate a timing clock synchronized with the transmitter clock. Similarly, a PLL can synchronize the oscillators in the transmitter and receiver to remove carrier frequency offsets. In fast fading situations we can directly adjust the timing error using an equalizer. However, MLE is the only technique that works in a fast fading environment, as it does not require symbol integration. This is evident from the graphs in Figure 7.23, wherein MLE estimates give low error variance without averaging, especially at SNR ≥ 6 dBs.

It is interesting to note the relationship between time and frequency synchronization. If frequency synchronization is a problem, lowering the number of subcarriers, which will increase the subcarrier spacing, can reduce it. This will, however, increase the demands on the time synchronization, since the symbol length gets shorter (i.e., a larger relative timing error will occur). Thus, the synchronizations in time and frequency are closely related to each other.

7.11.1   The Correlation Algorithm Using the Guard Interval

The magnitude of $G(j_{max})$ is maximum if $z(j_{max})$ is the last sample of the current OFDM symbol, since then the guard samples constituting the cyclic extension and their copies in the current OFDM symbol are perfectly aligned in the summation windows. The simulated accuracy of the OFDM symbol [7] synchronization in an AWGN channel is shown in Figure 7.24.

Observe in the figure that for SNRs in excess of about 7 dB, the histogram is tightly concentrated around the perfect estimate, typically resulting in OFDM symbol timing estimation errors below $\pm 20 T_s$. However, as even slightly misaligned time domain FFT windows cause phase errors in the frequency domain, this estimation accuracy is not sufficient. To improve the OFDM symbol timing synchronization, the estimates must be lowpass filtered and averaged over a large number.

Assuming perfect estimation of the position $j_{\max}$ of the correlation peak $G(j_{\max})$, the performance of the fine frequency error estimation in an AWGN environment is shown in Figure 7.25.

Observe that for AWGN SNR values above 10 dB, the estimation error histogram is concentrated to errors below about 0.02 Δf, where Δf is the subcarrier spacing. It can, therefore, be concluded that this approach performs well for acquisition provided we can estimate the peak of the correlation function correctly. This means high averaging. However, frequency tracking, especially for low SNRs, is poor. Hence, as this method requires high averaging, it is useless for bursty channels.

7.12   Carrier Synchronization

7.12.1   Pilots

In this concept, the authors [11] have developed an algorithm that traces the maximum of a sync-shaped function. For any OFDM system, the output $z_{n,l}$ of the lth carrier at time nT_{sym} can be written as

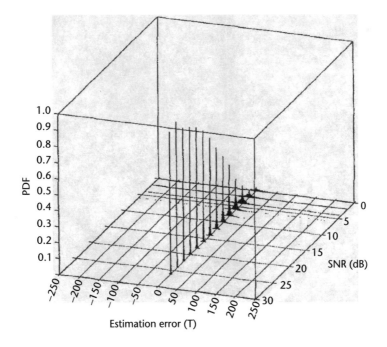

Figure 7.24   Histogram of the symbol timing estimation errors normalized to the sample interval T_s in an AWGN channel for $N = 512$ and $N_g = 50$ with no lowpass filtering of the estimates. (*From:* [7]. © 1996, IEEE.)

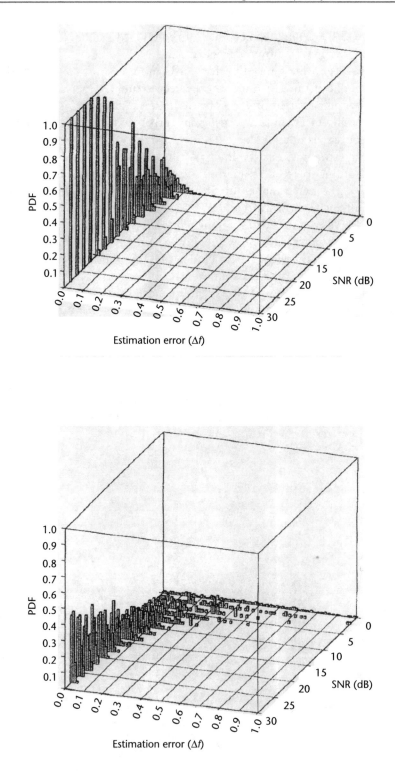

Figure 7.25 Histogram of the simulated frequency estimation error for the frequency acquisition and frequency tracking algorithms using $N = 512$ and $N_g = 50$ over AWGN channels. (*From:* [7]. © 1996, IEEE.)

$$z_{n,l} = a_{n,l} H(\Omega_1) + n_{n,l} \qquad (7.35)$$

where $T_{sym} = T_G + T_{sub}$, T_G being the guard time, and T_{sub} being the time separation between the subchannels; $H(\Omega_1) = H\left(\dfrac{2\pi l}{T_{sub}}\right)$ is the Fourier coefficient of the overall channel impulse response of the transmission channel, including the transmitter filter, the channel, and the receiver. It is assumed that the guard time is larger than the channel impulse response, thereby ensuring that there is no ISI. The complex noise process corrupting the lth subchannel is represented by $n_{n,l}$. Equation (7.35) is valid only in case of perfect synchronization. Otherwise the impact of crosstalk is given by (as an additional noise component)

$$z_{n,l} = e^{j2\pi f_0 n T_{sym}} a_{n,l} \, si(\pi f_0 T_{sub}) H(f_l) + n_{n,l} + \tilde{n}_{n,l} \qquad (7.36)$$

The exponential term results from the phase drift caused by the frequency offset and it represents the impact of the crosstalk. The phase of this complex valued noise process can be shown to be uniformly distributed between $[-\pi, \pi]$ and the power of this term is given by

$$P_{CT} = P_{crosstalk} = \sum_{l \neq v} si^2(\pi(v - l - f_0 T_{sub})) \qquad (7.37)$$

with $f_0 T_{sub} < 1/2$.

In practice, the underlying principle of the frequency estimation algorithm can be reduced to a phase shift between two subsequent subchannel samples (e.g., $z_{n,l}$ and $z_{n+1,l}$) without the need to generate an estimate of the channel Fourier coefficient $H(\Omega_1)$ as in (7.35). Hence, the assumption that the frequency offset is less than half the subcarrier spacing (7.37) yields a situation where there is a one-to-one correspondence between the phase rotations and the frequency offset. This is the core of the argument. The influence of the modulation is removed in the data-aided (DA) case by a multiplication with the conjugate complex value of the transmitted symbols. The generalized estimator is given by,

$$\hat{f} 2\pi T_{sym} = \frac{1}{D} \arg\left\{ \sum_{j=1}^{L_F-1} \left(z_{n+D,p(j)}(\hat{f}_{acq}) z_{n,p(j)}^*(\hat{f}_{acq})\right)(c_{1,j}^* c_{0,j}) \right\} \qquad (7.38)$$

where the function $p(j)$ gives the position of the jth sync-subchannel, which carries one of the L_F known training symbol pairs $(c_{1,j}^* c_{0,j})$. Note that $c_{1,j}^*$ and $c_{0,j}$ are transmitted over the same subchannel and the symbols $\{c_{0,j}\}$ belong to the nth time period and $\{c_{1,j}\}$ to the $(n + D)$th time period, respectively. D is an integer. The overall required amount of training symbols equals $2L_F$. These L_F subchannels should be spread uniformly over the $N - 2N_G$ subchannels, where N_G is the number of unmodulated subchannels to avoid aliasing effects at the receiver. The acquisition algorithm is based on the fact that the magnitude of the expression within the arg function of (9) reaches its maximum if $\hat{f}_{acq}$ coincides with f_0, because this expression obeys a $si^2(2\pi(f_0 - f_{\text{trial}})T_{\text{sym}})$ law for $|f_0 - f_{\text{trial}}|T_{\text{sub}} < 0.5$. Therefore, we consider the following maximum search procedure:

$$\hat{f}2\pi T_{\text{sub}} = \max_{f_{\text{trial}}}\left\{\left|\sum_{j=0}^{L_F-1}\left(z_{n+D,p(j)}(f_{\text{trial}})z^*_{n,p(j)}(f_{\text{trial}})(c^*_{1,j}c_{0,j})\right)\right|\right\}$$

(7.39)

where f_{trial} is the trial frequency and $z_{n,p(j)}(f_{\text{trial}})$ is the output of the FFT unit if its N input samples are frequency corrected by f_{trial}. In practice, the authors found that it was sufficient to space the trial parameters $0.1/T_{\text{sub}}$ apart from each other. Figure 7.26 shows the plot of (7.39) for various SNR values and $L_F = 50$ ($N = 1,024$) drawn over $f_0 - f_{\text{trial}}$.

The authors found that for SNRs > 5 dB, the acquisition behavior was satisfactory for AWGN channels as well as for frequency selective fading channels. In this analysis, it is assumed that timing synchronization is achieved.

7.13   Sampling-Frequency Synchronization

The aspects of timing synchronization have already been covered. The behavior of a nonsynchronized OFDM system need not interest us, as we are fundamentally interested in synchronized systems.

7.14   Observations

We have now analyzed the various schemes involved in synchronization. The basic conclusion is that no single method is efficient. Hence, we need to integrate the

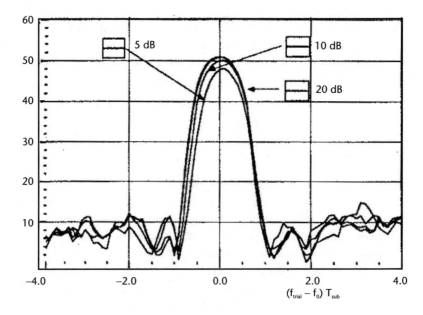

Figure 7.26   Carrier synchronization. (*From:* [11]. © 1994, IEEE.)

various techniques to find an effective and efficient solution. Bearing this in mind, we can make the following observations:

1. ML estimation can provide us the rough time offsets, which we can use in a PLL to synchronize the sampling frequencies of the transmitter and receiver. Similarly, the frequency offset estimate can be used to synchronize the carrier oscillators of the transmitter and receiver. Better still we can link the sampling oscillator and the carrier oscillator to a common source, thereby reducing the complexity.

2. The correlation algorithm technique [6] offers better results (being more accurate provided there is high averaging) than ML estimation [7], but ML estimation offers the additional advantage of being able to work in fast fading channels. Hence, logically we should combine the two techniques using an algorithm.

3. The algorithm we will need to use can be, most likely, rule based.

4. In view of the fact that, individually, the correlation algorithm estimation technique or the ML estimators are inferior to the approach suggested by Harada, the circuit is considerably simplified. When we look at the integrated techniques, however, the circuit will turn out to be a superior system.

7.15   Suggested Solution to the Synchronization Problem

The suggested block diagram is shown in Figure 7.27.

In Figure 7.27, the ML estimator determines the timing offset and passes the data to an algorithm. The algorithm also receives timing offsets acquired by alternate techniques from the subchannels. The requirement here need not be stringent, because we are not depending on any one method to give us the final answer. We

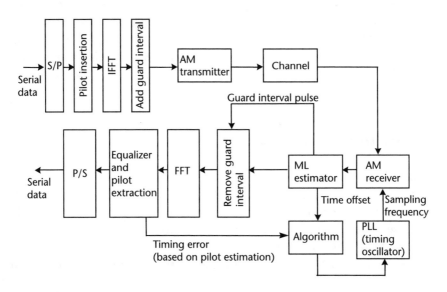

Figure 7.27   Algorithmic synchronization.

can use specially inserted pilot symbols in the subchannels and then carry out MLE on these, as suggested by Kapoor et al. [19] or we can carry out correlation estimation based on cyclic prefixes and so on. The ML estimator simultaneously determines the guard interval location and gives this information as a guard interval pulse to guard interval removal circuit. This circuit has a serial-to-parallel converter, which then gives the frames to the FFT. The algorithm determines the timing offset and gives it to the VCO of the PLL. In a fast fading situation, the PLL is practically ineffective, being essentially controlled by the VCO, which, in turn, is controlled by the algorithm output. However, in a slow fading environment, the PLL will help in correcting the sampling frequency. Hence, in such a situation, the algorithm and the PLL will act in tandem to accurately track the sampling frequency. If desired, we can use the timing offsets for acquisition and the frequency offsets for tracking. The algorithm needs to be optimized for speed for on-line performance.

We have seen in the preceding sections that there is a need for a linking technique, which links the pilot-based estimates with the MLE estimates, toward the common goal of synchronizing the sampling frequency. The question now facing us is how to implement the algorithm [20]. The details are beyond the scope of this work.

7.16   Channel Estimation

OFDM basically consists of subcarriers or tones, which are narrowband in nature. These tones lend themselves to channel estimation using techniques discussed in Chapter 4. In the synchronization methods previously discussed, MLE techniques give us channel estimates as a byproduct. The more popular ones are the data-aided techniques using pilots. In packet transmission systems (Chapter 8), these pilots are inserted as training symbols, appended to the individual packets. These aspects have already been discussed in Chapter 4. Generally, for SISO systems, least squares estimation has proved extremely effective for determination of CSI. However, it is noted that the topic of channel estimation for OFDM in general and MIMO systems in particular is a subject of ongoing research and has resulted in many patents. We have barely discussed a few major and essential ones in this work. The reader is advised to study the many ideas available in journals and conference proceedings.

7.17   Peak to Average Power Ratio

Based on our discussions until now, it can be seen that OFDM has many advantages, which make it an attractive scheme for high-speed transmission links. However, one major difficulty is its large peak to average power ratio (PAPR). These are caused due to the coherent summation of the OFDM subcarriers. When N signals are added *with the same phase,* they produce a peak power that is N times the average power. These large peaks cause saturation in the power amplifiers, leading to intermodulation products among the subcarriers and disturbing out of band energy. Hence, it becomes worth our while to reduce PAPR. Toward this end, there are several proposals such as clipping, coding, peak windowing and so on. Unfortunately, reduction of PAPR comes at a price of performance degradation.

Consider a complex envelope of the OFDM signal, consisting of N carriers and given by,

$$S(t) = \sum_{k=-\infty}^{\infty} \sum_{n=0}^{N-1} a_{n,k} g(t - kT) e^{jn\frac{2\pi}{T}t} \tag{7.40}$$

where $g(t)$ is a rectangular pulse of duration T and T is the OFDM symbol duration.

PAPR is defined by

$$PAPR = \frac{\max_{t \in [0,T]} |S(t)|^2}{\epsilon\{|S(t)|^2\}} \tag{7.41}$$

where $\epsilon\{\cdot\}$ is the expectation operator.

From the central limit theorem, for large values of N, the real and imaginary values of $S(t)$ become Gaussian distributed. The amplitude of the OFDM signal, therefore, has a Rayleigh distribution with zero mean and a variance of N times the variance of one complex sinusoid. Assuming the samples to be mutually uncorrelated, the cumulative distribution function for the peak power per OFDM symbol is given by [21]

$$P\{PAPR > \gamma\} = 1 - (1 - e^{-\gamma})^N \tag{7.42}$$

This tells us that PAPR occurs infrequently. Sometimes PAPR is described by peak envelope and crest factor (CF), which is the square root of PAPR. We shall, however, follow the definition of PAPR.

7.17.1   Schemes for Reduction of PAPR

7.17.1.1   Block Coding

This scheme [22] involves finding code words with minimum PAPR from a given set of code words and to map the input data blocks of these selected code words. Thus it avoids transmitting those code words, which exhibit high peak envelope power. This reduction in PAPR, however, comes at the expense of a decrease in coding rate. Obviously this is not suitable for a large number of subcarriers and high order bit rates.

7.17.1.2   M Sequences

This was proposed by [23]. The idea here is to map a block of m input bits to an m sequence $[C_0, \ldots, C_{N-1}]$ of length $N = 2^m - 1$. This results in a code rate of $(m/2^m - 1)$. The m sequences are a class of $(2^m - 1, m)$ cyclic codes obtained from a primitive polynomial of degree m over a GF(2). It was shown in [24] that the achievable PAPR is only 5 dB to 7.3 dB for m between 3 and 10. The problem with this approach is the extremely low rate for large values of m.

7.17.1.3 Clipping and Peak Windowing

A simple technique to reduce PAPR is to clip the signal such that the peak amplitude becomes clipped to some desired maximum level. Since large peaks occur with a very low probability, clipping could be an effective technique for the reduction of PAPR. However, clipping is a nonlinear process and may cause significant inband distortion, which degrades the bit error rate performance and out-of-band noise, which reduces spectral efficiency. Filtering after clipping can reduce the spectral splatter but may also cause some peak regrowth [25]. If digital signals are clipped directly, the resulting clipping noise will all fall inband and cannot be reduced by filtering. Hence, in such cases, it is wiser to pad the OFDM block with zeros and then take a longer IFFT. Filtering after clipping is required to reduce the out-of-band clipping noise.

A different approach is to multiply the large signal peaks with a Gaussian-shaped window [26]. In fact, any window can be used provided it has good spectral properties [27]. Since the OFDM time signal is multiplied with the spectrum of the chosen window, the resulting spectrum is a convolution of the original OFDM spectrum with the spectrum of the applied window. So, ideally, the window should be as narrowband as possible. On the other hand, the window should not be too long in the time domain, because that implies that many signal samples are affected, which increases the bit error rate. Examples of suitable window functions are Cosine, Kaiser, and Hamming windows. It was shown in [27] that PAPR could be achieved independent of the number of subcarriers and at the cost of a slight increased ion BER and out-of-band radiation. The windowing of an OFDM signal is shown in Figure 7.28 [1].

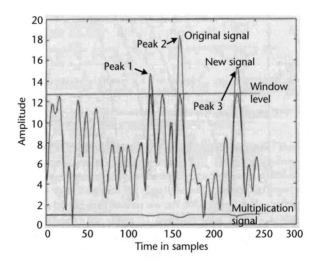

Figure 7.28 Windowing an OFDM time signal. (*From:* [1]. © 2000, Artech House. Reprinted with permission.)

In Figure 7.29 [1] we note the difference in spectrum caused due to clipping and peak windowing. Making the window width wider can decrease the shape of the "skirts." But this encompasses more signals that just the peak signal, leading to higher bit error rates.

Every power amplifier has a power transfer characteristic, as shown in Figure 7.30.

The 1-dB compression point is defined as the point where the output power falls by one dB due to saturation. Hence, the linear region is the region below that 1-dB compression point. If we wish to avoid nonlinear distortion, we need to operate in this region. If the power spike crosses this 1 dB level then, due to

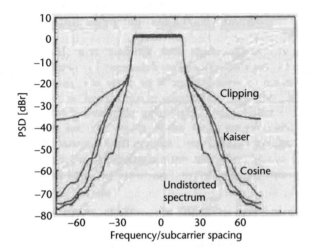

Figure 7.29   Frequency spectrum of an OFDM signal with 32 subcarriers with clipping and peak windowing at a threshold level of 3 dB above rms amplitude. (*From:* [1]. © 2000, Artech House. Reprinted with permission.)

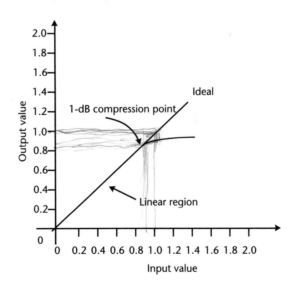

Figure 7.30   Power amplifier transfer characteristic.

saturation, the signal gets "clipped." Hence, we need to "backoff" the power amplifiers. The backoff relative to the maximum output power is determined such that any significant distortion of the spectrum caused due to signal peaks will be manifest at a level far below the inband spectral density. Obviously the more stringent the specification (level) the more the backoff is required. For example, if we choose a level of −50 dB, the required backoff will be higher than would be necessary at, say, −30 dB below the inband spectral density. At the chosen level, the spectrum will widen, compared with the width of the spectrum at −3 dB due to this backoff. The amount of backoff is usually so adjusted that at the chosen level, the spectrum becomes twice that at −3 dB. By controlling the amount of backoff, we control the amount of nonlinearities due to large peaks driving the amplifier into saturation. This, in turn, widens the spectrum. Hence, if we reduce the PAPR, the demand on the amount of backoff reduces for that same chosen level. The difference in backoff is much less than the difference in PAPR at the input of the power amplifier (e.g., without peak windowing the PAPR is about 18 dB for the OFDM signal with 64 subcarriers [1]). With peak windowing this is reduced to approximately 5 dB. Hence, we need to backoff the amplifier by a value slightly greater than 5 dB to achieve minimum spectral distortion. Assume we were not using peak windowing. In such a case, the backoff need not be in excess of 18 dB since there is relatively little energy in signals parts that have a large PAPR. Hence, the spectrum is not unduly affected. We can even clip the peak signal to 10 dB below the maximum without unduly affecting the spectrum, simply because there is less energy in the peak signal [1]. However, once the PAPR has been reduced to, say, 5 dB using any reduction technique, signals close to this value (say, a dB below) cannot tolerate any distortion because there is more energy in them than was originally in the basic unreduced peak signal. Hence, the lower the PAPR is rendered by PAPR reduction techniques, the more stringent the signal becomes against nonlinearities in the area of the maximum PAP ratio. The obvious conclusion from this discussion is that clipping is the preferred mode, compared with peak windowing, because there is not much energy in the clipped signal. Hence, the packet error rates are much less with clipping [1], though the spectrum in the case of peak windowing is better with steeper "skirts."

7.17.1.4 PAP Reduction Codes

A sequence x of length N is said to be complementary to another sequence y if the following condition holds on the sum of both autocorrelation functions [1]:

$$\sum_{k=0}^{N-1} (x_k x_{k+i} + y_k y_{k+i}) = 2N, \, i = 0 \tag{7.43}$$

$$= 0, \, i \neq 0$$

This is called a *Golay complementary sequence*.
Taking Fourier transforms of both sides,

$$|X(f)|^2 + |Y(f)|^2 = 2N \tag{7.44}$$

where $|X(f)|^2$ is the power spectrum of x and is the Fourier transform of its autocorrelation function. $X(f)$ is defined as

$$X(f) = \sum_{k=0}^{N-1} x_k e^{-j2\pi kfT_s} \tag{7.45}$$

where T_s is the sampling interval of the sequence x. From (7.44), the maximum value of the power spectrum is upper-bounded by $2N$:

$$|X(f)|^2 \le 2N \tag{7.46}$$

The average power of $X(f)$ [from (7.45)] is N if we assume that the power of the sequence x is equal to 1. The PAPR is given by

$$PAPR \le \frac{2N}{N} = 2 \tag{7.47}$$

Now IFFT is nothing but conjugated FFT scaled by $1/N$. Hence, (7.47) is also valid for IFFT of the sequence x. Hence, by using a complementary code as input to generate an OFDM signal, it is guaranteed that PAPR cannot exceed 3 dB.

This technique can also be modified to incorporate FEC capability for fading channels without having to incorporate a separate FEC coding block. The disadvantage of Golay complementary codes is that if there are a large number of subcarriers, it becomes unfeasible to generate a sufficient number of complementary codes with a length equal to the number of channels.

7.17.1.5   Symbol Scrambling

In this method for each OFDM symbol, the input sequence is scrambled by a certain number of scrambling sequences. Then the output signal with the smallest PAP is transmitted. If the scrambling sequences are uncorrelated, the resulting OFDM signals and corresponding PAPR will be uncorrelated. In such an event, if the PAPR for one OFDM symbol has a probability p of exceeding a certain level without scrambling, the probability is decreased to p^k by using k scrambling codes. This implies that symbol scrambling does not guarantee a PAPR below some level, but rather it decreases the probability that high PAPRs will occur. *Selected mapping* and *partial transmit sequences* are scrambling techniques. In the former case, independent scrambling rotations are applied to all subcarriers, whereas in the latter case scrambling rotations are applied to groups of subcarriers. Further details can be found in [28, 29]. This technique produces moderate results, as can be seen in Figure 7.31. In this case the figure shows that at −50-dB bandwidth (at this level, the bandwidth is twice that at −3 dB) one scrambling code requires a 7.2-dB backoff, whereas 10 scrambling codes require 7.5-dB backoff. Compare these figures to a backoff requirement of 8.5 dB without any scrambling.

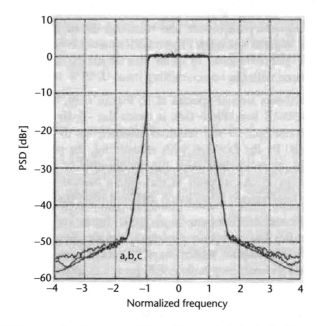

Figure 7.31   OFDM spectra for (a) no scrambling with an 8.5-dB backoff, (b) one scrambling code
with a 7.2-dB backoff, and (c) 10 scrambling codes with a 6.5-dB backoff. (*From:* [1].
© 2000, Artech House. Reprinted with permission.)

7.18   Application to Packet Transmission Systems

One of the basic reasons for the popularity of OFDM systems is the growth of
packet transmission systems like HIPERLAN and IEEE 802.11a. These systems
are destined to play a key role in wireless communications in the coming years.
The most popular route is to use OFDM for high-throughput systems. This is
because the multipath problems are adequately addressed in OFDM, unlike in
other types of communication systems. Hence, OFDM has become a vital partner
in the drive toward high bit-rate systems, the so-called (4G). This standard envisages
attaining throughputs as high as 155 mbit/s, though it has yet to be formally
defined. It is expected to be defined by 2007. In defining standards for packet
transmission systems, many ideas on synchronization from the topics discussed in
this chapter have been incorporated. This also holds true for channel estimation
techniques. We shall discuss one such standard, IEEE 802.11a, in Chapter 8.

7.19   Conclusions

In this chapter, we carried out a brief survey of existing techniques in OFDM
toward synchronization. Based on this survey, the following conclusions can be
made:

1. Each technique has its advantages and disadvantages. Among these, the
   MLE approach works best in a fast fading environment.

2. There is a need to combine the best features of each approach using software.

3. The topics of synchronization and channel estimation are very important and are the subject of ongoing research and patents. The ideas discussed in this chapter constitute the initial effort in this area and have been considerably improved since the time when they were first published. The interested reader is directed to look into the considerable literature available on these topics.

7.20 Simulation Exercises

Study the coded orthogonal frequency division multiplexing (COFDM) simulink model for coded OFDM modulation. The program carries out 16-QAM modulation based on a simple coded OFDM system, which forms the basis for the IEEE 802.11a system to be discussed in Chapter 8. The reader can see the performance of the LSE channel estimator in real time.

References

[1] Van Nee, R., and R. Prasad, *OFDM for Wireless Multimedia Communications,* Norwood, MA: Artech House, 2000.

[2] Prasad, R., *Universal Wireless Personal Communications,* Norwood, MA: Artech House, 1998.

[3] Speth, M., et al., "Optimum Receiver Design for Wireless Broadband Systems Using OFDM: Part 1," *IEEE Trans. Comm.,* Vol. 47, November 1999, pp. 1668–1677.

[4] Jankiraman, M., "Wideband Multimedia Solution Using Hybrid OFDM/CDMA/SFH Techniques," *Ph.D. thesis,* University of Aalborg, Aalborg, Denmark, September 2000.

[5] Warner, W. D., and C. Leung, "OFDM/FM Frame Synchronization for Mobile Radio Data Communication," *IEEE Trans. Vehicular Tech.,* Vol. 42, No. 3, August 1993, pp. 302–313.

[6] Schmidl, T. M., and D. C. Cox, "Robust Frequency and Timing Synchronization for OFDM," *IEEE Trans. Comm.,* Vol. 45, April 1996, pp. 1613–1621.

[7] Keller, T., and L. Hanzo, "OFDM Synchronization Techniques for Wireless Local Area Networks," *IEEE PIMRC '96,* October 1996, pp. 963–967.

[8] Sandell, M., et al., "On Synchronization in OFDM Systems Using the Cyclic Prefix," Div. of Signal Processing, Lulea University of Technology, Sweden.

[9] Harada, H., and R. Prasad, "Performance Analysis of an OFDM-Based ATM Communication System," *Proc. PIMRC 1997,* Helsinki, Finland, September 1997, pp. 1095–1099.

[10] Moose, P. H., "A Technique for OFDM Frequency Offset Correction," *IEEE Trans. Comm.,* Vol. 42, No. 10, October 1994, pp. 2908–2914.

[11] Classen, F., and H. Meyr, "Frequency Synchronization Algorithms for OFDM Systems Suitable for Communication Over Frequency-Selective Fading Channels," *IEEE Vehic. Technol. Conf.,* Vol. 3, Stockholm, Sweden, June 1994, pp. 1655–1659.

[12] Daffara, F., and A. Chouly, "Maximum Likelihood Frequency Detectors for Orthogonal Multicarrier Systems," *Proc. Intern. Conf. Commun.,* Switzerland, May 1993, pp. 766–771.

[13] Proakis, J., *Digital Communications,* New York: Prentice-Hall, 3rd edition, 1995.

[14]   Pollet, T., P. Spruyt, and M. Moeneclaey, "The BER Performance of OFDM Systems Using Nonsynchronized Sampling," *Proc Globecom,* Vol. 1, San Francisco, CA, November 1994, pp. 253–257.

[15]   Sandell, M., et al., "Timing Frequency Synchronization in OFDM Systems Using the Cyclic Prefix," Div. of Signal Processing, Lulea University of Technology, Sweden.

[16]   Van de Beek, et al., "Low-Complex Frame Synchronization in OFDM Systems," *Fourth IEEE International Conference on Universal Personal Communications,* November 1995, pp. 982–986.

[17]   Sandell, M., et al., "ML Estimation of Timing and Frequency Offset in Multicarrier Systems," *Research Report,* Div. of Signal Processing, Lulea University of Technology, Sweden.

[18]   Sandell, M., et al., "ML Estimation of Timing and Frequency Offset in OFDM Systems," *IEEE Trans. on Signal Processing,* Vol. 45, No. 7, July 1997, pp. 1800–1805.

[19]   Kapoor, S., D. J. Marchok, and Y.-F. Huang, "Pilot Assisted Synchronization for Wireless OFDM Over Fast Time Varying Fading Channels," *VTC-98 Proceedings,* May 1998, pp. 2077.

[20]   Jankiraman, M., and R. Prasad, "A Novel Algorithmic Synchronization Technique for OFDM-Based Wireless Multimedia Communications," *ICC '99 Proceedings,* Vancouver, Canada.

[21]   Muller, S. H., and J. B. Huber, "A Novel Peak Power Reduction Scheme for OFDM," *PIMRC 1997,* Vol. 3, pp. 1090–1094.

[22]   Jones, A. E., T. A. Wilkinson, and S. K. Barton, "Block Coding Scheme for Reduction of Peak to Mean Envelope Power Ratio of Multicarrier Transmission Schemes," *Electronics Letters,* Vol. 30, No. 25, December 1994, pp. 2098–2099.

[23]   Li, X., and J. A. Ritcey, "M-sequences for OFDM PAPR Reduction and Error Correction," *Electronic Letters,* Vol. 33, No. 7, March 1997, pp. 554–555.

[24]   Tellambura, C., "Use of M-sequences for OFDM Peak to Average Power Ratio Reduction," *Electronic Letters,* Vol. 33, No. 15, May 1997, pp. 1300–1301.

[25]   Li, X., and L. J. Cimini, "Effects of Clipping and Filtering on Performance of OFDM," *Letters,* Vol. 2, No. 5, May 1998, pp. 131–133.

[26]   Pauli, M., and H. P. Kuchenbecker, "On the Reduction of the Out of Band Radiation of OFDM Signals," *ICC 1998,* Vol. 3, pp. 1304–1308.

[27]   Van Nee, R., and A. Wild, "Reducing the Peak to Average Power Ratio of OFDM," *VTC 1998,* pp. 2072–2077.

[28]   Muller, S. H., et al., "OFDM With Reduced Peak-to-Average Power Ratio by Multiple Signal Representation," *Annals of Telecommunications,* Vol. 52, No. 1–2, February 1997, pp. 58–67.

[29]   Muller, S. H., and J. B. Huber, "OFDM With Reduced Peak-to-Average Power Ratio by Optimum Combination of Partial Transmit Sequences," *Electronics Letters,* Vol. 33, No. 5, February 1997, pp. 368–369.

CHAPTER 8

IEEE 802.11a Packet Transmission System

8.1 Introduction

IEEE 802.11a packet transmission system is a WLAN system. WLAN is a flexible data communication system that can either replace or extend a wired LAN to provide added functionality. Using RF technology or infrared (IR), WLANs transmit and receive data over the air, through wall, ceilings, and even cement structures, without wired cabling. A WLAN provides all the features and benefits of traditional LAN technologies like Ethernet and Token Ring, but without the limitations of being connected by a cable. This provides greatly increased freedom and flexibility [1, 2].

WLANs have been used increasingly in many critical applications over the past few years, particularly since 1997 when the first IEEE 802.11 WLAN standard was issued, followed by its European competitor standard, HIPERLAN. In certain locations, the use of WLANs could save millions of dollars in cost and deployment time when compared with permanent wired networks. In other locations, WLAN services are complimentary to existing wired LANs, adding the advantage of user mobility.

An enormous number of WLAN standards exist from different standardization organizations and they are competing with each other in terms of quality of performance. The main problem is that there is not one unique standard like Ethernet with a guaranteed compatibility between all standards and devices, but rather many proprietary standards pushed by many independent organizations and incompatible between themselves. The positive spin is that this brings some good aspects like lower product price but, at the same time, it introduces some challenges concerning lack of compatibility and interoperability.

Currently, there are two major WLAN standards in operation—IEEE 802.11 and HIPERLAN/2. Both are OFDM-based systems. This chapter attempts to cover certain basic aspects of the IEEE 802.11a system without delving into too much detail to enable the reader to appreciate the use of such a packet transmission system for MIMO. Toward this end, we shall emphasize the so-called physical layer. This is distinct from the packet transmission system protocol manager called medium access control (MAC). The interested reader is advised to study the many excellent references for a deeper understanding of the IEEE 802.11a system, since such details are beyond the scope of this book.

8.2 Background

Over the past 10 years or so an alternative to wired LAN structures has evolved in the form of the WLAN. The 1G WLAN products operated in the unlicensed 900- to 928-MHz industrial scientific and medical (ISM) band, with low range and throughput offering (500 kbit/s). These were subject to a lot of extraneous interference and consequently did not perform well. But they enjoyed a reputation of being inexpensive due to advances in the development of semiconductor technologies. However, this band became extremely crowded with other products, leaving no room for further development.

The second generation in 2.40- to 2.483-GHz ISM band WLAN products boosted by the development of semiconductor technology was developed by a huge number of manufacturers. Using spread-spectrum technology and modem modulation schemes, this generation of products was able to provide data rates up to 2 mbit/s, but again the band became crowded, since the most widely used product in 2.4 GHz is the microwave oven, which caused interference.

Third generation products assembled with more complex modulation techniques in 2.4-GHz band allow 11 mbit/s data rate. In June 1997, the IEEE finalized the initial standard for wireless LANs: IEEE 802.11. The first fourth generation standard, HIPERLAN, came as a specification from the European Telecommunication Standard Institute (ETSI) Broadband Radio Access Network (BRAN) in 1996, operating at 5-GHz band. Unlike the lower frequency bands used in prior generations of WLAN products, the 5-GHz bands do not have a large "indigenous population" of potential interferers like microwave ovens or industrial heating systems, as was true in 900-MHz and 2.4-GHz bands [2]. In late 1999, IEEE published two supplements to the 802.11—802.11b and 802.11a—following the predecessor successes and interest from the industry [2]. ETSI's next generation HIPERLAN family, HIPERLAN/2, proposed in 1999, operates in the same band as its predecessor and is still under development. The goal is to provide high-speed (raw bit rate 54 mbit/s) communications access to different broadband core networks and moving terminals [1]. It is expected that 802.11b will compete with HIPERLAN/1 and 802.11a will compete with HIPERLAN/2 in the near future.

8.3 Wireless LAN Topology

The infrastructure mode and the ad-hoc mode are the two most common topologies supported by WLAN. The infrastructure mode is sometimes called basic service set (BSS), which relies on an access point (AP) that acts as a controller in each radio cell or channel. If station A wants to communicate with station B, it goes through the AP. This mode of operation is suitable for business applications, both indoors and outdoors, where an area much larger than a radio cell has to be covered. The access point performs several tasks, like connecting to a wired network or as a bridging function to connect multiple WLAN cells or channels.

Ad-hoc modes are known as "peer-to-peer" modes in some literature. In this mode, mobile nodes can network among themselves without the help of any fixed or wireless infrastructure like AP. It is principally used to build a network quickly and easily where no infrastructure is available. Users in the military who want to share files, or perhaps event attendees at a convention center who need to share information, are two examples of who could use this mode. Another mode sometimes referred to in the IEEE 802.11 standard is extended service support (ESS), where multiple BSSs are joined together to use the same channel to boost the aggregate throughput. Basically this mode is a set of BSSs working together.

8.4   IEEE 802.11 Standard Family

8.4.1   802.11

The 802.11 standard for WLAN operates at data rates up to 2 mbit/s in the 2.4-GHz ISM band. The goal of this standard was to serve the same purpose as IEEE 802.3 for wired Ethernet to define an open standard for wireless networks so that consumers were no longer tied to a single vendor with proprietary technologies [2]. This standard describes the specification of one MAC layer and three physical layers: frequency hopping, direct sequence, and diffused infrared. The MAC has two main standards of operation, a distributed mode (CSMA/CA) and a coordinated mode (polling mode). 802.11 uses MAC level retransmissions, as well as request-to-send/clear-to-send (RTS/CTS) and fragmentation. This standard includes optional, but quite complex, power management features, which support two separate modes: active mode and power save mode. Power management features define functionality relating to how stations can enter into a power mode and the functionality relating to when another station desires to communicate with it during power saving state. But the standard does not define when to enter or leave a low-power operating state. This is why power management features are considered complex.

The standard also includes optional authentication (open system and shared key) and encryption using wired equivalence privacy (WEP) [3, 4]. With the WEP enabled, the body of the data frame, not the header, is encrypted (RC4 symmetrical stream cipher 40-bit key) with a common key used for both encryption and decryption.

8.4.2   802.11b

The 802.11b is the standard for WLAN operations at data rates up to 11 mbit/s in the 2.4- (2.4 to 2.4835) GHz ISM band, which provides for an 83-MHz spectrum. This is the same RF band of wireless spectrum used by cordless phones and microwave ovens. It is an expansion and, much like the IEEE 802.11 standard, supports transmission using DSSS modulation. It allows transmission at such a rate and at a distance of several hundred feet. The distance depends on impediments, materials, environment and the line of sight for IR-based networks [5]. It was the first widely available WLAN technology to provide data rates similar to wired LAN. Organizations were quick to realize that the technology, operating at speeds of 11 mbit/s, could very easily address most mainstream and enterprise-wide applications such as e-mail messaging, database and Internet access, and traditional office applications. Although 11 channels are available throughout this band, only three are nonoverlapping or clear channels. The occupied bandwidth of the spread-spectrum channel is 22 MHz, spaced 25 MHz apart. The available bandwidth decreases if users roam more than 400 ft from an access point.

802.11b's physical layer is slightly different compared with its predecessor, while using the same type of MAC layer. A high-rate extension of 802.11, called HR PHY, is implemented in 802.11b to achieve higher bit rates. This uses complementary code keying (CCK) [5]. CCK is a set of 64 eight-bit code words used to encode data. It has unique mathematical properties that allow them to be correctly distinguished from one another by a receiver even in the presence of substantial noise and multipath interference. The 802.11b standard is considered a competitor to HIPERLAN/1 technology.

8.4.3   802.11a

The 802.11a is the standard for WLAN operations at data rates up to 54 mbit/s in the 5- (5.15 to 5.825) GHz unlicensed national information infrastructure (UNII) band, which is designed to provide short-range, high-speed wireless networking communication. The MAC layer is the same as in 802.11 and 802.11b, but it does not use spread-spectrum technique in physical layers. Instead it uses OFDM, which is the basis for such high data rates. This standard uses 300 MHz of bandwidth and the spectrum is divided into three sections, each section being subjected to a restriction on maximum allowed power. The first 100 MHz in the lower frequency portion is restricted to a maximum power output of 50 mW, the second 100 MHz has a higher 250-mW limit, and the third 100 MHz has a maximum of 1.0W output power intended for outdoor applications [5]. In each section there are four channels available, which yields a total of 12 channels. This is three times more than that offered by 802.11b. The 802.11a standard has a wide variety of high-speed data rates available: 6, 9, 12, 18, 24, 36, 48, and 54 mbit/s; it is mandatory for all products to have 6 mbit/s, 12 mbit/s, and 24 mbit/s rates [5]. This standard is considered a competitor to HIPERLAN/2 technology.

8.4.4   Others

802.11g takes the best features of 802.11a and 802.11b. It will operate in 2.4-GHz band but will use OFDM as the modulation scheme in the physical layer. The original idea in using this standard is to maintain backward compatibility with 802.11b products. 802.11e, 802.11f, 802.11h, 802.11i, and 802.11n are the future 802.11 standard variants and are still awaiting ratification. The reader is advised to check the Web site for some more variants.

8.5   WLAN Protocol Layer Architecture

Wireless LAN protocols are seen as logical layered architecture to comply with the ISO model. As mentioned earlier, 802.11 standard families mostly deal with two lower layers of OSI architecture: the data link layer (DLL) and the physical layer (PHY). In the layered concept, one lower layer is considered a service provider and the upper layer is the so-called service user. DLL is further divided into sub-layers: logical layer control (LLC) and MAC. The initial idea was to use the same LLC developed for an 802-compliant system and use upper-layer protocols without much concern that they differ significantly. For example, one uses an unreliable air medium and the other uses reliable wire media. The physical layer and MAC layer of IEEE 802.11a wireless LAN standard will be covered next.

8.6   Medium Access Control

Basically, the MAC layer is a program that turns on a processor; it manages and maintains communications between the radio network interface card (NIC) and AP by coordinating access to a shared radio channel. The goal of the MAC layer is to provide access control functions such as address coordination, frame check sequence generation, and checking for shared-medium PHYs [5]. The main purpose of the MAC protocol is to regulate the use of the medium, and this is done through a channel access mechanism, the core of the MAC, which is a way to divide the main resource between nodes and the radio channel, by regulating the use of it.

An ideal MAC layer should provide the following features [5]:

- Good throughput, since the spectrum is a scarce resource.
- Less delay due to the fact that there will be more and more time-bound multimedia applications.
- Transparency to different PHY layers.
- Fairness to access because of unequal received power in fading channels.
- Low battery power consumption since the portable and mobile devices will be battery powered.
- Maximum number of nodes in a coverage area
- Less channel interference.
- Security to an acceptable level.

8.6.1   IEEE 802.11 MAC Layer

IEEE 802.11 uses distributed MAC protocol based on CSMA/CA as the channel access mechanism. CSMA/CA is used by most wireless LANs in the ISM bands. It specifies how the node uses the medium (i.e., when to listen, when to transmit [5]). It is extremely unusual for a wireless device to be able to receive and transmit simultaneously. This is the reason why IEEE 802.11 uses collision avoidance (CA) rather than collision detection (CD), as is used in wired LANs. Since it is impractical for wireless devices to communicate with all other devices directly, IEEE 802.11 implements a network allocation vector (NAV), a value that indicates to a station the amount of time that remains before the medium will become available. In that sense, NAV can be considered a virtual carrier sense mechanism. By combining physical carrier sense and virtual carrier sense mechanism, the MAC protocol implements the CA portion of CSMA/CA access mechanism [6].

As mentioned earlier, IEEE 802.11 supports one mandatory and two optional coordination function schemes: distributed coordinated function (DCF), which is based on CSMA with collision avoidance (CSMA/CA) protocol; DCF with hand-shaking (CTS/RTS); and point coordinating function (PCF) for time-bound multi-media services [5]. DCF, the fundamental access method, is used to support asynchronous message passing mechanism, delivering the best-effort services but no bandwidth and latency guarantees. The main advantages of DCF are its suitabil-ity for network protocols such as TCP/IP, because it adapts quite well to variable conditions of traffic and is quite robust against interference. With DCF, 802.11 stations contend for access and attempt to send frames when there is no other station transmitting. The protocol starts by listening on the channel; stations deliver MAC service data units (MSDU) of arbitrary length after detecting that there is no other transmission in progress on the wireless medium. However, if two stations detect the channel as free at the same time, a collision occurs. Therefore, a CA mechanism, defined in 802.11, takes care of reducing the probability of such collisions. As a part of CA, before starting a transmission, a station performs a backoff procedure that states that it has to keep sensing the channel for an additional random time after detecting the channel is free, for a minimum duration called DCF interframe space (DIFS). Only if the channel is idle for this additional random period is the station allowed to initiate transmission. This ensures that multiple stations wanting to send data do not transmit at the same time [5].

As mentioned earlier, with radio-based LANs, a transmitting station cannot listen for collisions while sending data, mainly because the station cannot have its receiver on while transmitting the frame. As a result, the receiving station needs to send an acknowledgement (ACK) by checking the CRC of the received packet if it detects no error in the received frame. If the sending station does not receive an ACK after a specified period of time, the sending station will assume that there was a collision and retransmit the frame or fragment.

As an optional feature, the 802.11 standard includes Request-to-Send/Clear-to-Send (RTS/CTS) mechanisms to reduce so-called "hidden station" problems—where a station, believing the channel to be idle, begins transmitting without successfully detecting the presence of a transmission already in progress and causes collisions. If RTS/CTS is enabled, a station will refrain from sending data frame until the station completes an RTS/CTS handshake with another station, such as an access point. A station initiates the process by sending an RTS frame. The access point receives the RTS and responds with the CTS frame. The station must receive a CTS frame before sending the data frame. The CTS also contains a time value that alerts other stations to hold off from accessing the medium while the station initiating the RTS transmits its data. DCF with handshaking is an overhead to the protocol; [7] has presented, based on the study performed by K. C. Chen, that throughput reduces by 63% due to CTS/RTS overhead, compared with the case when there is no hidden station problem.

Priority-based access is another way to gain access to the medium. This is a contention-free access protocol usable on infrastructure network configuration containing a controller called point coordinator with access point; this mode is referred to as the PCF [5]. For supporting time-bound delivery of data frames, the 802.11 standard defines the optional PCF where the access point grants access to individual stations to the medium by polling the station, according to a polling list, during the contention-free period and then switches to DCF mode. Stations cannot transmit frames unless the access point polls them first. PCF has higher priority than DCF, because it may start transmission after a shorter duration than DIFS; this time space is called PCF interframe space (PIFS). With PCF, a contention-free period (CFP) and contention period (CP) alternate over time, in which a CFP and the following CP form a superframe. During the CFP, the PCF is used to access the media, while the DCF is used during the CP. We now describe the general MAC frame format and how it forms data unit. MAC accepts MSDU from higher layers and adds headers and trailers to create MAC protocol data unit (MPDU), or PDU for short. The MAC may fragment MDSU into several frames (fragmentation), increasing the probability of each individual frame being delivered successfully [7]. The header, MSDU, and trailer contain information like address information, IEEE 802.11-specific protocol information, information for setting the NAV, and frame check sequence for verifying the integrity of the frame.

Further details on the MAC layer are beyond the scope of this book.

8.7   Physical Layer

The PHY is the interface between the MAC and wireless media, which transmit and receive data frames over a shared wireless media. The PHY provides three levels of functionality: First, it provides a frame exchange between the MAC and PHY under the control of the physical layer convergence procedure (PLCP), a sublayer between the MAC and physical medium dependent layer (PMD). Second, the PHY uses signal carrier and spread-spectrum modulation to transmit data frames over the media under the control of PMD. Third, the PHY provides a carrier sense indication back to the MAC to verify activity on the medium [7].

The IEEE 802.11 standard actually specifies a choice of three different PHY layers, any of which can underline a single MAC layer. Specifically, the standard provides for an optical-based PHY that uses infrared light to transmit data and two RF-based PHYs that leverage different types of spread-spectrum radio communications. The IR PHY will typically be limited in range and most practically implemented within a single room. The RF-based PHYs, meanwhile, can be used to cover significant areas and indeed entire campuses when deployed in cellular like configurations.

The infrared PHY provides for 1-mbit/s-peak data rates with a 2-mbit/s rate optional and relies on pulse position modulation (PPM). The IR technology is cheaper, simpler, and widely used but the problem is that it cannot penetrate obstacles and needs a direct line of sight of communications. The RF PHYs includes direct sequence spread spectrum (DSSS) and frequency hopping spread spectrum (FHSS) choices. Both of these techniques use spread-spectrum technology, which trades bandwidth for reliability The goal is to use more bandwidth than the system really needs for transmission to reduce the impact of localized interference on the system by artificially spreading the transmission band so that the transmitted signal can be accurately received and decoded in face of noise. But they differ significantly in the way they work. In the 2.4-GHz band, the regulation specifies that systems have to use one of the two main spread-spectrum techniques: FHSS or DSSS [5].

8.7.1 Frequency Hopping Spread Spectrum

In frequency hopping spread-spectrum systems, the carrier frequency of the transmitter abruptly changes in accordance with a pseudorandom code sequence. The FHSS method works by dividing the 2.4-GHz bandwidth into 75 subchannels, each having 1-MHz bandwidth. The sender and receiver agree on a subchannel hopping pattern and the data is sent. Each sender/receiver pair in the network medium selects a different frequency-hopping pattern, minimizing the chance of two pairs using the same subchannel. A minimum hop rate of 2.5 hops per second is specified for the United States. The limitation of this method is introduced by the (1-MHz) bandwidth of each of the subchannels, which allows a maximum throughput of 2 mbit/s. This situation is made worst by the hopping overhead, limiting this method to a small throughput [5].

8.7.2 Direct Sequence Spread Spectrum

The DSSS seems to be the most promising physical layer in the IEEE 802.11 standard and it is relatively simple to implement. In this scheme a narrowband carrier is modulated by a code sequence. The carrier phase of the transmitted signal is abruptly changed in accordance with this code sequence. This method divides the 2.4-GHz band into 14 22-MHz subchannels, with no hopping between subchannels. Data is sent through one 22-MHz channel and special technique "chipping" is used to compensate for channel noise. Chipping simply converts raw bit data into redundant bit patterns called "chips," which provide a form of error checking and correction at the receiver side, minimizing the need for retransmission. The resulting data is then modulated onto the carrier using either differential binary phase shift keying or differential quadrature phase shift keying. By spreading the data bandwidth over a much wider frequency band, the power spectral density of the signal is reduced by the ratio of the data bandwidth to the total spread bandwidth. In a DSSS receiver the incoming spread-spectrum data is fed to a correlator, where it is correlated with a copy of the pseudorandom spreading code used at the transmitter. Since noise and interference are by definition decorrelated from the desired signal, the desired signal is then extracted from a noisy channel. The usual implementation of DSSS in the 2.4-GHz band employs a 13-MHz wide channel to carry a 1-MHz signal. Channels are centered at 5-MHz spacing, giving significant overlap [2]. The advantage of this technique is that it reduces the effect of narrowband sources of interference.

A comparison of DSSS and FSSS is necessary to have a better understanding of the 802.11 PHY technologies. In terms of complexity, the DSSS is more complicated than FSSS, which allows lower implementation cost. In terms of bandwidth sharing, the two technologies differ. The same is true regarding resistance to interference. DSSS seems to have a lower overhead on the air. Transmission time in DSSS is shorter, since it does not require spending time to change frequency of the channel, unlike FSSS. The IEEE 802.11 standard does not strictly express which PHY to use and hence it leaves the issue open for manufacturers to come up with different incompatible PHYs in products [7].

8.7.3 Orthogonal Frequency Division Multiplexing and 5-GHz WLAN Physical Layer

The OFDM physical layer delivers up to 54-mbit/s data rates in a 5-MHz band. The OFDM physical layer, commonly referred to as 802.11a and HIPERLAN/2, will likely become the basis for high-speed wireless LANs. It is worth mentioning that OFDM is not really a modulation scheme. Rather it is a coding or transport scheme. OFDM divides a single digital signal across 1,000 or more signal carriers simultaneously. The signals are sent at right angles (orthogonal) to each other so they do not interfere with each other. The benefits of OFDM are high spectral efficiency, resilience to RF interference, and lower multipath distortion. The orthogonal nature of OFDM allows subchannels to overlap, having a positive effect on spectral efficiency. The subcarriers transporting information are just far enough apart to avoid interference with each other, theoretically. These aspects were covered in sufficient detail in Chapter 7.

OFDM has been selected as the modulation scheme for HIPERLAN/2 and 802.11a due to good performance on highly dispersive channels. The key feature of the physical layer is to provide modes with different code rates and modulation schemes, which are selected by link adaptation. The interleaved data is subsequently mapped to data symbols according to either a BPSK, QPSK, 16-QAM, or 64-QAM scheme. The OFDM modulation is implemented by means of inverse FFT. Approximately 48 data symbols and four pilots are transmitted in parallel in the form of one OFDM symbol [6]. We shall examine the PHY layer in more detail.

A list of the key parameters of the systems is shown in Table 8.1.

Selection of the transmission rate is determined by a *link adaptation scheme,* wherein we select the best coding rate and modulation scheme based on channel conditions. The WLAN standard does not, however, explicitly specify the scheme. Data for transmission is supplied to the physical layer via a PDU train. This train is a binary sequence of 1s and 0s. The schematic is shown in Figure 8.1.

The binary input data is initially sent to a length 127 pseudorandom sequence scrambler. The purpose of the scrambler is to prevent a long sequence of 1s or 0s. This helps with the timing recovery at the receiver. Since, during timing recovery, we resort to edge detection of packets, a long series of 1s or 0s could be detrimental to its efficient functioning. The signal is then sent to a convolution encoder. The

Table 8.1 Performance Specifications

IEEE 802.11a Standard	
Data rate	6, 9, 12, 18, 24, 36, 48, 54 Mbit/s
Modulation	BPSK, QPSK, 16 QAM, 64 QAM
Channel coding rates	1/2, 9/16, 2/3, 3/4
Number of subcarriers	52
Number of pilot tones	4
OFDM symbol duration	4 μsec
Guard interval	800 nsec
	400 nsec
Subcarrier spacing	312.5 kHz
Signal bandwidth	16.66 MHz
Channel spacing	20 MHz

From: [8]. © 1999, IEEE. Reprinted with permission. All rights reserved.

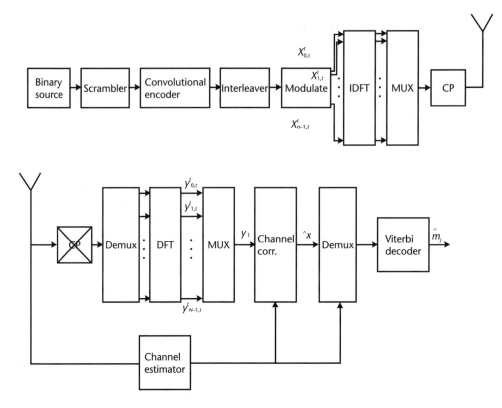

Figure 8.1   A block diagram of the IEEE 802.11a transceiver. (*From:* [8]. © 1999, IEEE. Reprinted with permission. All rights reserved.)

de facto standard for this encoder is (2,1,7). The other rates shown in Table 8.1 are achieved by puncturing the output of this encoder. Puncturing involves deleting coded bits from output data sequence, such that the ratio of uncoded bits to coded bits is greater than the mother code. For example, to achieve a 2/3 code rate, one bit out of every four bits is deleted from the coded sequence. The signal is then sent to an interleaver. The idea of interleaving is to disperse a block of data in frequency so that the entire block does not experience deep fade in the channel. This prevents burst errors at the receiver. Otherwise the convolution decoder in the receiver will not perform very well in the presence of burst errors. This interleaving is carried out at block level (i.e., the interleaving operates on one block of bits at a time). The number of bits in the block is called *interleaving depth,* which defines the delay introduced by interleaving. A block interleaver can be described as a matrix to which data is written in columns and read in rows, or vice versa. Block interleaver is simple to implement using random access memory (RAM). In this case we use an 8 × 6 block interleaver, making the interleaving depth as 48. The reason for this will be explained shortly. The interleaved coded bits are grouped together to form symbols. The symbols are then modulated using one of the schemes listed in Table 8.1. Hence, BPSK uses one bit at a time, QPSK uses 2 bits at a time, 16 QAM uses 4 bits at a time, and 64 QAM uses 6 bits at a time per symbol. The modulation symbols are then mapped to the subcarrier of the 64-point IDFT, thereby creating an OFDM symbol. In the IEEE 802.11a standard, the bandwidth is typically restricted to 20 MHz. It is important that there should be no spectral leakage outside this bandwidth. This can occur owing to two reasons:

- Prior to transmission, the OFDM symbol is subjected to windowing. If the window used is a rectangular one, then there will be an expansion of the spectrum due to the window edges. This causes leakage. We mitigate this (it cannot be avoided) by using a shaped window like a raised cosine window.
- Even then, there will be a certain amount of leakage. This is reduced by not using the edge carriers.

Due to this reason, we cannot use all of the 64 subcarriers. Hence, we use only 48 for data, four for pilots, and the remaining 12 are not used. Out of these 12, 11 are on the edges and one is in the center. This center subcarrier is not used because it is dc arising out of the IDFT operation. Hence, it is for this reason we have only 48 data subcarriers out of a possible 64, leading to the use of an 8×6 block interleaver, as discussed earlier. The four pilot subcarriers are necessary to determine the phase shift suffered by the carrier signal during its passage through the channel. This information will then be used in the receiver to exactly match the carrier frequencies, which is so essential for synchronization. The output of the IDFT is converted to a serial sequence and a guard interval or cyclic prefix (CP) is added. This is per the OFDM transmission theory discussed in Chapter 7. Thus the total duration of the OFDM symbol is the sum of the CP plus the useful symbol duration. Obviously the cyclic prefix is considered overhead in the OFDM frame along with the preamble. The preamble is basically a frame comprising training symbols used for synchronization and channel estimation. We shall discuss it later in this chapter. After the CP has been added, the entire OFDM symbol is transmitted across the channel. This constitutes one packet. It should be noted that as long as the duration of the CP is longer than the channel impulse response, ISI is eliminated, as discussed in Chapter 7. This is the greatest advantage of OFDM in that the multipaths are minimal. This is the reason why IEEE 802.11a achieves a high throughput of 54 mbit/s, which is much higher than what is achievable by other non-OFDM systems.

The process at the receiver is just the reverse. However, the first thing the receiver needs to do is achieve timing synchronization. This implies that the system clock in the receiver needs to be synchronized with that of the transmitter, allowing for the delay across the channel. Various algorithms are used for this purpose, but they are beyond the scope of this book. The interested reader is advised to examine the references.

In addition to timing recovery, the receiver must also compute automatic gain control (AGC) for the A/D converter. The purpose of the AGC is to maintain a fixed signal power to the A/D converter to prevent signals from saturating or clipping the output of the A/D converter. OFDM is a frequency domain modulation technique. Hence, it is essential that we have an accurate estimate of the frequency offset, caused by oscillator instability, at the receiver. Furthermore, we need to estimate the channel as accurately as possible. These tasks are achieved by incorporating training sequences as a preamble to the actual OFDM symbol. To reduce the uncertainty in the channel estimation, two OFDM symbols containing training sequences are provided: short training and long training. The short training is used to provide coarse and long training fine estimation of time and frequency errors. The algorithms used for these are defined in the standard but will not be discussed here, as it is beyond the scope of this chapter. The long training sequence comprising two OFDM symbols is used to estimate the CSI. Knowing the CSI, the receiver can then demodulate, deinterleave, and feed the signal to a Viterbi algorithm for decoding. The channel estimation is carried out using the least squares estimation technique discussed in Chapter 4 and based on the two long training symbols. Though it is sufficient for SISO systems, it is insufficient for MIMO systems. These aspects will be examined in Chapter 9.

To give the reader a deeper insight into the process of OFDM modulation/demodulation in IEEE 802.11a systems, we will now discuss some figures.

In Figure 8.2, the input bits are encoded, interleaved, and mapped onto a constellation, in this case 16 QAM. Thereafter, the complex values from this constellation are loaded onto frequency bins prior to IFFT. There are 52 carriers shown in the figure, starting from −26 to +26 (this includes four pilots). Once the complex numbers are loaded, we carry out IFFT and convert the signal from frequency domain to time domain. We then append the guard interval. Remember that there are 64 subcarriers (i.e., 64 data samples). If the cyclic prefix is of length 16 samples, then there will be a total of 80 data samples. From Table 8.1, we know that one OFDM symbol is, as per standard, 4 μsec. This implies that these 80 data samples (constituting one OFDM symbol) are of length 4 μsec. Hence, the OFDM symbols need to come out at

$$\frac{4 \times 10^{-6}}{80} = 0.05 \times 10^{-6} \; \mu \text{ sec rate, (i.e. 20 MHz)}$$

Hence, in Figure 8.2, the clocking out of these symbols is at a 20-MHz rate. Since the constellation is 16 QAM and there are 48 data carriers, we are transmitting

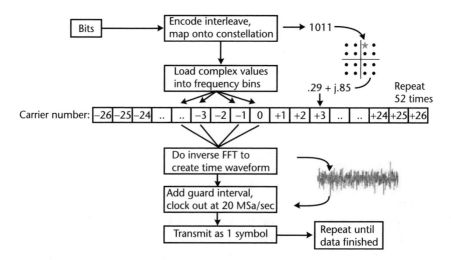

Figure 8.2 Generating OFDM.

48×4 bits every 4 μsec (i.e., 48 mbit/s). This figure is calculated assuming that there is no coding. But recall that 1/2 rate convolution coding is the minimum allowed. If we assume this value, then the data transmission rate is actually 24 mbit/s, which is the standard listed in Table 8.1. The other data rates listed arise from different combinations of modulation and convolution coding rates. This is shown in Table 8.2 [5].

Figure 8.3 shows the transmitted spectrum for one OFDM symbol. Note the sharp wedge in the center corresponding to the dc subcarrier. Note also that beyond ± 10 MHz, the spectrum attenuates drastically (i.e., there is no leakage outside the 20-MHz bandwidth specified in Table 8.1).

Figure 8.4 shows the complete situation when the transmitter is transmitting using 16-QAM modulation. The channel is a dispersive fading channel. There are two types of channel classifications for Rayleigh channels—dispersive and flat fading. By dispersive, we mean that some subcarriers fade differently compared with other subcarriers. Obviously, this is a function of the coherent bandwidth. By flat fading, we mean that all the subcarriers fade together (i.e., they rise and fall together). This occurs when the coherent bandwidth encompasses the entire lot of subcarriers. In Figure 8.4 the dispersive nature of the channel is obvious because the spectrum is not flat compared with the one in Figure 8.3, which is a

Table 8.2   Different Throughput Combinations

Data Rate (Mbit/s)	Modulation	Coding Rate (R)	Coded Bits Per Subcarrier (N_{BPSC})	Coded Bits Per OFDM Symbol (N_{CBPS})	Data Bits Per OFDM Symbol (N_{DBPS})
6	BPSK	1/2	1	48	24
9	BPSK	3/4	1	48	36
12	QPSK	1/2	2	96	48
18	QPSK	3/4	2	96	72
24	16 QAM	1/2	4	192	96
36	16 QAM	3/4	4	192	144
48	64 QAM	2/3	6	288	192
54	64 QAM	3/4	6	288	216

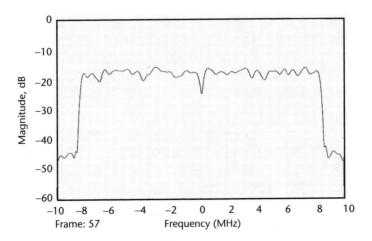

Figure 8.3   Transmitted OFDM spectrum.

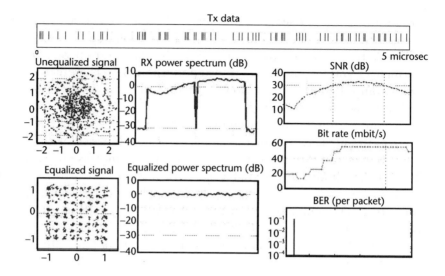

Figure 8.4 Composite situation for dispersive channel.

case of flat fading. Note the quality of the signal received (i.e., the 16-QAM constellation) before and after equalization (i.e., after correcting for the channel). Also note in the SNR graph how the bit rate improves when the SNR is high enough. In fact the curve flattens out at 48 mbit/s (there is no convolution coding in this demo). Conversely, when the SNR is low, not only is the bit rate poor but we notice packet errors also.

The reader is advised to try out these simulations using MATLAB® SIMULINK® and the IEEE 802.11a SIMULINK zip file accompanying this book. The zip file describes an IEEE 802.11a system but without convolution coding and synchronization algorithms (i.e., we assume perfect synchronization). The interesting aspect here is that the code employs link adaptation scheme. The reader can ascertain that as the channel quality becomes poor, the type of modulation becomes robust (i.e., constellation size comes down).

Figure 8.5 shows the preamble that is appended to the OFDM symbol. The preamble carries the training symbols discussed earlier. The parts from A_1 to A_{10} are short training symbols that are identical and 16 samples long. CP is a 32-sample cyclic prefix that protects the long training symbols C_1 and C_2 from ISI caused by the short training symbols. The long training symbols are identical, 64 samples

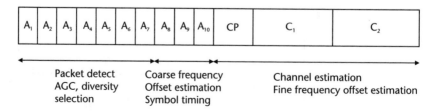

Figure 8.5 The preamble. (*From:* [8]. © 1999, IEEE. Reprinted with permission. All rights reserved.)

long OFDM symbols. However, the guidelines are not binding requirements of the standard. The design engineer has the freedom to use any other available method or develop new algorithms. The quest for suitable training symbols for synchronization and channel estimation is a continuous process and there are many patents in this regard [9–11]. The structure of this preamble enables the receiver to use very simple and efficient detection algorithms to detect the packet. In fact, the Schmidl and Cox algorithm discussed in Chapter 7 is one.

The preamble of Figure 8.5 is then appended to one signal packet, followed by a series of data packets, each 4 μsec long, as shown in Figure 8.6. There are a few salient points to be observed regarding the timings. The short training symbols are each 800 ns long, constituting totally for 10 symbols a total allotted time period of 8 μsec. The long training symbol is 3.2 μsec long per symbol. The two long training symbols are preceded by a cyclic prefix 32 samples long. As per earlier discussions, our clocking rate is 20 MHz. This means 16 samples take $(1/20 \times 10^{-6}) \times 16 = 0.8$ μsec. Hence, the cyclic prefix for the long training symbols will be 32 samples (i.e., 1.6 μsec long). This makes this the long training symbol set, taking a time duration of 8 μsec. This is followed by a signal set with its cyclic prefix. This signal set is part of the MAC protocol and tells the receiver about the rate and length of data to follow. This has the length of one OFDM symbol, which is the basic symbol of 3.2 μsec length preceded by a 16 sample cyclic prefix of 0.8 μsec length, which totals a length of 4 μsec. Note that the cyclic prefix of the OFDM symbol is 16 samples, compared with the cyclic prefix of the long training symbols, which is 32 samples. This is because it is extremely important that there be no ISI of the long training symbols with the short training symbols. This is critical since the long training symbols are used for channel estimation plus fine frequency offset correction, both extremely vital for correct detection. Finally, the signal set is followed by the data set with their respective cyclic prefixes. In Figure 8.6, the data set is shown open ended (i.e., they can be of any length, though in the figure only two are shown). The total number of data sets actually transmitted

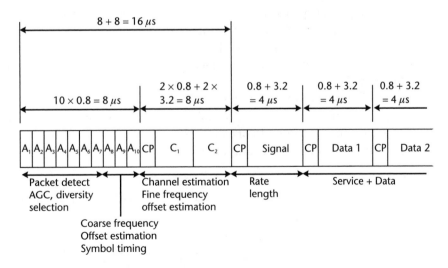

Figure 8.6 OFDM training structure. (*From:* [8]. © 1999, IEEE. Reprinted with permission. All rights reserved.)

depends on the time allowed by the MAC protocol, which in turn is dependant on the channel condition, number of users, traffic density, and so on. Figure 8.6 constitutes one transmitted packet. This packet is transmitted at 4 μsec rate (i.e., 20 MHz as per the standard). It is noted that although it is left to the user to design a particular training packet depending on the algorithm used, the timings are defined by the standard and need to be adhered to.

8.8   Synchronization and Packet Detection Algorithms

The main assumption usually made when WLAN systems are designed is that the channel impulse response does not change significantly during one data burst. This assumption is justified by the short time duration of the transmitted packets, usually a couple of milliseconds, and because the transmitter and receiver in most applications move very slowly relative to each other. Under this assumption, most of the synchronization for WLAN receivers is done during the preamble and need not be changed during the packet [6].

The timing estimation problem comprises two main tasks: packet synchronization and symbol synchronization. The IEEE 802.11a MAC protocol is essentially a random access network. Hence, the receiver does not know when the packet starts. This is the core issue and every other function of the receiver depends on this one basic fact.

8.8.1   Packet Detection

Packet detection is the task of finding an approximate estimate of the start of the preamble of an incoming data. It is, therefore, the first synchronization algorithm that is performed, so the rest of the synchronization process is dependent on good packet detection performance. There are many packet detection algorithms. We shall only discuss one algorithm. This is an application of the Schmidl algorithm used for acquiring symbol timing discussed in Chapter 7. We modify this approach for packet detection. We take advantage of the inherent periodicity of the short training symbols at the start of the preamble. The algorithm is called the delay and correlate algorithm [6]. The signal flow structure of this algorithm is shown in Figure 8.7.

The figure shows two sliding windows, C and P. The C window is a cross-correlation between the received signal and a delayed version of the received signal. Hence, the name delay and correlate. The delay Z^D is equal to the period of the start of the preamble; for example $D = 16$ for IEEE 802.11a systems, the period

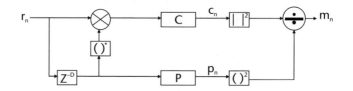

Figure 8.7   Signal flow of the delay and correlate algorithm.

of the short training symbols. The P window calculates the received signal energy during the cross-correlation window. The value of the P window is used to normalize the decision statistic, so that it is not dependent on absolute received power level. The value of c_c is calculated according to (8.1) and the value of p_n according to (8.2).

$$c_n = \sum_{k=0}^{L-1} r_{n+k} r_{n+k+D}^* \qquad (8.1)$$

$$p_n = \sum_{k=0}^{L-1} r_{n+k+D} r_{n+k+D}^* = \sum_{k=0}^{L-1} |r_{n+k+D}|^2 \qquad (8.2)$$

Then the decision statistic m_n is calculated from (8.3)

$$m_n = \frac{|c_n|^2}{(p_n)^2} \qquad (8.3)$$

c_n and p_n are again sliding windows, so the general recursive procedure can be used to reduce computational workload. Figure 8.8 shows the decision statistic m_n for IEEE 802.11a preamble in 10 dB SNR [6]. The overall response is restricted between [0, 1] and the step at the start of the packet is prominent. Initially the received signal consists of only noise. This causes the output c_n of the delayed cross-correlation to be a zero-mean random variable, since the cross-correlation

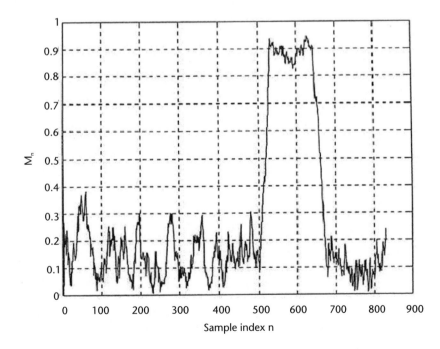

Figure 8.8   Response of the delay and correlate packet detection. (*From:* [6]. © 2002. Reprinted by permission of Pearson Education Inc., Upper Saddle River, NJ.)

of noise samples is zero. This explains the low level of m_n before the start of the packet. Once the start of the packet is received, c_n is a cross-correlation of the identical short training symbols, which causes m_n to jump quickly to its maximum value. This is then a good indicator of the start of the packet.

8.8.2   Symbol Timing

Symbol timing is the task of finding the precise moment when individual OFDM symbols start and end. This defines the FFT window (i.e., the set of samples used to calculate the FFT of each received symbol). The output of the FFT is then used to demodulate the subcarriers of the symbol. The packet detector detects the packet and the symbol timing algorithm refines the estimate to sample-level precision. This is carried out by cross-correlating the received signal r_n and a known reference c_k. This known reference can be the start of the long training symbol. This cross-correlation is defined by

$$\hat{c}_s = \arg\max_n \left| \sum_{k=0}^{L-1} r_{n+k} \, c_k^* \right|^2 \tag{8.4}$$

The value of n that corresponds to the maximum absolute value of the cross-correlation is the symbol timing estimate. In (8.4) the length L of the cross-correlation determines the performance of the algorithm. The larger the better, but it involves more computation. Figure 8.9 shows the output of a cross-correlator that uses the first 64 samples of the long training symbols of the IEEE 802.11a standard as the reference signal. The simulation was run in AWGN channel with 10 dB SNR. The high peak at $n = 77$ clearly shows the correct symbol timing point [6].

Ideally the timing point should be exactly at the end of the cyclic prefix and at the start of the OFDM symbol. This is impossible to realize in practice. The

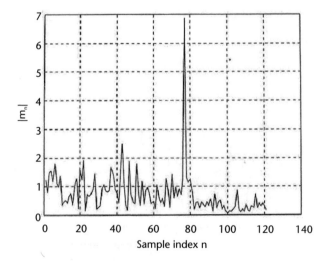

Figure 8.9   Response of the symbol timing cross-correlation. (*From:* [6]. © 2002. Reprinted by permission of Pearson Education Inc., Upper Saddle River, NJ.)

best one can hope for is that the DFT should start somewhere within the cyclic prefix and end at the last sample of the OFDM symbol, as shown in Figure 8.10. Due to the circular convolution properties of the cyclic prefix, since it contains the last samples of the symbol, this does not inconvenience us.

8.8.3   Sampling Clock Frequency Error

The sampling clock frequency in the receiver is extremely critical to proper synchronization. There is always a mutual drift between the clock in the transmitter and the clock in the receiver. This causes the digital to analog converter (DAC) in the transmitter and the analog to digital converter (ADC) in the receiver to be at variance with each other. Ideally both should sample at the same time, but in reality the sampling instants of these clocks slowly shift relative to each other. This drift in clock rates rotates the subcarriers (maximum rotation being suffered by the outermost carriers) and causes a loss of SNR due to ICI generated by the slightly incorrect sampling instants, which causes loss of orthogonality of the subcarriers. This aspect was extensively discussed in Chapter 7. We reproduce (7.30) here for convenience.

$$y_{i,k} = x_{i,k} \cdot h_{i,k} \cdot \operatorname{sinc}(\delta f \cdot T_{FFT}) \cdot e^{j\psi_{i,k}} + n_{i,k}$$

where

$y_{i,k}$ is the received sequence

$x_{i,k}$ is the transmitted sequence

$h_{i,k}$ is the channel impulse response

$n_{i,k}$ is the independent noise

$$\psi_{i,k} = \theta_0 + 2\pi\delta f \left(kT_s + \frac{T_{FFT}}{2} + \delta t \right) + 2\pi\delta t \left(\frac{i}{T_{FFT}} \right) \qquad (8.5)$$

where

θ_0 is the carrier phase offset

T_s is the symbol time

i is the subcarrier number

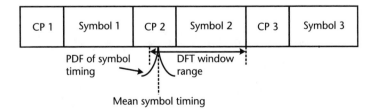

Figure 8.10   Symbol timing variations.

The evaluation of the expression in (8.5) shows that there is a common phase rotation and attenuation of all subcarriers due to the carrier frequency offset (δf) and carrier phase offset (θ_0) and a progressive phase rotation (proportional to i) due to the symbol timing offset (δt).

We need to correct this rotation caused by the sampling frequency offset (symbol timing offset δt). The correction techniques for carrier frequency offset and carrier phase offset will be discussed next.

8.8.3.1 Correcting the Sampling Frequency Error

There are two main approaches.

- The problem can be corrected by adjusting the sampling frequency of the receiver ADC, as shown in Figure 8.11 (upper part of figure).
- The rotation can be corrected after the DFT processing by derotating the subcarriers, as shown in Figure 8.11 (lower part of figure).

These solutions have been analyzed by Pollet et al. [12], wherein the first method is called synchronized sampling and the second method nonsynchronized sampling.

The adjustment of the clock of the ADC in Figure 8.11(a) seems very logical. However, the option shown in Figure 8.11(b) is more popular, wherein instead of a VCXO we use a fixed crystal to control the ADC. This is motivated by the desire to simplify the analog part of the receiver, as analog components are relatively costly compared with digital gates. Hence, by using a fixed crystal, the number of analog components can be reduced, thereby lowering the cost. Figure 8.11(b) shows an additional block called "rob/stuff" just after the ADC. This block is required because the drift in sampling instant will eventually be larger than the sampling period. When this happens, the "rob/stuff" block will either "stuff" a duplicate sample or "rob" one sample from the signal, depending on whether the receiver clock is faster or slower than the transmitter clock. This process prevents the receiver sampling instant from drifting so much that the symbol timing would be

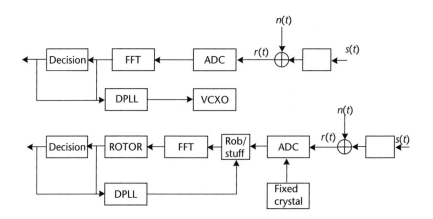

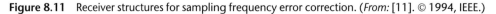

Figure 8.11 Receiver structures for sampling frequency error correction. (*From:* [11]. © 1994, IEEE.)

incorrect. The "ROTOR" block performs the required phase corrections with the information provided by the digital phase locked loop (DPLL) that estimates the sampling frequency error.

8.8.4   Carrier Frequency Synchronization

The effect of carrier frequency offset was discussed in Chapter 7. This offset results in loss of amplitude of the subcarriers and ICI caused by the neighboring carriers. The amplitude loss occurs because the desired subcarrier is no longer sampled at the peak of the sinc-function of DFT. Adjacent carriers cause interference because they are not sampled precisely at their zero crossings (i.e., orthogonality is lost). The overall effect on SNR is analyzed by Pollet et al. [13] and for relatively small frequency errors, the degradation in dB was approximated by

$$SNR_{loss} = \frac{10}{3 \ln 10} (\pi T f_\Delta)^2 \frac{E_s}{N_0} \text{ dB} \tag{8.6}$$

where f_Δ is the frequency error as a fraction of the subcarrier spacing and T is the sampling period. The performance effect varies strongly with the modulation used because smaller constellations are more tolerant of frequency errors compared with larger constellations. This is shown in Figure 8.12 [6].

In Figure 8.12, 64 QAM cannot tolerate more than 1% error in the carrier frequency for a negligible SNR loss of 0.5 dB, whereas QPSK can tolerate up to 5% error for the same SNR loss. This does not imply that we should only operate

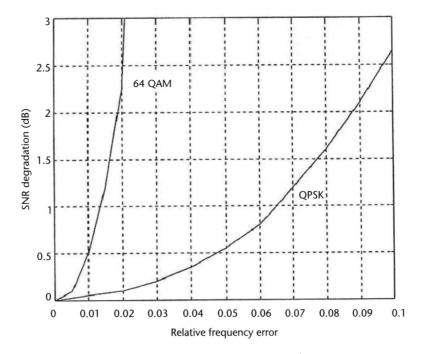

Figure 8.12   SER degradation due to frequency offset at SER = 10^{-4}. (*From:* [6]. © 2002. Reprinted by permission of Pearson Education Inc., Upper Saddle River, NJ.)

with small constellations, since large constellations operate at higher SNRs compared with small constellations. This directly improves the performance of the frequency error estimates. There are three techniques to solve this problem [6]:

- Data-aided algorithms based on training information embedded in the transmitted signal.
- Nondata-aided algorithms that analyze the received signal in frequency domain.
- Cyclic prefix-based algorithms that use the inherent structure of the OFDM signal provided by the cyclic prefix.

The first option is best suited for WLAN applications, since the preamble allows the receiver to use efficient maximum likelihood algorithms to estimate and correct the frequency offset before the actual information portion of the packet starts. The other two options are more popular with broadcast or continuous transmission systems.

8.8.4.1 Data-Aided Maximum Likelihood Estimator

This estimator operates on the received time domain signal (i.e., before the FFT [6]).

Let the transmitted signal be s_n, then the complex baseband model of the passband signal y_n is

$$y_n = s_n e^{j2\pi f_{tx} nT_s} \tag{8.7}$$

where f_{tx} is the transmitter carrier frequency and T_s the sampling period. After the receiver downconverts the signal with a carrier frequency f_{rx}, the received complex baseband signal r_n, neglecting noise, is

$$
\begin{aligned}
r_n &= s_n e^{j2\pi f_{tx} nT_s} e^{-j2\pi f_{rx} nT_s} \\
&= s_n e^{j2\pi (f_{tx} - f_{rx}) nT_s} \\
&= s_n e^{j2\pi f_\Delta nT_s}
\end{aligned}
\tag{8.8}
$$

where $f_\Delta = f_{tx} - f_{rx}$ is the difference between the transmitter and the receiver carrier frequencies. Let D be the delay between the identical samples of two repeated symbols. Then the frequency offset estimate is given by

$$
\begin{aligned}
z &= \sum_{n=0}^{L-1} r_n r_{n+D}^* \\
&= s_n e^{j2\pi f_\Delta nT_s} \left(s_{n+D} e^{j2\pi f_\Delta (n+D)T_s} \right)^* \\
&= \sum_{n=0}^{L-1} s_n s_{n+D}^* e^{j2\pi f_\Delta nT_s} e^{-j2\pi f_\Delta (n+D)T_s} \\
&= e^{j2\pi f_\Delta DT_s} \sum_{n=0}^{L-1} |s_n|^2
\end{aligned}
\tag{8.9}
$$

The final expression in (8.9) is a sum of complex variables with an angle proportional to the frequency offset. The frequency error estimator is given as

$$\hat{f}_\Delta = -\frac{1}{2\pi DT_s} \, \measuredangle z \tag{8.10}$$

where $\measuredangle z$ operator takes the angle of its argument.

Range of Operation
We now determine the range of operation of the frequency synchronization algorithm. We need to determine how large an offset can be estimated by this algorithm. It can be seen from (8.9) and (8.10) that the range is directly related to the length of the repeated symbols. The angle of z is of the form $-2\pi f_\Delta DT_s$ which is unambiguously defined only in the range $[-\pi \ \pi]$. Thus, if the absolute value of the frequency error is larger than the following limit

$$|f_\Delta| \geq \frac{\pi}{2\pi DT_s} = \frac{1}{2DT_s} \tag{8.11}$$

the estimate will be incorrect, since z has rotated an angle larger than π. This maximum allowable frequency error is usually normalized with the subcarrier spacing f_s. If the delay D is equal to the symbol length, then

$$\frac{1}{2DT_s} = \frac{1}{2} f_s \tag{8.12}$$

Thus the frequency error can be, at most, a half of the subcarrier spacing. It should be noted that if the repeated symbols include a cyclic prefix, the delay is longer than the symbol length and, hence, the range of the estimator is reduced.

As an example, we can calculate the value of this limit for the IEEE 802.11a system for both the short and long training symbols. For the short training symbols, the sample time is 50 ns and the delay $D = 16$. Thus, the maximum frequency error that can be estimated is

$$f_{\Delta\max} = \frac{1}{2DT_s}$$

$$= \frac{1}{2 \times 16 \times 50 \times 10^{-9}} \tag{8.13}$$

$$= 625 \text{ kHz}$$

This should be compared with the maximum possible frequency error in an IEEE 802.11a system. The carrier frequency is approximately 5.3 GHz, and the standard specifies a maximum oscillator error of 20 parts per million (ppm). Thus, if the transmitter and receiver clocks have the maximum allowed error but with opposite signs, the total observed error would be 40 ppm. This amounts to a frequency error of $f_\Delta = 40 \times 10^{-6} \times 5.3 \times 10^9 = 212$ KHz. Hence, the maximum possible frequency error is well within the range of the algorithm. Now consider the long training symbols. The only significant difference is that the delay $D = 64$ is four times longer. Hence the range is

$$f_{\Delta\max} = \frac{1}{4 \times 2DT_s} = 156.25 \text{ kHz} \qquad (8.14)$$

This is less than the maximum possible error defined in the standard. Thus, this estimator would not be reliable if only the long training symbols were used. Schmidl and Cox [14] have shown that at high SNR, the variance $\sigma_{f_\Delta}^2$ of the estimator is proportional to

$$\sigma_{f_\Delta}^2 = -\frac{1}{L \times SNR} \qquad (8.15)$$

This implies that the more the samples in the sum, the better the quality of the estimator.

Based on the preceding discussion, it is clear that we require a two-step frequency estimation process with a coarse frequency estimate performed from the short training symbols and fine frequency synchronization from the long training symbols.

8.8.5 Carrier Phase Tracking

Frequency estimation is never accurate. Hence, there will always be some residual frequency error. This residual frequency error causes constellation rotation (see Section 8.8.3). This constellation rotation is the same for all subcarriers, unlike the case of timing offset estimate, wherein it varies depending on the position of the subcarrier, being maximum at the edges and decreasing toward the center. Initially, the effect on bit error rates is small, but as the number of demodulated symbols increase, the amount of rotation increases symbol after symbol, until finally after 10 symbols, the demodulated symbols will cross the decision boundaries, resulting in errors. This is shown in Figure 8.13 for QPSK [6].

There is, therefore, a need to track the carrier phase throughout to ensure that the demodulated symbols never cross the decision boundaries. This is accomplished by data-aided tracking of the carrier phase, using four predefined subcarriers among the transmitted data. These special subcarriers are called *pilot* subcarriers. They are meant for tracking the carrier phase. After the FFT operation on the nth received symbol, the pilot subcarriers $R_{n,k}$ are equal to the product of the channel frequency response H_k and the known pilot symbol $P_{n,k}$, rotated by the residual frequency error.

$$R_{n,k} = H_k P_{n,k} e^{j2\pi n f_\Delta} \qquad (8.16)$$

If the channel estimate $\hat{H}_k$ is available, the phase estimate is given by

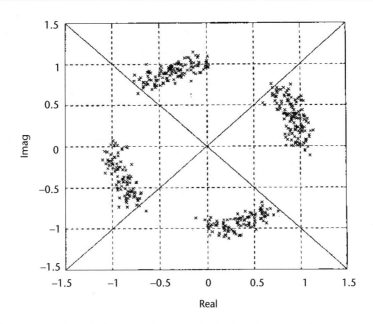

Figure 8.13 Constellation rotation with 3 KHz (1% of subcarrier spacing) frequency error during 10 symbols. (*From:* [6]. © 2002. Reprinted by permission of Pearson Education Inc., Upper Saddle River, NJ.)

$$\hat{\Phi}_n = \measuredangle \left[\sum_{k=1}^{N_P} R_{n,k} \left(\hat{H}_k P_{n,k} \right)^* \right] \tag{8.17}$$

$$= \measuredangle \left[\sum_{k=1}^{N_P} H_k P_{n,k} e^{j2\pi n f_\Delta} \left(\hat{H}_k P_{n,k} \right)^* \right]$$

If $\hat{H}_k = H_k$, (i.e., we have perfect channel state information), then

$$\hat{\Phi}_n = \measuredangle \left[\sum_{k=1}^{N_P} |H_k|^2 |P_{n,k}|^2 e^{j2\pi n f_\Delta} \right] \tag{8.18}$$

$$= \measuredangle \left[e^{j2\pi n f_\Delta} \sum_{k=1}^{N_P} |H_k|^2 \right]$$

The amplitude of the pilots is unity. We need not worry about the $[-\pi\ \pi]$ phase range, because the pilot data are known and, therefore, the phase ambiguity is automatically resolved correctly. It should be noted that if the channel estimates are not accurate, as is usually the case, they will contribute to the noise in the estimate.

8.9 Channel Estimation

Channel estimation is the technique of estimating the frequency response of the radio channel through which the transmitted signal travels before reaching the receiver antenna. Channel estimation can be carried out in the frequency domain or in the time domain. We shall examine the frequency domain approach. In this approach, there are two methods—preamble or pilots (see Chapter 4). The pilot method is usually used in broadcast OFDM systems. To estimate the channel, we use the long training symbols in the WLAN preamble. In IEEE 802.11a, there are 64 subcarriers, as there are 64 symbols in each frame of long training symbols, wherein each symbol corresponds to one subcarrier. Therefore, if we use the long training symbols, we can obtain an efficient estimate of the channel frequency response for all the subcarriers. The contents of the two long training symbols are identical, so averaging them can be used to improve the quality of the channel estimate. After the DFT processing, the received training symbols $R_{1,k}$ and $R_{2,k}$ are a product of the training symbols X_k and the channel H_k plus additive noise $W_{l,k}$,

$$R_{l,k} = H_k X_k + W_{l,k}$$

We then calculate the estimate as,

$$\hat{H}_k = \frac{1}{2}(R_{1,k} + R_{2,k})X_k^*$$

$$= \frac{1}{2}(H_k X_k + W_{1,k} + H_k X_k + W_{2,k})X_k^* \qquad (8.19)$$

$$= H_k |X_k|^2 + \frac{1}{2}(W_{1,k} + W_{2,k})X_k^*$$

$$= H_k + \frac{1}{2}(W_{1,k} + W_{2,k})X_k^*$$

where the training data amplitudes have been selected to unity amplitudes. The noise samples $W_{1,k}$ and $W_{2,k}$ are statistically independent. Therefore, the variance of their sum divided by two is half of the variance of the individual noise samples.

This concludes our study of the basic IEEE 802.11a system. The reader is encouraged to delve deeper into this topic from the references, since this type of system promises to be a vehicle for future high bit rate systems. The reader is also encouraged to utilize the software on IEEE 802.11a supplied with [6]. Some simulation results using this software will be discussed in the next chapter.

References

[1] Zahed, I., "Wireless LAN Technology: Current State and Future Trends," *Research Seminar on Telecommunications Software*, Helsinki University of Technology, Autumn 2002.

[2] *Wireless Local Area Networks: Issues in Technology and Standards*, www.radiolan.com/ds/WP-WLAN.

[3] Birkeland, T. A., and F. F. Nilsosson, *Limitations in Performance for WLAN Technologies*, Agder University of College, 2002, www.swing.hia.no/ikt02/ikt6400/g07/files/Poster.

[4] Tourrilhes, J., Hewlett Packard, *Wireless Overview: Some Wireless LAN Standards*, HP Labs Technical Report, www.hpl.hp.com/cgi-bin/AT-Tech_Reportssearch.cgi.

[5] Geier, J., *Wireless LANs: Implementing Interoperable Networks,* Macmillan Technical Publishing, 1999.

[6] Heiskala, J., and J. Terry, *OFDM Wireless LANs: A Theoretical and Practical Guide,* Indianapolis, IN: Sams Publishing, 2002.

[7] Santamaria, A., and F. J. Lopez-Hernandez, *Wireless LAN Standard and Applications,* Norwood, MA: Artech House, Autumn 2002, pp. 3–7, 45–77, 93–94, 109–112, 186–188.

[8] *IEEE Std. ISO/IEC 8802-11/And 1.* Copyright 1999, IEEE. All rights reserved.

[9] Mody, A. N., and G. L. Stuber, "Synchronization for MIMO-OFDM," *IEEE Global Commun. Conference,* San Antonio, TX, 2001.

[10] Mody, A. N., and G. L. Stuber, "Parameter Estimation for MIMO-OFDM," *IEEE Vehicular Technology Conference,* Rhodes, Greece, 2001.

[11] Mody, A. N., and G. L. Stuber, "Receiver Implementation for a MIMO-OFDM System," *IEEE Global Commun. Conference,* Taiwan, 2002.

[12] Pollet, T., P. Spruyt, and M. Moeneclaey, "The BER Performance of OFDM Systems Using Non-Synchronized Sampling," *IEEE Global Telecommun. Conference,* December 1994, San Francisco, CA, pp. 253–257.

[13] Pollet, T., M. van Bladel, and M. Moeneclaey, "BER Sensitivity of OFDM Systems to Carrier Frequency Offset and Wiener Phase Noise," *IEEE Trans. on Commun.,* Vol. 43, Issue 2, Part 3, February–April 1995, pp. 191–193.

[14] Schmidl, T. M., and D. C. Cox, "Low-Overhead, Low-Complexity [Burst] Synchronization for OFDM," *IEEE International Conference on Commun.,* Vol. 3, June 1996, Dallas, TX, pp. 1301–1306.

CHAPTER 9

Space-Time Coding for Broadband Channel

9.1 Introduction

In the preceding chapters, we concentrated on space-time coding for frequency-nonselective flat fading channels pertaining to narrowband systems. However, it will be appreciated that such systems are not readily found in nature. In the real world we have what are called frequency-selective fading channels. In wideband wireless communications, the coherent bandwidth is narrower than the signal bandwidth. This implies that we will experience dispersive fading or independent fading, wherein certain frequencies will fade more than other frequencies (i.e., if you observe the spectrum, they will rise and fall differently with respect to each other). We studied this phenomenon in Chapter 8. Recently, there has been an increasing interest in providing high data rate services like videoconference, multimedia and so on over wideband channels. Space-time coding techniques, by their very nature, readily lend themselves to high data rate situations. Unfortunately, they require a flat fading channel to function correctly. Therefore, it becomes extremely necessary to harness their power for wideband systems. We now investigate the effect of frequency-selective fading on space-time code performance.

9.2 Performance of Space-Time Coding on Frequency-Selective Fading Channels

We now consider the performance of space-time codes in frequency-selective fading channels. We consider the case of multiple transmit antennas ($M_T > 1$) and one receive antenna ($M_R = 1$) in an ST-coded system. The results can be extended to more general cases. We assume that the fading parameters between pairs of transmit antennas and receive antennas are identically distributed and independent.

Let $u_i(t)$ denote the signal transmitted from the ith antenna and let $h^i(t, \tau_i)$ denote the channel impulse response between the ith transmit antenna and receiver. Let t_n denote the sampling time $t = nT_s$, where T_s is the symbol duration. After matched filtering, the received signal at t_n is given by [1],

$$
r_n = \frac{1}{T_s} \int\limits_{nT_s}^{(n+1)T_s} \left[\sum_{i=1}^{M_T} \int\limits_0^\infty u_i(t - \tau_i) h^i(t, \tau_i) \, d\tau_i \right] dt + n_n \tag{9.1}
$$

where n_n is an independent sample of a zero-mean complex Gaussian random process with variance $N_0/2$ per dimension. After some manipulations, the received signal can be expressed as comprising the following three terms [1],

$$r_n = \alpha \sum_{i=1}^{M_T} \sum_{\ell=1}^{L} h_\ell^j(t_n) s_n^i + I_n + n_n \tag{9.2}$$

where s_n^i is the message for the ith antenna at the nth symbol period, I_n is the term representing the ISI, and α is a constant dependent on the channel power delay profile, which can be shown as,

$$\alpha = \frac{1}{T_s} \int_{-T_s}^{T_s} p(\tau + \tau_d)(T_s - |\tau|)\, d\tau \tag{9.3}$$

The values of α for different power delay profiles are given by [1]

$$\alpha = \begin{cases} 1 - d & \text{Exponential or 2-ray equal-gain profile} \\ 1 - \sqrt{2/\pi}d & \text{Gaussian profile} \end{cases} \tag{9.4}$$

The mean value of the ISI term is zero and the variance is given by [1]

$$\sigma_{I_n}^2 = \begin{cases} 3M_T d^2 E_s & \text{Exponential or 2-ray equal-gain profile} \\ 2M_T(1 - 1/\pi)d^2 E_s & \text{Gaussian profile} \end{cases}$$

where E_s is energy per symbol. Using these results, if we approximate the ISI term by a Gaussian random variable with zero mean and variance $\sigma_{I_n}^2$, it follows that

$$r_n = \alpha \sum_{i=1}^{M_T} \sum_{\ell=1}^{L} h_\ell^j(t_n) s_n^i + \tilde{n}_n \tag{9.5}$$

where $\tilde{n}_n$ is an additive zero-mean complex Gaussian random variable with variance $\sigma_{I_n}^2 + N_0$, which is *uncorrelated with the signal term*. The PEP under this approximation is given by [1]

$$P(s \to e) \leq \left[\prod_{i=1}^{r} \lambda_i \frac{\alpha^2}{\sigma_{I_n}^2/N_0 + 1} \right]^{-M_R} \left(\frac{E_s}{4N_0} \right)^{-rM_R} \tag{9.6}$$

where $r \leq M_T$ is the rank of the code word distance matrix and λ_i, $i = 1, 2, \ldots, r$ are the nonzero eigenvalues of the matrix. From (9.6) we note that a diversity gain of rM_R is achieved. This is the same as was derived in Chapter 5 for frequency nonselective fading channels. The coding gain is

$$G_{\text{coding}} = \frac{\left(\displaystyle\prod_{i=1}^{r} \lambda_i\right)^{1/r} \dfrac{\alpha^2}{\sigma_{I_n}^2/N_0 + 1}}{d_u^2} \tag{9.7}$$

The coding gain is reduced by a factor of $\left(\dfrac{\alpha^2}{\sigma_{I_n}^2/N_0 + 1}\right)$ compared with the one in frequency flat channels. Furthermore, at high SNRs, there exists an irreducible error rate "floor," as shown in Figure 9.1 [1].

This irreducible error rate "floor" persists even if we increase the number of states. This is clearly due to the existence of delay spread. The ISI due to the multipaths is causing this "floor" to appear. We can mitigate the effects of ISI by resorting to adaptive equalization, but in such an environment it will be too complex to implement. The other option is to resort to OFDM. OFDM, by its nature, converts a frequency-selective fading channel into a frequency-nonselective fading channel. The subcarriers (tones) in an OFDM symbol are essentially narrowband signals. Consequently they combat multipath effects in the channel. This aspect was discussed extensively in Chapter 7. These tones readily lend themselves to use as vehicles for space-time codes. Therefore, OFDM is an enabler for space-time coding in real life.

9.3   Space-Time Coding in Wideband OFDM Systems

We now examine the behavior of STC in OFDM systems. Specifically we single out one well-known OFDM-based wideband system—IEEE 802.11a packet transmission system. In Chapter 8, we briefly examined the design of such a system. If we can incorporate space-time coding into this system, we will obtain a high-rate

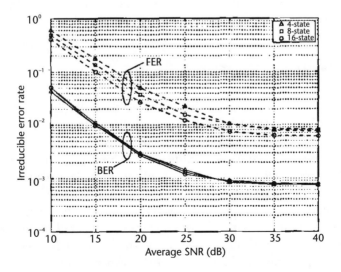

Figure 9.1   Performance of ST codes in multipath fading channels. (*From:* [1]. © 2000, IEEE.)

packet transmission system suitable for high-throughput applications like videoconferencing and multimedia. Toward this end, we will need to modify the system block diagram (Figure 8.1), as shown in Figure 9.2.

It can be seen in Figure 9.2, that there are two new blocks added, as shown in dotted squares. These are "Tx. Diversity Encoder" in the transmitter chain and "diversity combiner" in the receiver chain. The encoder converts the modulated symbols into space-time coded signals. In the figure, we are considering a 2×2 system. The diversity combiner in the receiver takes the output from the demultiplexers and carries out space-time decoding on the two inputs. The Viterbi decoder then finally gets the original modulated stream for convolution decoding. There is a demodulator and deinterleaver before the decoder (not shown). By this modification, we are now in a position to fully exploit the IEEE 802.11a system for MIMO, with all its advantages as an established packet transmission system. These modifications are by no means sufficient, but it will enable us to determine how an established packet transmission system behaves if adapted for MIMO. To explain the configuration further, let us take an example of Alamouti's code. The Tx. diversity encoder splits the data into two orthogonal streams. The first stream is fed to the top IFFT modulator and the bottom to the lower IFFT modulator. Remember from Chapter 8 that the modulator is 64 points long, out of which only 48 are data. Hence, the data stream in each case is split into blocks of 48 each, to which the four pilot signals for carrier phase locking are appended. These 52 subcarriers are then padded with zeros to make the full OFDM symbol of 64 subcarriers. We shall now refer to these 64 subcarriers as 64 symbols. We then convert this from parallel to serial in a multiplexer (mux) and then append the CP. Hence, the incoming data stream from the modulator is split into two orthogonal streams comprising 64 symbols (this makes one OFDM symbol) and 16 cyclic prefix symbols, making a total of 80 symbols in each packet. In this manner the information is transmitted in packets of 80 symbols each. Hence, we have a block matrix of NM_T at the transmitter where in our case $N = 64$. Consequently, the channel matrix $\mathbf{H}$ will be of size $NM_T \times NM_R$. The receiver is the exact reverse process after the incoming packets are stripped of their cyclic prefixes. The performance of this scheme will be discussed in simulations in Section 9.5.

9.4 Capacity of MIMO-OFDM Systems

We investigate the capacity behavior of wireless OFDM based *spatial multiplexing systems* (discussed in Chapter 6) in broadband fading environments for the case where the channel is unknown at the transmitter and perfectly known at the receiver. We will examine the influence of physical parameters such as the amount of *delay spread, cluster angle spread*, and *total angle spread* and system parameters such as the number of antennas and antenna spacing on ergodic capacity and outage capacity. Before proceeding further, it will be useful to recapitulate what we have learned in Chapter 6. In OFDM-based spatial multiplexing, each antenna transmits statistically independent data symbols from different antennas and different tones. If the channel is unknown at the transmitter we allocate the total available power uniformly across all the antennas and OFDM tones. Once transmitted, these

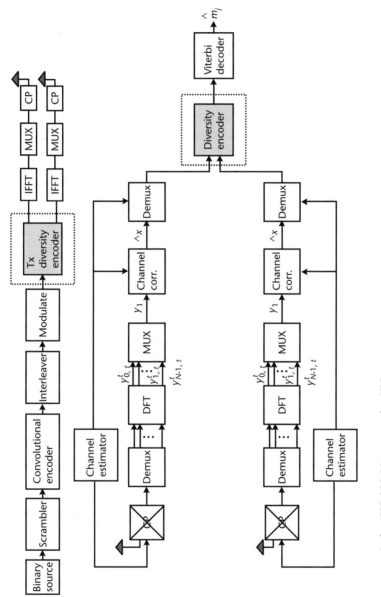

Figure 9.2   Integration scenario for IEEE 802.11a system for STC.

symbols, depending on their frequency, will choose a delay path across the channel. These paths are also called *data pipes*. Remember that our precondition for space-time coding is that the environment is ideally a rich scattering environment. In such an environment there are bound to be a number of delay paths. The symbols propagate along these paths and reach the receiver. The orthogonality between the data streams is dependent on the fact that these paths are mutually independent. Hence, each symbol stream does not "see" the other symbol streams. Intuitively, the more the streams or delay paths the greater the capacity because there are more data pipes. We will now mathematically prove this assumption. Section 9.4 is reproduced from an excellent paper by Bolcskei et al. ([2]. © 2002, IEEE. Reprinted with permission). In this analysis, L is the number of delay paths across the channel, with each path being considered a scatterer cluster and each of the paths emanating from within the same scatterer cluster experiencing the same delay. Microdelay variations within the scatterers will be neglected.

9.4.1 Assumptions

Propagation scenario: We assume that the subscriber unit (SU) is surrounded by local scatterers so that fading at the SU antennas is spatially uncorrelated. The BTS is high enough so that it is unobstructed and no local scattering occurs. Therefore, spatial fading at the BTS will be correlated depending on the BTS antenna spacing, the angle spread observed at the BTS array, and coherent distance for that channel. We incorporate the power delay profile of the channel but neglect shadowing. These assumptions are typical for a cellular environment, wherein the BTS is on a tower or on the roof of a tall building and the terminal is on the street level and experiences local scattering. We restrict ourselves to the uplink case, the results being similar for downlink. In the following derivation, boldface letters signify vector-matrix notation, thus **H**.

Channel: Assume that there are L distinct scatter clusters and that each of the paths emanating from within the same scatterer cluster experiences the same delay. Micro delay variations within the scatterers will be neglected. Let $s[n]$ be the $M_T \times 1$ transmitted signal vector and $r[n]$ the $M_R \times 1$ received signal vector. Then,

$$\mathbf{r}[n] = \sum_{\ell=0}^{L-1} \mathbf{H}_\ell \mathbf{s}[n - \ell] \tag{9.8}$$

where the $M_R \times M_T$ complex valued random matrix $\mathbf{H}_\ell$ represents the ℓth tap of the discrete-time MIMO fading channel impulse response. In general in a wideband channel, there will be a continuum of delays. The channel model (9.8) is based on the assumption that there are L resolvable paths, where $L = [B\tau]$ with B and τ denoting the signal bandwidth and delay spread, respectively. The individual $\mathbf{H}_\ell$ are (possibly correlated) ZMCSCG. Different *scatterer clusters* are uncorrelated, that is,

$$\epsilon\left[vec\{\mathbf{H}_\ell\} vec^H\{\mathbf{H}_{\ell'}\} \right] = \mathbf{0}_{M_R M_T} \text{ for } \ell \neq \ell' \tag{9.9}$$

where ϵ denotes the expectation operator and superscript H stands for conjugate transposition and where

$$vec\{\mathbf{H}_\ell\} = \begin{bmatrix} \mathbf{h}_{\ell,0}^T & \mathbf{h}_{\ell,1}^T & \cdots & \mathbf{h}_{\ell,M_T-1}^T \end{bmatrix}^T$$

with $\mathbf{h}_{\ell,k}$ being column vectors of the matrix $\mathbf{H}_\ell$ and $\mathbf{0}_{M_R M_T}$ denoting the all zero matrix of size $M_R M_T \times M_R M_T$. Each scatterer cluster has a mean angle of arrival at the BTS denoted as $\bar{\theta}_\ell$, a cluster angle spread δ_ℓ (proportional to the scattering radius of the cluster), and a path gain σ_ℓ^2 (derived from the power delay profile of the channel).

Array geometry: We assume a uniform linear array (ULA) at both the BTS and SU with identical antenna elements. We can also extend out results to nonuniform arrays. The relative antenna spacing is denoted as $\Delta = d/\lambda$ where d is the absolute antenna spacing and $\lambda = c/f_c$ is the wavelength of a narrowband signal with center frequency f_c.

Fading statistics: We assume that the $\mathbf{h}_{\ell,k}$ ($\ell = 0, 1, \ldots, L-1$; $k = 0, 1, \ldots, M_T - 1$) have zero mean (i.e., pure Rayleigh fading) and that the $M_R \times M_R$ correlation matrix $\mathbf{R}_\ell = \epsilon\{\mathbf{h}_{\ell,k}\ \mathbf{h}_{\ell,k}^H\}$ is independent of k or, equivalently, the fading statistics are the same for all transmit antennas. We define $\rho_\ell(s\Delta, \bar{\theta}_\ell, \delta_\ell) = \epsilon\{h_{\ell,k}^r (h_{\ell,k}^{(r+s)})^*\}$ for $\ell = 0, 1, \ldots, L-1$; $k = 0, 1, \ldots, M_T - 1$ to be the fading correlation between two BTS antenna elements spaced $s\Delta$ wavelengths apart. The correlation matrix $\mathbf{R}_\ell$ can then be written as

$$[\mathbf{R}_\ell]_{m,n} = \sigma_\ell^2 \rho_\ell\big((n-m)\Delta, \bar{\theta}_\ell, \delta_\ell\big) \tag{9.10}$$

The correlation matrixes already take into account the power delay profile of the channel.

We factor the $M_R \times M_R$ correlation matrix $\mathbf{R}_\ell$ according to $\mathbf{R}_\ell = \mathbf{R}_\ell^{1/2}\mathbf{R}_\ell^{1/2}$, where $\mathbf{R}_\ell^{1/2}$ is of size $M_R \times M_R$, the $M_R \times M_T$ matrix $\mathbf{H}_\ell$ can be written as

$$\mathbf{H}_\ell = \mathbf{R}_\ell^{1/2}\mathbf{H}_{\omega,\ell}, \ \ell = 0, 1, \ldots, L-1 \tag{9.11}$$

where $\mathbf{H}_{\omega,\ell}$ is an uncorrelated matrix of size $M_R \times M_T$ with i.i.d $\mathcal{CN}(0, 1)$ entries. We have essentially decomposed the ℓth tap of the stochastic MIMO channel impulse response into the product of a deterministic matrix $\mathbf{R}_\ell^{1/2}$, taking into account the spatial fading correlation at the BTS and a stochastic matrix of i.i.d complex Gaussian random variables $\mathbf{H}_{\omega,\ell}$.

We assume that the angle of arrival for the ℓth ($\ell = 0, 1, \ldots, L-1$) path cluster at the BTS is Gaussian distributed around the mean angle of arrival $\bar{\theta}_\ell$ [i.e., the actual angle of arrival is given by $\theta_\ell = \bar{\theta}_\ell + \hat{\theta}_\ell$ with $\hat{\theta}_\ell \sim \mathcal{N}(0, \sigma_{\theta_\ell}^2)$]. The variance $\sigma_{\theta_\ell}^2$ is proportional to the angular spread δ_ℓ and, hence, the scattering radius of the ℓth path cluster. For small angular spread the correlation function can be approximated as [2]

$$\rho_\ell\left(s\Delta, \bar{\theta}_\ell, \delta_\ell\right) \approx e^{-j2\pi s\Delta \cos\left(\bar{\theta}_\ell\right)} \; e^{-\frac{1}{2}\left(2\pi s\Delta \sin\left(\bar{\theta}_\ell\right)\sigma_{\theta_\ell}\right)^2} \tag{9.12}$$

This approximation is accurate for small angular spread, but it does indicate a trend for large angular spreads, such as uncorrelated spatial fading. Note that if $\sigma_{\theta_\ell} = 0$, the correlation matrix $\mathbf{R}_\ell$ collapses to a rank-1 matrix and can be written as $\mathbf{R}_\ell = \sigma_\ell^2 \mathbf{a}\left(\bar{\theta}_\ell\right)\mathbf{a}^H\left(\bar{\theta}_\ell\right)$ with the array response vector of the ULA given by

$$\mathbf{a}(\theta) = \begin{bmatrix} 1 & e^{j2\pi\Delta \cos(\theta)} & \cdots & e^{j2\pi(M_R-1)\Delta \cos(\theta)} \end{bmatrix}^T \tag{9.13}$$

9.4.2 Mutual Information

We now derive the expression for mutual information of OFDM-based spatial multiplexing system. We will thereafter use it to compute the ergodic capacity and study the outage properties of the system.

Spatial multiplexing, as we already know from Chapter 6, has the potential to drastically increase the capacity of wireless radio links with no additional power or bandwidth consumption. The gain in terms of ergodic capacity over SISO systems resulting from the use of multiple antennas is called *multiplexing gain.* OFDM turns a frequency-selective channel into a frequency-nonselective one. This allows for simple equalization in that for each OFDM tone, we need to only invert the constant matrix. In OFDM-based spatial multiplexing, the data streams are passed through the OFDM modulators, as is shown in Figure 9.2, and then launched from the individual antennas. This takes place simultaneously from all the M_T antennas. In the receiver, the individual signals are passed through OFDM demodulators, separated, and then decoded. We assume that the length of the cyclic prefix is more than the maximum delay spread in the channel. This will obviate ISI. This assumption guarantees that the frequency-selective fading channel decouples into a set of parallel frequency flat fading channels [3]. The transmitted data can be shown as frequency vectors $\mathbf{s}_k = \begin{bmatrix} s_k^0 & s_k^1 & \cdots & s_k^{(M_T-1)} \end{bmatrix}^T$ with s_k^i denoting the data symbol transmitted from the ith antenna on the kth tone and defining $\mathbf{H}\left(e^{j2\pi\theta}\right) = \sum_{\ell=0}^{L-1} \mathbf{H}_\ell e^{-j2\pi\ell\theta}$ $(0 \le \theta < 1)$, it can be shown that

$$\hat{\mathbf{s}}_k = \mathbf{H}\left(e^{j2\pi\frac{k}{N}}\right)\mathbf{s}_k + \mathbf{n}_k \tag{9.14}$$

where $\hat{\mathbf{s}}_k$ is the received data vector for the kth tone, N is the total number of OFDM tones, and $\mathbf{n}_k$ is additive white Gaussian noise satisfying

$$\epsilon\left\{\mathbf{n}_k \mathbf{n}_\ell^H\right\} = \sigma_n^2 \mathbf{I}_{M_R} \delta[k - \ell] \tag{9.15}$$

where $\mathbf{I}_{M_R}$ is the identity matrix of size M_R. We observe from (9.14) that equalization requires the inversion of a constant matrix for each tone $k = 0, 1, \ldots, N-1$.

We stack the vectors $\hat{\mathbf{s}}_k$, $\mathbf{s}_k$ and $\mathbf{n}_k$ according to [2]

$$\hat{s} = \left[\hat{s}_0^T \ \ \hat{s}_1^T \ \ \dots \ \ \hat{s}_{N-1}^T \right]^T, \ s = \left[s_0^T \ \ s_1^T \ \ \dots \ \ s_{N-1}^T \right]^T, \ n = \left[n_0^T \ \ n_1^T \ \ \dots \ \ n_{N-1}^T \right]^T$$

where $\hat{s}$ and n are $M_R N \times 1$ vectors and s is an $M_T N \times 1$ vector. We note that based on (9.15) we can infer that the noise vector n is white, that is,

$$\epsilon\left\{ nn^H \right\} = \sigma_n^2 I_{M_R N}$$

H is now a block-diagonal matrix of size $NM_R \times NM_T$, that is,

$$H = \text{diag}\left\{ H\left(e^{j2\pi \frac{k}{N}} \right) \right\}_{k=0}^{N-1}$$

We now rewrite (9.14) as

$$\hat{s} = Hs + n \tag{9.16}$$

We assume that each of the OFDM symbols transmitted uses a new realization of the channel impulse response matrix H_ℓ and that this matrix remains constant throughout the duration of the OFDM symbol (quasi-static channel). Using (9.16), the mutual information (in bit/s/Hz) of the OFDM-based spatial multiplexing system under an average transmitter power constraint is given by [4, 5]

$$I = \frac{1}{N} \log_2 \left[\det\left(I_{M_R N} + \frac{1}{\sigma_n^2} H\Sigma H^H \right) \right] \tag{9.17}$$

where Σ with $\text{Tr}(\Sigma) \leq P$ is the covariance matrix of the Gaussian input vector s and P is the maximum overall transmit power. The symbol Tr stands for trace of a matrix. We are familiar with this symbol from Chapter 5. We note that in (9.17), the mutual information is normalized by N, since N data symbols are transmitted in one OFDM symbol. We ignore the loss in spectral efficiency due to the cyclic prefix. The $NM_T \times NM_T$ matrix Σ is a block-diagonal matrix given by

$$\Sigma = \text{diag}\left\{ \Sigma_k \right\}_{k=0}^{N-1}$$

where the $M_T \times M_T$ matrixes Σ_k are the covariance matrixes of the Gaussian vectors s_k, and as such determine the power allocation across the transmit antennas and across the OFDM tones. Remember, this implies that the power is not only divided across the transmit antennas, as is usual, but also *within* each antenna across the OFDM tones. If we knew the channel, then we would resort to the water-pouring algorithm (see Chapter 2) to distribute this power for maximum capacity. Since, in our case, we have no knowledge of the channel at the transmitter, the total power is allocated uniformly across all antennas and OFDM tones. Therefore, we set $\Sigma_k = \frac{P}{M_T N} I_{M_T}$ $(k = 0, 1, \dots, N-1)$, which is easily verified to result in $\text{Tr}(\Sigma) = P$. Using (9.17) we obtain,

$$I = \frac{1}{N} \sum_{k=0}^{N-1} I_k = \frac{1}{N} \sum_{k=0}^{N-1} \log_2 \left[\det \left(\mathbf{I}_{M_R} + \rho \mathbf{H} \left(e^{j2\pi\frac{k}{N}} \right) \mathbf{H}^H \left(e^{j2\pi\frac{k}{N}} \right) \right) \right]$$

(9.18)

where $\rho = \dfrac{P}{M_T N \sigma_n^2}$. We call I_k the mutual information of the kth MIMO OFDM subchannel. Since $\mathbf{H}\left(e^{j2\pi(k/N)}\right)$ is random, I_k is also random.

Proposition 1:   The distribution of I_k ($k = 0, 1, \ldots, N-1$) is independent of k and given by

$$I_k \sim \log_2 \left[\det \left(\mathbf{I}_{M_R} + \rho \mathbf{\Lambda} \mathbf{H}_\omega \mathbf{H}_\omega^H \right) \right] \text{ for } k = 0, 1, \ldots, N-1 \qquad (9.19)$$

where $\mathbf{\Lambda}$ defines the eigenvalues of the *sum correlation matrix* $\mathbf{R}$ $\left(\text{defined as } \mathbf{R} = \sum_{\ell=0}^{L-1} \mathbf{R}_\ell\right)$ and is given by $\mathbf{\Lambda} = \text{diag}\left\{\lambda_i(\mathbf{R})\right\}_{i=0}^{M_R-1}$, and $\mathbf{H}_\omega$ is an $M_R \times M_T$ i.i.d random matrix with $\mathcal{CN}(0, 1)$ entries. Finally, $\lambda_i(\mathbf{R})$ denotes the ith eigenvalue of $\mathbf{R}$.

We give this proposition without proof. The interested reader is referred to [2].

I apologize to the reader for all this heavy mathematics, but it is necessary to see what is behind (9.18), as it forms the basis for what is to follow.

9.4.3   Ergodic Capacity and Outage Capacity

There are basically two cases that we need to investigate.

- *Ergodic case:* The assumption here is that the transmission time is long enough to reveal the long-term ergodic properties of the fading channel. From probability theory, we know that the term ergodic implies that the process is such that one sample from it is representative of that class. In this case the Shannon capacity (the maximum possible capacity for a channel) is given by $C = \epsilon\{I\}$ with I defined by (9.18). For a good code, at rates lower than C the error probability decays exponentially with the transmission length. The assumption here is that the fading process is ergodic, coding and interleaving are performed across OFDM symbols, and the number of fading blocks spanned by a code word goes to infinity, whereas the block size (NM_T) remains constant and finite. Capacity can be achieved in principle by transmitting a code word over a very large number of independently fading blocks. We note that the capacity obtained for an OFDM-based spatial multiplexing system is a lower bound for the capacity of the underlying broadband MIMO channel. This means that an OFDM-based system attains the same capacity as the minimum possible capacity that can be attained by the actual MIMO channel through which the transmission propagates. This is determined by not only M_T and M_R but also the size (rank) of the channel impulse response matrix $\mathbf{H}$, which, in turn, is determined by the number of delay paths in the channel, which, in turn, determines the number of eigenvalues in the channel sum correlation matrix $\mathbf{R}$. We already examined the mechanics of this in Chapter 6.

- *Nonergodic case:* In this case, we assume that the code word spans an arbitrary but fixed number of blocks while the block size goes to infinity. This occurs when there is a finite time limit imposed for the transmission, like in the case of speech transmission over wireless channels. Unlike in the previous case, these assumptions give rise to error probabilities that do not decay with an increase of block length. Therefore, a capacity in the Shannon sense does not exist since with nonzero probability, which is independent of the code length, the mutual information I in (9.18) falls below any positive rate, as small as it may be. Thus we invoke the concept of outage. Assuming that the code words extend over a single block, the outage (or failure) probability for a given rate is the probability that I falls below that rate. In this case capacity is viewed as a random entity since it depends on the instantaneous random channel parameters.

9.4.4 Influence of Channel and System Parameters on Capacity

We now study the influence of channel and system parameters on ergodic capacity and outage capacity.

9.4.4.1 The Ergodic Case

The ergodic capacity is obtained from (9.18) as

$$C = \epsilon\left\{ \frac{1}{N} \sum_{k=0}^{N-1} I_k \right\}$$

Using Proposition 1, which says that the I_k $(k = 0, 1, \ldots, N-1)$ all have the same distribution given by (9.19), the ergodic capacity is obtained as

$$C = \epsilon\left\{ \log_2 \left[\det\left(\mathbf{I}_{M_R} + \rho \mathbf{\Lambda} \mathbf{H}_\omega \mathbf{H}_\omega^H \right) \right] \right\} \tag{9.20}$$

where the expectation is taken with respect to $\mathbf{H}_\omega$. We carry out an asymptotic analysis by assuming M_T is large. Then for a fixed M_R, $\frac{1}{M_T} \mathbf{H}_\omega \mathbf{H}_\omega^H \to \mathbf{I}_{M_R}$. Hence, in the large M_T limit, we get

$$C = \log_2 \left[\det\left(\mathbf{I}_{M_R} + \overline{\rho} \mathbf{\Lambda} \right) \right] \tag{9.21}$$

where $\overline{\rho} = M_T \rho = \dfrac{P}{N \sigma_n^2}$. In the low SNR regime (i.e., for small $\overline{\rho}$ and large M_T limit), we obtain from (9.21)

$$C = \log_2 \left(\prod_{i=0}^{M_R-1} (1 + \overline{\rho} \lambda_i(\mathbf{R})) \right) \approx \log_2 (1 + \overline{\rho} \mathrm{Tr}(\mathbf{R}))$$

where all the higher order terms in $\overline{\rho}$ have been neglected. Thus in the low SNR regime, the ergodic capacity is determined by the trace of the sum correlation matrix $\mathbf{R}$. Now since the transmitted power is a constant value, the $\text{Tr}(\mathbf{R})$ is fixed, leading us to the conclusion that the quantum of delay spread in the channel has no impact on ergodic capacity when the SNR is low.

In the high SNR case, we obtain

$$C = \sum_{i=0}^{M_R-1} \log_2(1 + \overline{\rho}\lambda_i(\mathbf{R})) \tag{9.22}$$

This implies that the eigenvalue spread of the sum correlation matrix $\mathbf{R} = \sum_{\ell=0}^{L-1} \mathbf{R}_\ell$, therefore critically determining the ergodic capacity. This is obvious when we look at the problem differently. The more the eigenvalues, the more are the delay paths through the channel. The more the delay paths, the more routes are available from the transmitter to the receiver. Remember that the number of eigenvalues determines the rank of the sum correlation matrix. Each delay path contributes to one eigenvalue. Hence, the more the delay paths, the higher the capacity of the channel. It can be shown that for $\text{Tr}(\mathbf{R}) = 1$, (9.20) maximizes if $\lambda_i(\mathbf{R}) = \dfrac{1}{M_R}$ $(i = 0, 1, \ldots, M_R - 1)$ and M_T is finite. The proof is beyond the scope of this book and the interested reader is referred to [2]. The implication here is that a deviation of $\lambda_i(\mathbf{R})$ as a function of I from a constant function will, therefore, result in a loss in terms of ergodic capacity or equivalently reduced multiplexing gain.

Impact of Cluster Angle Spread and Antenna Spacing
If the cluster angle spread is small and/or the antennas spacing is small (this occurs when there a lot of separation between the BTS and SU), then the delay paths will be correlated and less vice versa. This becomes obvious when we look at the eigenvalue distribution. A correlated situation gives rise to a peaky distribution, whereas a relatively uncorrelated situation gives rise to a broader eigenvalue distribution. These cases are shown in Figure 9.3(a) and (b), respectively [2].

A corollary is that the higher the delay spread the more the total angle spread and the more eigenvalues "raked" in the flatter the eigenvalue distribution will be. This is obvious if we consider that when the total angle is small, the paths will be more correlated at the antenna. Consequently, the eigenvalue count will drop and the eigenvalue distribution will be peaky and the capacity reduces. If the total angle spread is large, the eigenvalue distribution becomes flat and, therefore, the capacity increases. Hence, there is a lot of interaction between the eigenvalue distribution, angle spread (both total and cluster angles), and capacity. These aspects are proved mathematically in [2] and the proof is beyond the scope of this book. Figure 9.4(a) and (b) show an example limiting distributions for a three-path channel with a total angle spread of 22.5° and 90°, respectively [2]. In the former case the capacity is 7.3 bit/s/Hz and in the latter case 9.87 bit/s/Hz [2]. Remember, this argument stems from the assumption that delay spreads determine the total angle spread. This assumption has been proved in [2].

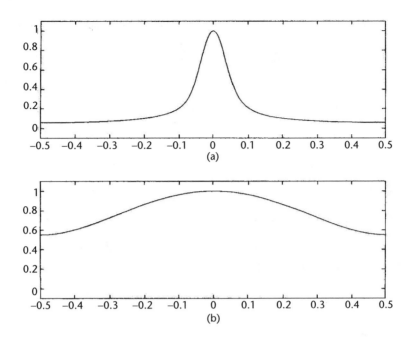

Figure 9.3   Eigenvalue distribution of the correlation matrix $\mathbf{R}_\ell$ for the case of (a) high spatial fading correlation and (b) low spatial fading correlation. (*From:* [2]. © 2002, IEEE.)

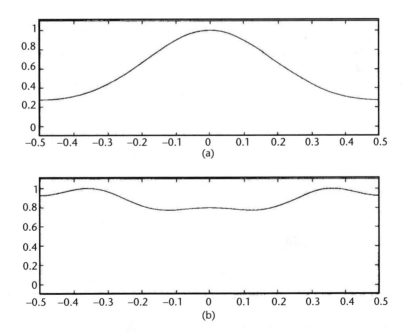

Figure 9.4   Eigenvalue distribution of the sum correlation matrix $\mathbf{R} = \sum_{\ell=0}^{L-1} \mathbf{R}_\ell$ for fixed cluster angle spread and for the cases of (a) small total angle spread and (b) large total angle spread. (*From:* [2]. © 2002, IEEE.)

Suppose we fix $\mathrm{Tr}\,(\mathbf{R})$ and take a flat fading scenario with a small cluster angle spread. In such a case, $\mathbf{R} = \mathbf{R}_\ell$ with rank 1. In this case the matrix $\mathbf{\Lambda H}_\omega \mathbf{H}_\omega^H$ has rank 1 with only one data pipe (i.e., there is no multiplexing gain). Now compare this with a delay spread scenario where $L \geq M_R$ and each of the R_ℓ ($\ell = 0, 1, \ldots, L - 1$) has rank 1, but the sum correlation matrix $\mathbf{R}$ has full rank. For this to happen, a sufficiently large total angle spread is necessary. Clearly, in this case M_R spatial data pipes will open up and we will get a higher ergodic capacity, because the rank of $\mathbf{R}$ is higher than in the flat fading case. We can therefore conclude that, in practice, MIMO delay spread channels offer an advantage over MIMO flat fading channels in terms of ergodic capacity.

9.4.4.2 The Nonergodic Case

This is the case that pertains to IEEE 802.11a and similar systems. In [6] it has been demonstrated that SISO delay-spread channels offer significant advantages over flat fading channels in terms of outage probabilities or outage capacity. The outage properties are determined by the number of diversity degrees of freedom in the channel. In our case, we have both spatial diversity and frequency diversity available. We can therefore expect that both diversity sources will contribute to the outage capacity of the system. Assuming that a code word spans one block, we recall that the outage probability for a given rate R is the probability that $I = \frac{1}{N} \sum_{k=0}^{N-1} I_k$ falls below that rate. The distribution of I is hard to compute analytically. We therefore resort to simulations. The individual I_k all have the same distribution. The correlation between the I_k, however, depends on the amount of delay spread in the system. To analyze this statement, we will consider two extreme cases—flat fading and high delay spread. In the high delay spread case, the correlation between I_k is small, as there are a large number of independently fading taps in the channel. The mean of I is independent of the correlation between the I_k and is given by (9.20). The variance of I, however, significantly depends on the amount of space-frequency diversity. Consequently, this determines the outage properties. If we denote the variance of I_k as σ_I^2 (recall that the distribution of I_k is independent of k), in the flat fading case we have $\mathrm{var}\,(I) = \sigma_I^2$, whereas in the high delay spread case (under the idealistic assumption of full decorrelation of the I_k) we obtain $\mathrm{var}\,(I) \approx \frac{1}{N} \sigma_I^2$. Figure 9.5(a) and (b) illustrate example histograms of I for a 64-tone OFDM system (IEEE 802.11a) in the flat fading case and in the high delay spread case, respectively. It can be seen that in the high delay spread case the distribution is significantly more concentrated around the mean. Take a rate of 7.5 bit/s/Hz. For this rate, clearly from Figure 9.5 the outage probability will be much lower for the high delay spread case than for the flat fading case.

Since the rank of the individual correlation matrixes R_ℓ ($\ell = 0, 1, \ldots, L - 1$) determines the number of spatial degrees of freedom in each path of the MIMO channel (and hence the diversity gain), it is to be expected that the rank of the individual correlation matrixes and not the rank of the sum correlation matrix $\mathbf{R}$ determines the outage properties. This can be illustrated by assuming a simple example where all the $\mathbf{R}_\ell$ have rank 1 but are such that the sum correlation matrix

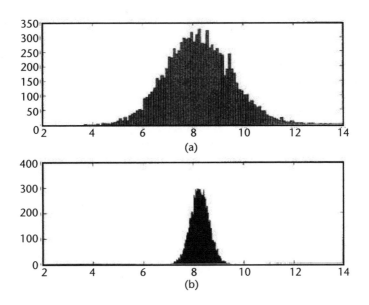

Figure 9.5   Histograms of I in bit/s/Hz in the (a) flat fading case and (b) the high delay spread case. (*From:* [2]. © 2002, IEEE.)

has full rank. In this case, it readily follows from (9.11) that the number of degrees of freedom in each path is M_T and, hence, the total number of degrees of freedom in the channel is LM_T, irrespective of the rank of the sum correlation matrix **R**. In the case where the individual correlation matrixes $\mathbf{R}_\ell$ are full rank ($M_R \times M_R$) and the sum correlation matrix **R** is also full rank, the number of degrees of freedom in the channel will be $LM_T M_R$ and hence significantly better outage properties than in the fully correlated case can be expected.

9.4.5   Simulations

These aspects are proved by Bolcskei et al. for ergodic and nonergodic cases in the following simulations, reproduced here for inspection [2]. Approximately 1,000 Monte Carlo runs were performed. The power delay profile was exponential. The number of tones in the OFDM system was $N = 512$, the CP length was 64, and the relative antenna spacing was set to $\Delta = 0.5$. Uniform tap spacing was assumed and $SNR = M_T \rho = \dfrac{P}{N\sigma_n^2}$. In Figure 9.6, the system was a 4×4 system. The power was normalized to 1 ($\mathrm{Tr}(\mathbf{R}) = 1$). The cluster angle spread was assumed to be $\sigma_{\theta,\ell} = 0$ ($\ell = 0, 1, \ldots, L - 1$). In the flat fading case the mean angle of arrival was set to $\bar{\theta}_0 = \pi/2$. In the delay spread case a total angle spread of 90° was assumed. Figure 9.6(a) shows that the ergodic capacity increases with rising delay spread L. This increase is up to four. It does not improve beyond four because this system has a maximum of $M_T = M_R = 4$. Since the number of antennas on each side of the channel is four, we have no means of using the higher number of available paths L. Remember the golden rule that $C = \min(L, M_T, M_R)$. Figure 9.6(b) shows the ergodic capacity for the same parameters as above, except for

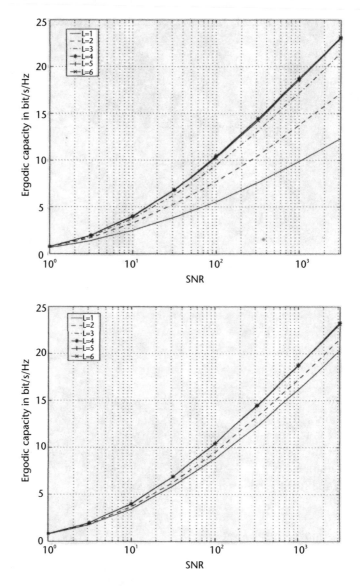

Figure 9.6 Ergodic capacity (in bit/s/Hz) as a function of SNR for various values of L and (a) small cluster spread and (b) large cluster spread. (*From:* [2]. © 2002, IEEE.)

the cluster angle spread, which was increased to $\sigma_{\theta,\ell} = 0.25$ ($\ell = 0, 1, \ldots,$ $L - 1$). In this case the rank of the individual correlation matrixes $\mathbf{R}_\ell$ is higher than 1 (since the cluster spread angle was increased from the earlier 0) and the improvement in terms of ergodic capacity resulting from the presence of multiple taps is less pronounced, but the improvement is there nevertheless. Once again we remind the reader that this result is a consequence of the assumption that delayed paths tend to increase the total angle spread.

In Figure 9.7 we investigate the impact of delay spread on the outage properties of the system. Again, for fixed $\text{Tr}(\mathbf{R}) = 1$, the figure shows the outage probability for $L = 1$, 5 and 16 and an SNR of 10 dB. Here we assume that there is no

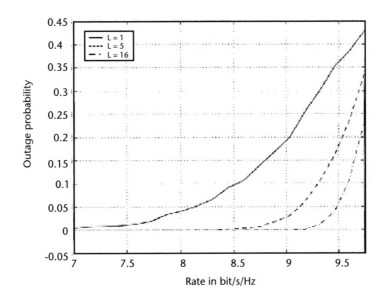

Figure 9.7   Outage probability for L = 1, 5 and 16 at an SNR of 10 dB. (*From:* [2]. © 2002, IEEE.)

spatial fading correlation. It is clearly seen that the outage probability decreases significantly with increasing delay spread.

In Figure 9.8, we investigate the impact of spatial fading correlation on outage probability. For $M_T = M_R = 4$, $L = 10$ and $s = \sigma_{\theta,\ell} = 0.25, 0.5, 0.7$, the figure shows the outage probability as a function of rate for an SNR of 10 dB. In all

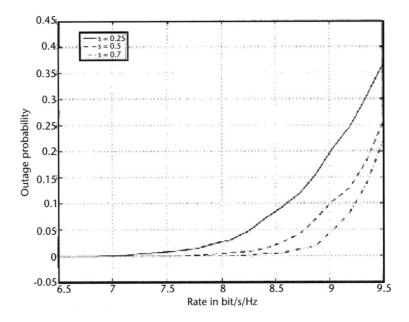

Figure 9.8   Outage probability for various values of $s = \sigma_{\theta,\ell}$ at an SNR of 10 dB. (*From:* [2]. © 2002, IEEE.)

three simulations the mean angles of arrival were chosen such that the sum correlation matrix **R** was full rank. In all cases $\mathrm{Tr}(\mathbf{R}) = 1$. It can be seen that even though the sum correlation matrix **R** had full rank, the outage probability critically depends on the individual cluster angle spread and, hence, the rank of the individual correlation matrixes $\mathbf{R}_\ell$.

9.4.6 Summary

In this section we derived expressions for the ergodic and outage capacities of OFDM-based spatial multiplexing systems for the case where the channel is unknown at the transmitter and perfectly known at the receiver. We studied the influence of propagation parameters and system parameters on ergodic capacity and outage probability and demonstrated the beneficial impact of delay spread and angle spread on capacity. Specifically we showed that in the MIMO case, as opposed to the SISO case, delay spread channels may provide advantage over flat fading channels not only in terms of outage capacity, but also in terms of ergodic capacity (provided the assumption that delayed paths tend to increase the total angle spread is true). We furthermore found that while the multiplexing gain is governed by the rank of the sum correlation matrix **R**, the diversity gain is governed by the rank of the individual correlation matrix $\mathbf{R}_\ell$. Finally, the capacity of an OFDM-based spatial multiplexing system is defined by

$$C = \min(L, M_T, M_R) \tag{9.23}$$

9.5 Performance Analysis of MIMO-OFDM Systems

We now analyze the performance of the scheme in Figure 9.2. We assume a Rayleigh channel with a maximum delay spread of 75 ns. Unless otherwise stated, we assume perfect channel knowledge at the receiver and perfect synchronization. We have no knowledge of the channel at the transmitter. We employ interleaving, which gives us a typically 5 dB advantage (this varies depending on the channel and the type of modulation used). The modulation employed is 16 QAM and we introduce a carrier phase noise of 10 Hz. The carrier phase noise problem was discussed in Chapters 7 and 8. We recap here. The bandwidth of an IEEE 802.11a system is 20 MHz. There are 64 subcarriers in each OFDM symbol. These make for an intercarrier spacing Δf of $\dfrac{20 \times 10^6}{64} = 312.5$ KHz. We use carrier oscillators for shifting the baseband signal to carrier frequencies in the UNII band. Ideally, these carrier oscillators should have line spectrum. In reality, this is not attainable and they consequently have a spectral width of typically 30 KHz with a specified noise floor, typically −130 dBc/Hz (dBc is the power in dB relative to the carrier); this width varies from vendor to vendor and the cost of the oscillator. This carrier phase noise, if it is sufficiently broad enough, will cause ICI between subcarriers, resulting in degradation of performance. The quantum of degradation depends on the type of constellation used and the spectral width of the phase noise of the carrier oscillator. For example, for BPSK, there is no appreciable degradation, but in the case of 16 QAM there can be about 0.5 dB degradation, for a phase noise of 40 KHz. Further discussion on this topic is beyond the scope of this book. The reader is referred to [7] for further details. Hence, the 10-Hz phase noise in this simulation is considered negligible.

In Figure 9.9, we have also used the channel estimation method (LSE) discussed in Chapter 8. The number of packets transmitted at each SNR is 1,000. The phase noise curve is obtained using channel estimation plus phase noise. This is because if we introduce phase noise, we need to use channel estimation techniques to recover the signal. The estimated curves are without any phase noise. We note a 3.5-dB gap in performance at a BER of 10^{-2} between the perfect curve and the estimated curve for the Alamouti-MIMO case. The perfect curve follows the curves plotted in Chapter 4 and this consequently shows us that there is no degradation in performance because of the OFDM modulation. It also shows that OFDM indeed does convert a wideband channel into a set of parallel narrowband channels; otherwise, the STBC will not work. However, in the SISO mode, the performance gap between perfect and estimated is around 1.2 dB at the same BER. This is much better than the performance of the Alamouti case. This shows that in the MIMO case, the estimation algorithm is insufficient. This is not due to the loss of orthogonality in the MIMO case as, in this simulator, the data streams are orthogonal in time, but due to its relatively poor quality of estimate. We note that in both the MIMO and SISO case, the estimated and phase noise curves coincide, which is as it should be, since the phase noise is negligible. The reader is encouraged to try higher phase noise like 100 KHz and see the degradation in performance, using the WLAN simulator supplied with [7].

We now examine the performance of MRC and STBC, using the accompanying software, assuming perfect channel knowledge at the receiver and perfect synchronization. This is shown in Figure 9.10.

From Figure 9.10, we note the following:

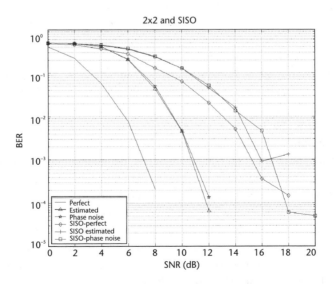

Figure 9.9 Performance of 2×2 STBC and SISO in an IEEE 802.11a environment.

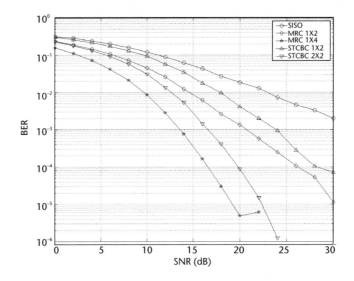

Figure 9.10   Comparison of performance—STBC, MRC, and SISO.

- There is a 3-dB gap between the 1×2 (MRC) curve and 2×1 (STC) curve. This is because the power is equally divided at the transmitter, unlike in the MRC case, where there is only one antenna (full power).
- For similar reasons 1×4 (MRC) is superior to 2×2 (STC).
- Even with this lower power the diversity gain between SISO and 2×1 (STC) is 5 dB and with 2×2 (STC) is 10 dB. This is only 3 dB worse than 1×4 (MRC).

In Figure 9.11 we compare the performance of V-BLAST spatial multiplexing algorithm with STBC, using the accompanying software.

In this simulation, the V-BLAST is based on a zero-forcing receiver with OSUC. In this case, as discussed in Chapter 6, the diversity order at the receiver is more than $M_R - M_T + 1$ and less than M_R. We note that as the diversity order increases, the performance of V-BLAST improves, which is to be expected. The interesting point to note here is that STBC 2×4 is superior to V-BLAST 2×4. The difference in performance is due to the diversity gain of the STBC compared with V-BLAST. The diversity order of STBC is 8. Whereas, in the V-BLAST, the diversity order is only 3. We did not use any interleaving in these simulations. In the accompanying simulator, the V-BLAST option disables the interleaver. This has been done for the sake of simplicity. If we incorporate interleaving, then we would need to deinterleave at the receiver before demodulation. In the case of the V-BLAST algorithm, this makes the simulator design complex.

In Figure 9.12, we note the performance of the FFT method discussed in Chapter 4 for channel estimation using QPSK modulation. This was done with the accompanying software, which assumes perfect synchronization. The channel is a Rayleigh slow fading channel. It will be recalled that in the FFT method we truncate the window in the time domain at the cyclic prefix. If we are sure of the

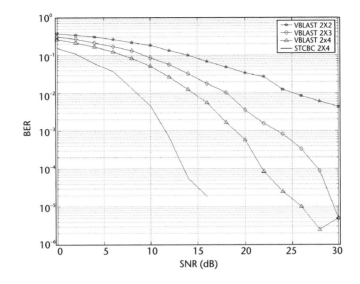

Figure 9.11   Comparison of performance, V-BLAST, and STBC in MIMO-OFDM environment.

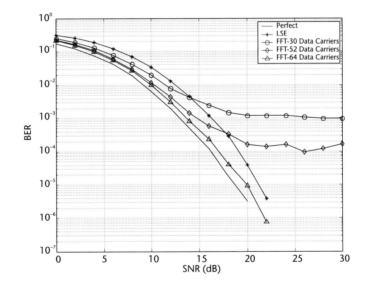

Figure 9.12   Performance comparison of FFT method in a 2 × 2 system.

delay spread in the channel, we can truncate to the nearest channel tap. In these simulations we truncated to the 10th sample when the cyclic prefix is 16 samples corresponding to 800 ns with a bandwidth of 20 MHz. The accompanying software is a generic OFDM-based packet transmission system with a facility for varying the OFDM size (FFT size) as well as a facility for varying the number of data carriers in an OFDM packet while padding the remaining carriers, including the central dc carrier with zeros. Some aspects of the software like bandwidth and channel estimation are similar to the IEEE 802.11a standard. The IEEE 802.11a standard incorporates, however, only the LSE algorithm. This software also incorporates the FFT estimation approach. Figure 9.12 shows the result of varying the number of data carriers on the performance of the FFT estimation algorithm. As usual, at a BER of 10^{-4} there is a performance gap of around 3 dB from the ideal if we use the LSE method. But if we use all the data carriers of the 64-point FFT, as in this simulation, the performance gap is only 0.7 dB. However, if we use only 30 data carriers out of the available 64, the curve levels off at around 20 dB SNR at a BER of 10^{-3}. This phenomenon occurs because we truncate across the available 30 data carriers at the 10th sample but spread the result across the entire 64 carriers of the full FFT. This introduces errors due to the limited number of data carriers. It therefore stands to reason that if we increase the number of data carriers, the adverse effect of the truncation step will be reduced. This is borne out in the figure for the case of 52 data carriers, wherein the leveling off occurs at a BER of 10^{-4} at the same SNR. Finally if we use all the 64 data carriers there is no such leveling-off phenomenon. Hence, we can conclude that in the FFT type of estimation we would ideally like to use all the available subcarriers. However, this is not possible in practice because we need to set the subcarriers on either end to zero to avoid ICI, as discussed in Chapter 7. Furthermore, since the center carrier is dc, it also needs to be set to zero. This implies that for a 64-point OFDM system, we have a serious "floor" problem to contend with if we decide to use the FFT method of channel estimation. What if we increase the size of the OFDM system?

Figure 9.13 shows the situation when we increase the size of the OFDM system to 256 and 512 points. It can be seen that with a full system, both 64 and 256 points yield the same result. In the case of 256 points, we find the same improvement in performance, as the number of data carriers is increased from 52 to 200. However, 256/200 is inferior to 64/52 in terms of performance, as can be seen from Figure 9.12. In terms of percentage of used carriers, both are nearly similar. But this argument loses ground when we look at the performance of 512/400 (same

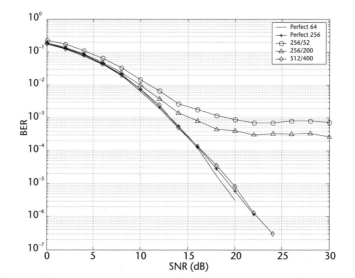

Figure 9.13   Performance of FFT estimation with higher order OFDM systems.

percentage). It is nearly perfect. Hence, it is a good idea to use as large a system as possible when using this method of estimation. In a packet system, it is essential that the channel be estimated as correctly as possible in *one packet*. The reader is referred to [8] for additional interesting analysis of this method.

In Figure 9.14, we plot the behavior of the system using the synchronization algorithms discussed in Chapter 8. The software used here is the one supplied with [7]. This software pertaining to an IEEE 802.11a system was modified to also carry out FFT estimation. We only discuss the results here. The reader is referred to [7] for further details on the software and other aspects. Remember that synchronization algorithms require channel estimation. In this plot, we used two types of estimation techniques—LSE and FFT. We note that there is an error "floor" with the FFT algorithm. This occurs due to the truncation phenomenon. Recall that the FFT algorithm is designed to reduce noise so as to make the channel estimates more accurate. This is done by converting to time domain and truncating at the end of the cyclic prefix. The argument here is that if the cyclic prefix is correctly designed, the entire impulse response of the channel will be confined to *within* the cyclic prefix. Anything outside this is noise and can be discarded. When we truncate the time window, and then take the FFT of the result, we retrieve this channel information but with a much reduced noise level. This works fine, as we have seen in Figure 9.13. We now examine the performance during synchronization. During synchronization our endeavor is to synchronize to the start of the OFDM symbol. This process of synchronization is carried out based on channel estimation. In IEEE 802.11a systems, we employ zero padding (64/52). Therefore, the performance is similar to the one in Figure 9.12, which is not surprising. Therefore, there is an error "floor" at high SNRs. We cannot get rid of this "floor" because of the nature of the phenomenon and the need for zero padding of the edge subcarriers to avoid ICI. However, we can mitigate it by increasing the number of subcarriers to, say,

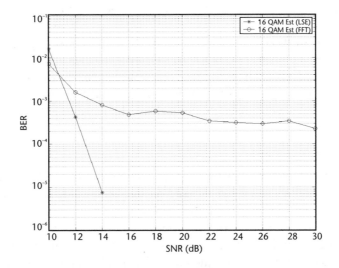

Figure 9.14   Performance of synchronization algorithms using LSE and FFT estimation for a 2 × 2 system.

512 subcarriers. In such an event we spread this error over a larger number of subcarriers. Another interesting aspect to note is that the SNR is shown varying from 10 dB to 30 dB. Below 10 dB there is no reliable synchronization with either algorithm. We require a basic level of SNR for both the algorithms to work correctly.

The algorithm for channel estimation using FFT does have certain disadvantages, but its importance cannot be denied. There are many papers on channel estimation for MIMO systems. But for these algorithms to have practicable value, it is essential that in packet transmission systems, the channel be estimated in *one packet* and we then use these estimates to recover the transmitted signal from the rest of the packet, assuming quasi-stationarity. Hence, from this point of view the FFT technique performs better than LSE or MMSE.

9.5.1 Analysis

Based on the preceding simulations, we can make the following observations:

- We need to improve on the channel estimation technique for MIMO purposes by resorting to an algorithm other than LSE. The FFT approach is one such.
- To obtain good results using the FFT approach, it is necessary to increase the number of subcarriers from the present 64 in IEEE 802.11a and similar systems. This will make for better accuracy of channel estimates due to the higher number of points.
- The side effect of the last point is that the data rate will also increase, bandwidth permitting. This is a welcome development.

9.6 CDMA-OFDM-MIMO

We now examine a novel approach to interface MIMO systems to code division multiple access (CDMA). In literature a lot of papers have been written in the problems encountered in interfacing CDMA to MIMO [9, 10]. The approach we are about to discuss is different from these. It was first proposed in [11]. In this approach we interface CDMA to OFDM. We then integrate this CDMA-OFDM system to MIMO. This proposal was originally given for operation in the 60-GHz band but without MIMO, if sufficient bandwidth is available. If the bandwidth is not available, we achieve the high throughput using MIMO (LST) techniques.

9.6.1 Introduction

Multicarrier systems have gained an increased interest during the last years. This has been fuelled by a large demand on frequency allocation, resulting in a crowded spectrum as well as a large number of users requiring simultaneous access. The quest received a fillip with the onset of CDMA systems. CDMA protocols do not achieve their multiple-access property by a division of the transmissions of different users in either time or frequency and it is already getting too crowded in these domains. Instead, make a division by assigning each user a different code. This code is used to transform a user's signal to a wideband signal (spread-spectrum signal). If a receiver receives multiple wideband signals, it will use the code assigned to a particular user to transform the wideband signal received from that user back to the original signal. All other code words will appear as noise due to decorrelation. Each user, therefore, operates independently with no knowledge of the other users. There is, however, a problem. The power of multiple users at a receiver determines the noise floor after decorrelation. If this power of a near user is not controlled compared with the power of a far user, then these signals will not appear equal at the base station. Then the stronger received signal levels raise the noise floor at the base station demodulators for the weaker signals, thereby decreasing the probability that weaker signals will be received. This is called the "near-far" problem. The second problem in CDMA systems is that the signals travel by different paths to the receiver. It is therefore preferred to use a RAKE receiver to use maximum ratio combining techniques to take advantage of all the multipath delays to get a strong signal. These RAKE receivers access the principal multipaths, but not all multipaths.

There was, however, an industrial demand for very high bit rates, irrespective of the type of access scheme used. This gave rise to OFDM systems. In such systems very high data rates are converted to very low parallel data rates using a series-to-parallel converter. This ensures flat fading for all the subcarriers (i.e., a wideband signal becomes a packet of narrowband signals). This will automatically combat multipath effects, removing the need for equalizers and RAKE receivers. A variant of this approach was introduced earlier as multicarrier-CDMA or MC-CDMA [12–14]. This proposal envisages interfacing a DS-CDMA system with a system of orthogonal coding using Walsh coding. Various multiple access approaches were developed over the years, each with advantages and disadvantages. It is especially noted that one of the methods of implementing MC-CDMA system is to adopt the OFDM/CDMA approach. This is still a complex procedure because it involves spreading each bit in a parallel manner using Walsh coding. We shall briefly examine the advantages and disadvantages of the prevailing systems before proceeding to examine the new proposal. It is pointed out that the advantages/disadvantages listed are not comprehensive, but only those relevant to this topic.

It can be seen in Table 9.1 that each multiple access approach has its advantages and disadvantages. It is especially noted that one of the methods of implementing an MC-CDMA system is to adopt the OFDM/CDMA approach. This is still a complex procedure because it involves spreading each bit in a parallel manner using Walsh coding.

There is an urgent need to develop a system that does the same thing but in an easily achievable manner. This section pertains to such a system. This topic proposes a comprehensive approach maximizing the merits and minimizing the demerits of the individual components of the new scheme. In view of the novelty of the idea, compared with MC-CDMA, we have chosen to call this approach the hybrid OFDM/CDMA/SFH approach or hybrid approach for short. This has essentially been developed for the 60-GHz frequency, but it is equally applicable at any other frequency, provided we have the necessary bandwidth.

This proposal pertains to the downlink as well as the uplink, the only difference being that for the synchronous downlink we can apply orthogonal Walsh-Hadamard sequences, leading to the well-known user separation for MC-CDMA systems.

Table 9.1 Types of Multicarrier Access Schemes

Type	Advantages	Disadvantages	References
DS-CDMA	i) Can address multiple users simultaneously and at same frequency. ii) Interference rejection.	i) Problems due to "near-far" effect. ii) Complex Time Domain RAKE receivers. iii) Synchronization within fraction of chip time becomes difficult.	[13]
SFH-CDMA	i) Reduces "near-far" effect. ii) Synchronization within fraction of hop time is easier. iii) No need for contiguous bandwidths.	Coherent demodulation difficult because of phase relationship during hops.	[13]
MC-CDMA	i) Higher number of users as full bandwidth is utilized unlike in DS-CDMA. ii) Effectively combines all the signal energy in the frequency domain, unlike CDMA	i) Peak-to-average ratio problem. ii) Synchronization problems. iii) Overcrowding of the spectrum as each bit is spread across the available bandwidth based on Walsh coding. iv) Complex Frequency Domain RAKE receivers.	[13, 14, 16]
OFDM	i) Robust against multipath effects. ii) Robust against narrowband interference. iii) Capable of single-frequency operation.	i) Sensitive to frequency offset and phase noise. ii) Synchronization problems. iii) Large peak-to-average power ratios.	[15]

On the other hand, in the asynchronous uplink scenario, PN sequences are used, with the drawback of high multiple-access-interference (MAI).

9.6.2 Overall System Concept

The overall concept is shown in the schematic in Figure 9.15.

9.6.2.1 Brief Description

It can be seen in the schematic that the transmitter and receiver are each divided into three subsections:

- Data modulation (demodulation) section.
- DS-CDMA section.
- SFH section.

The binary input data enters the data modulation section, where it is encoded by a forward error correction code. We can also use concatenated coding comprising

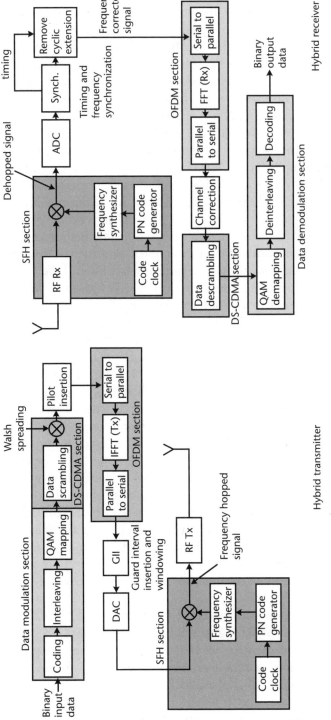

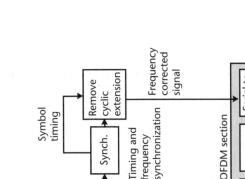

Figure 9.15 Overall system schematic.

a convolutional coding as an inner code followed by an outer coding as a block code (e.g., a Reed-Solomon code). This makes for a large coding gain with less implementation complexity compared with a single code. This coding is followed by interleaving to randomize the occurrence of bit errors due to deep fades across certain subcarriers. This is followed by QAM mapping. Thereafter, the data enters the DS-CDMA section. In this section, the data is subjected to scrambling based on PN sequences. This aspect is similar to the implementation in the IS-95 system. Thereafter, Walsh coding is used to provide orthogonal covering, because PN sequences by themselves are insufficient to ensure user separation. The Walsh function matrix will be an $N \times N$ matrix where N is the number of OFDM points. We then obtain what we can call a DS-CDMA signal. We insert pilot symbols after this step. In doing so, we must take care regarding the size of the Walsh matrix. For example, if our OFDM system has 32 points, we can have, at the most, a 24-length Walsh sequence. This leaves eight subcarriers for pilots. This is assuming that we use all the subcarriers. In practice, this is not possible, because we need to leave the edge subcarriers unused. This problem is explained as follows. We need to allow for the skirt of the lowpass antialiasing filter in the receiver. Subcarriers that lie beyond the bandwidth of interest should contain zero information because these subcarriers will lie along the slope of the lowpass filter. If they contain information, their amplitudes will not be uniform since they lie along the slope. It will be recalled that the fundamental assumption for orthogonality between subcarriers is that they have constant amplitude but differ only in phase. Hence, if any subcarrier of interest lies along the slope of the lowpass filter, we will have ICI. Therefore, we need to allow a safety zone, as it were, around each frame by inserting zeros to subcarriers around the edges of the OFDM symbol. During this process, we must ensure that all the subcarriers of interest lie within the passband of the lowpass filter. Furthermore, the communication spectrum is crowded. This means that no extraneous signal should exist beyond the pass band of the filter (i.e., along the slope). The zeros ensure this. The data sequence is then given to the OFDM section. This section is self-explanatory. One point to be noted here is that each user in the DS-CDMA section will share the entire lot of subcarriers with other users. The discrimination between users will only be possible in the DS-CDMA section of the receiver after descrambling and will be based on the orthogonality of the PN sequences and the Walsh coding. The analogue signal coming from the DAC is then frequency-hopped in the SFH section before being fed to the RF transmitter. The hop set for each user usually bears a definite relationship with the PN sequence of a particular user. During this process, it must be ensured that:

- The frequency synthesizer of the hopper and the carrier beat frequency oscillator of the RF transmitter are phase-locked.
- The frequency synthesizer of the dehopper and the beat frequency oscillator of the receiver (whose operating frequency is controlled by the synchronization circuit) are also phase-locked.

Failure to ensure these two aspects will result in ICI.

The hybrid receiver is exactly the reverse operation. The dehopped signal is given to the ADC and, thereafter, the digital signal processing starts with a training phase to determine the symbol timing and frequency offset. An FFT is used to demodulate all the subcarriers. The output of the OFDM section is the DS-CDMA sequence, which is then descrambled. The output of the DS-CDMA section is the QAM sequence, which are mapped onto binary values and decoded to produce binary output data. To successfully map the QAM values onto binary values, the reference phases and amplitudes of all subcarriers have to be acquired. Alternatively, differential techniques can be applied.

However, there are a few salient points to be noted:

1. *Bandwidth and other considerations:* We are assuming both Rayleigh and Rician fading conditions and AWGN. We are also assuming perfect OFDM synchronization with no carrier offset. Multimedia requirements of high bit rates, typically 155 mbit/s, require wide bandwidths of around 100 MHz or higher. In our case, however, we intend to operate at around 60 GHz, where larger bandwidths are available. This means Rician fading conditions and line-of-sight transmissions, conditions that are not so severe.

   The basic motivation for the scheme in Figure 9.15 is not so much having a robust design as having a design that can incorporate a lot of users.

   But we can expect a steep rise in the number of users when high data rates become realizable, especially with regard to video-telephones. It is the CDMA aspect (code diversity) that really gives rise to a lot of users. This is because frequency hopping has been introduced to obtain frequency diversity to reduce the "near-far" problem. This limits the number of users to avoid "collisions." The CDMA aspect makes up for this limitation by introducing a larger number of users due to code diversity. Interleaving and error correction coding may be dispensed with if the need so arises (i.e., if the channel is not severe). In case, in the foreseeable future, the channel does pose problems, we can increase the spread factor of the CDMA transmission (increase the bandwidth) and/or introduce FEC and interleaving. The OFDM aspect is required because it eliminates the need for RAKE receivers (there are no multipath delay effects) and allows us to use coherent modulation even when frequency hopping (because we will now hop on an OFDM-symbol basis) unlike most SFH systems, wherein maintaining phase continuity during hopping is difficult. It also helps reduce the burden of synchronization related to CDMA systems (see number 5 below). In this connection, it is noted that MC-CDMA also uses OFDM techniques but with RAKE receivers (in the frequency domain, due to Walsh spreading). In our style of signal processing, we do not use RAKE receivers. This crucial change from the MC-CDMA design results in a massive saving of hardware. We will accept the risk that some subcarriers will be in deep fade and correct for this eventuality using forward error correction coding (coded OFDM) and/or interleaving. This is different from the OFDM-CDMA approach discussed earlier, wherein RAKE combiners are used after OFDM demodulation to take advantage of the entire frequency spread of that particular bit. In reality, if we use an N-point OFDM system, we will need to use an N finger RAKE combiner. This is extremely costly and, hence, a compromise is achieved by using lesser number of fingers (e.g., a seven-finger combiner). This means that we do not take advantage of the entire frequency spread anyway. In a way, this is a waste of resources. Taking such matters into consideration, the hybrid system does not spread each bit in the frequency domain and does not, therefore, use RAKE combiners. Hence, we call this approach the hybrid OFDM/CDMA/SFH approach and not MC-CDMA.

2. *Coding:* In the CDMA transmitter (as in an IS-95 system), there are two levels of coding viz. Convolution encoding (for error correction) or concatenated coding and Walsh encoding (this is a spreading code and not an error correction coding). The latter is an orthogonal coverage, since PN sequences by themselves are insufficient to ensure channel isolation. The Walsh coding ensures orthogonality *between* users. The convolutional encoding ensures robustness of data.

3. *Modulation:* Unlike in a pure DS-CDMA system, in our case, the CDMA sequence after Walsh coding does not get converted to RF, but instead is fed as an input to the OFDM transmitter. In the OFDM transmitter it gets modulated as an OFDM signal and then via a P/S converter, it gets converted to RF.

4. *CDMA receiver:* Similarly, in the CDMA receiver, the OFDM receiver gives it a sequence at chip rate after OFDM demodulation. Thereafter, the CDMA signal processing is carried out, in that there is a *digital* correlator which ensures channel isolation based on PN sequences and Walsh coding. The output of the correlator is then given to a Viterbi decoder (for convolution decoding). The output from this decoder is the required data sequence. RAKE receivers are not required in this case unlike in DS-CDMA systems, since the OFDM system has no deleterious effects due to multipath.

5. *Synchronization:* Stringency of synchronization is, however, still required, as the PN sequences need to be synchronized. However, in such a hybrid system, the burden of synchronization is transferred to the OFDM system. The OFDM system has a more sophisticated synchronization system than does CDMA systems, as the OFDM system uses the cyclic prefixes for synchronization. Hence, the PN sequences emerging from the OFDM system and going to the CDMA system are already better synchronized than in a pure CDMA system. The reader will recall that synchronization is one of the limiting factors in CDMA systems for high data rates. It is expected that in our system, such problems will be considerably reduced, especially in the uplink, because the mobile receivers, thanks to OFDM, will be better synchronized to the transmitter and, consequently, with each other. This will reduce MAI compared with a pure DS-CDMA system.

6. *Bit error probabilities:* The proposed system is essentially a CDMA/ OFDM-FH system. This is because the transmission and reception is carried out by the OFDM-FH system. The CDMA aspect generates the data stream, but in a more complicated way.

7. *Trade-off between OFDM and CDMA:* The OFDM-FH system by itself does not solve the multimedia requirement. This is because multimedia requires very high bit rates, typically 155 mbit/s. This requires large

bandwidths of typically 100 MHz. By using FH among users to reduce the "near-far" effect suffered by CDMA systems, our number of users comes down drastically, being limited by the bandwidth available. By adding CDMA to this, we have rectified this problem by enhancing the number of users, since CDMA supports a larger number of users (being limited only by MAI) working at the same frequency. Hence, there is a trade-off.

We clarify this trade-off by using an example. Suppose in a hybrid system the CDMA end cannot handle more than 20 users due to MAI. These 20 users share one hop set. Therefore, among these users there will be adverse performance due to "near-far" effect. If we find that this "near-far" effect is intolerable, we reduce the number of users to, say, 10 and make the remaining 10 share another hop set. Due to bandwidth constraints suppose we can use only two hop sets, then we once again have a total of only 20 users for this hybrid system. But on the other hand, if the "near-far" effect is not too serious for 20 users, we can assign the other hop set to another 20 users, making 40 users in all. This trade-off between control of "near-far" effect and number of users depends on channel conditions. The hybrid system gives us this flexibility. Hence, there is eventually a trade-off between our desire to control the "near-far" effect and the number of users we desire.

Table 9.2 summarizes the overall system aspects.

9.6.3 Comparison with MC-CDMA

9.6.3.1 Basic Principles of MC-CDMA

This portion is based on the work done by Hara and Prasad [12, 13, 15]. MC-CDMA transmitter spreads the original signal using a given spreading code in the

Table 9.2 Hybrid System Overall Aspects

Problem	Solution
1. OFDM system only supports one user.	1. We use OFDM/CDMA which supports multiple users.
2. DS-CDMA does not permit very high data rates, owing to frequency selective fading at high data rates.	2. OFDM counters this by S/P coversion, allowing flat fading at subcarrier level.
3. DS-CDMA system suffers from "near-far" effect in the uplink.	3. This is solved in hybrid system using SFH.
4. CDMA systems use DS-SFH to control "near-far" effect. But DS-SFH can only support noncoherent modulation owing to hopping at bit level.	4. OFDM-FH system hops on frame basis allowing coherent modulation.
5. DS-CDMA systems cannot indefinitely support multiple users due to MAI and SI problems caused due to too many users.	5. Hybrid system allows any number of users by increasing the number of hops. Hence, Number of users = Number per CDMA system × Number of hops. Bandwidth should, however, be available.
6. DS-CDMA poses synchronization problems at very high chip rates.	6. OFDM systems have an easier synchronization problem due to using cyclic prefixes.

frequency domain. In other words, a fraction of the symbol corresponding to a chip of the spreading code is transmitted through a different subcarrier. For MC transmission, it is essential to have frequency nonselective fading over each subcarrier. Therefore, if the original symbol rate is high enough to become subject to frequency-selective fading, the signal needs to be first serial-to-parallel converted before spreading over the frequency domain. The basic transmitter structure of the MC-CDMA scheme is similar to that of the OFDM scheme, the main difference being that the MC-CDMA scheme transmits the *same symbol in parallel* through a lot of subcarriers, whereas the OFDM scheme transmits different symbols.

Figure 9.16 shows the MC-CDMA transmitter for the jth user with CBPSK format [15]. The input information sequence is first converted into P parallel data sequences $(a_{j,0}(i), a_{j,1}(i), \ldots, a_{j,P-1}(i))$, and then each serial/parallel converter output is multiplied with the spreading code with length K_{MC}. All the data in total $N = P \times K_{MC}$ (corresponding to the total number of subcarriers) are modulated in baseband by the IFFT and converted back into serial data. A guard interval is inserted between symbols to avoid ISI caused by multipath fading and finally the signal is transmitted after RF upconversion.

The received signal for MC-CDMA, is written as

$$r_{MC}(t) = \sum_{j=1}^{J} \int_{-\infty}^{+\infty} s_{MC}^{j}(t - \tau) \otimes h^{j}(\tau; t)\, d\tau + n(t)$$

$$= \sum_{i=-\infty}^{+\infty} \sum_{p=0}^{P-1} \sum_{m=0}^{K_{MC}-1} \sum_{j=1}^{J} z_{m,p}^{j}(t)\, a_{j,p}(i)\, d_{j}(m) \qquad (9.24)$$

$$\cdot p_{s}(t - iT_{s}')\, e^{j2\pi(Pm+p)\Delta f't} + n(t)$$

where $z_{m,p}^{j}(t)$ is the received complex envelope at the $(mP + p)$th subcarrier of the jth user, $\{dj(0), dj(1), \ldots, dj(K_{MC} - 1)\}$ is the spreading code, T_s is the symbol duration at subcarrier, Δf is the minimum subcarrier separation, and $p_s(t)$ is the rectangular symbol pulse waveform. Similarly, for the hybrid system, it can be shown that the received signal is given by,

$$r'(t) = \sum_{n=0}^{P-1} \sum_{m=1}^{N_{PG}} H_n b_n c_m p_c(t)\, e^{j2\pi \frac{n - P/2}{T_s'}} + n(t) \qquad (9.25)$$

where b_n is the data sequence and c_m is the time domain spreading code.

9.6.3.2   Comments on the MC-CDMA Technique

We can see that the MC-CDMA technique is very complex. However, mathematically the BER derived from this approach is no different from the OFDM-CDMA (hybrid) approach. This is apparent if you compare the equation structure of (9.24) with (9.25). It only differs in the manner in which the signal processing is implemented. The important distinction is that in MC-CDMA, if we have a set of

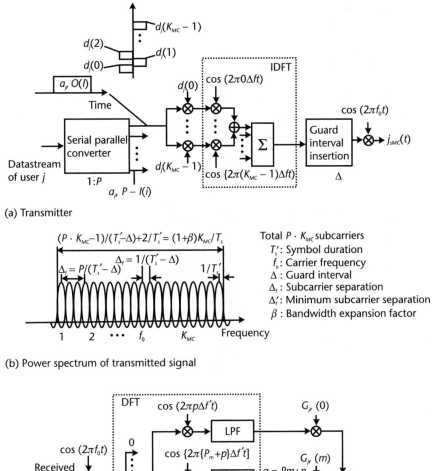

(a) Transmitter

(b) Power spectrum of transmitted signal

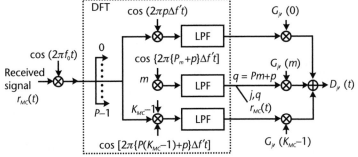

(c) Receiver

Figure 9.16 MC-CDMA system: (a) transmitter, (b) power spectrum of its transmitted signal, and (c) receiver. (*From:* [15]. © 2000, Artech House. Reprinted with permission.)

OFDM symbols, say, eight, we implement an eight-point OFDM system. This means these eight symbols comprise eight OFDM *frames*. One OFDM frame comprises eight subcarriers. Each symbol uses one complete OFDM frame. In contrast, the hybrid approach transmits all the eight symbols in one OFDM frame. Admittedly, the frequency diversity effect is applied more rigorously in MC-CDMA compared with the hybrid. In the former method, each symbol is transmitted over all the subcarriers.

This means that if there are certain carriers that do not perform well, we are still assured that the receiver gets that particular symbol through other carriers that operate well. The approach to simulation is also the same in hybrid as well as MC-CDMA. In both cases, each symbol in the set of symbols to be transmitted is spread per a code. In MC-CDMA the spreading code is based on the Walsh-Hadamard coding (i.e., each user is allotted one row of the Hadamard matrix). This ensures that users do not clash because the rows are mutually orthogonal. In the hybrid approach, the spreading is carried out per some PN sequence or Gold sequence. In both cases, depending on the spreading length, there will be processing gain. In the receiver, the received signal is once again multiplied by the *same* sequence. Thereafter there is an important distinction between the two systems. In the hybrid system, the received symbols are summed (integrated) after multiplication, whereas in the MC-CDMA approach, this summing is carried out per some combination law (e.g., maximal ratio combining, equal gain combining and so on) because the MC-CDMA system uses RAKE combiners. Hence, the results obtained using the OFDM/CDMA (hybrid) approach (i.e., the hybrid approach without SFH) in this chapter apply equally well to MC-CDMA systems provided that we do not lose any subcarriers due to deep fades.

There is also one positive factor in favor of the hybrid approach. We have seen that if we choose, we can also use Walsh-Hadamard coding in the hybrid approach as an additional precaution in addition to PN sequences. This is carried out before serial-to-parallel conversion.

There are also certain additional points in favor of the hybrid system. These are listed below:

1. In the MC-CDMA approach, the size of the OFDM modulator is dependent on the size of the Hadamard matrix. In the hybrid approach, we select the number of carriers and, consequently, the size of the OFDM system based on user requirements. The designer consequently has total flexibility in this respect.

2. The number of users depends on the size of the Hadamard matrix. We cannot have additional users because there will not be any more rows left in the Hadamard matrix. In the hybrid system, on the other hand, the number of users can be added indefinitely. This means that if the CDMA system cannot take more than say, 20 users, we simply add another frequency hop to the system. The total number of users equals the number of users per CDMA channel times the total number of frequency hops. Bandwidth should, however, permit this.

3. Frequency hopping is another major problem in MC-CDMA systems. Consider a case when we are transmitting a set of eight symbols based on a certain type of modulation like QPSK or 16 QAM. We use an eight-size Hadamard matrix and carry out an eight-point OFDM modulation. This means we need to transmit eight OFDM frames for the complete set. If we plan to frequency hop, then we need to delay the hopping until these eight frames are transmitted. If we do choose to hop with every frame as we do in hybrid systems, the task of compiling the eight symbols at the receiver becomes complex.

4. The data rate will become very slow in MC-CDMA systems compared with hybrid systems. Once again consider the case of an eight-symbol set based

on any type of modulation like QPSK or 16 QAM. We need to transmit eight frames before the symbol set is considered to have left the transmitter. In contrast, the hybrid system transmits just one frame.

To estimate the processing power required to implement a practical multimedia system, consider an example.

Total bandwidth: 150 MHz
User capacity: Single user (we shall hand over the multiuser problem to the CDMA part of the hybrid system)
Modulation used: 16 QAM
FFT size: 512
Guard period: 128 samples

If we assume the number of active carriers as 200, we obtain the following parameters:

Data rate: 600 mbit/s
Symbol duration: 1.3 μsecs
Total frame time: 1.7 μsecs $\approx$ 2 μsecs

We know that the number of complex calculation required for a 512-point FFT is 6,912. The maximum time that can be taken in performing the calculation is once every symbol (i.e., once every 2 μsecs). If we assume that the processor used requires two instructions to perform a single complex calculation, and that there is an overhead of 30% for scheduling of tasks and other processing, the minimum processing power required for this is then:

$$\text{MIPS} = \frac{6912 \times 2}{2 \times 10^{-6}} \times 1.3 \times 10^{-6} = 8,986$$

Thus, the transmitter requires > 9,000 MIPS to implement the transmitter. The receiver will require just as much. Thus, a full OFDM transceiver will require two boards of capability >9,000 MIPS each. This is beyond the range of current processors but within the capability of hardware-based systems. On the other hand, the MC-CDMA system will require 512 such IFFT boards in the modulator (for a 200-bit word + 312 zeros = 512 bits) and as many FFT boards in the receiver. These boards will need to be in parallel. Alternatively, these boards will need to be replaced by a single card in a modulator/demodulator, which is as fast. Such fast FFT boards are currently not available. This will make the system extremely expensive. Even in this case, the P/S converter will take the output of each FFT board in turn and serialize it. This means 512 OFDM symbols or 512 OFDM frames, if we include the guard bands, will have to be transmitted before the system is ready to look at the next word (the MC-CDMA system is transmitting a 512 × 512 matrix). This cannot be avoided. This will considerably slow down the data rate compared with a 512-point OFDM system, which is roughly 512 times faster, as it will transmit 512 bits at a time as one OFDM symbol. This hardware count for the MC-CDMA system is further increased by the use of RAKE receivers.

9.6.4 Interfacing with MIMO

We have briefly reviewed the technique required to interface CDMA with OFDM. We are already aware of interfacing OFDM with MIMO. The advantage in this proposal lies in the fact that we can still fully exploit the CDMA handover algorithms and so on without any major problems. The implications at higher layers need to be investigated. The performance aspects of this system in fading channels are maintained. If we require high throughputs we need to interface the OFDM part to MIMO. In such an event our system will be a CDMA-OFDM-MIMO system. This has a lot of potential, especially in existing CDMA 2000 cell phone systems, where requirement exists for high throughputs, as complexities pertaining to CDMA-MIMO interface are considerably mitigated using our proposed technique of inserting an OFDM interface between them.

9.7 Simulation Exercises

1. Implement the V-BLAST algorithm in the OFDM-based packet simulator using the MMSE method.
2. Simulate the V-BLAST algorithm using receiver correlation technique for a 2×4 system. Modify the GUI accordingly with a receive correlation checkbox. Why is the performance so poor?
3. Study the performance of the LSE as well as FFT estimation techniques using the simulator provided with this chapter. Modify it to incorporate block/convolutional coding and compare the performance. This will also give you an insight into MATLAB graphical user interface (GUI) techniques for simulation.

References

[1] Ging, Y., and K. B. Letaief, "Performance Evaluation and Analysis of Space-Time Coding in Unequalized Multipath Fading Links," *IEEE Trans. on Commun.*, Vol. 48, No. 11, November 2000.

[2] Bolcskei, H., D. Gesbert, and A. J. Paulraj, "On the Capacity of OFDM-Based Spatial Multiplexing Systems," *IEEE Trans. on Commun.*, Vol. 50, No. 2, February 2002.

[3] Peled, A., and A. Ruiz, "Frequency Domain Data Transmission Using Reduced Computational Complexity Algorithms," *Proc. IEEE ICASSP-80,* Denver, CO, 1980, pp. 964–967.

[4] Gallager, R. G., *Information Theory and Reliable Communication*, New York: John Wiley & Sons, 1969.

[5] Cover, T. M., and J. A. Thomas, *Elements of Information Theory*, New York: John Wiley & Sons, 1991.

[6] Ozarow, L. H., S. Shamai, and A. D. Wyner, "Information Theoretic Considerations for Cellular Mobile Radio," *IEEE Trans. Veh. Tech.*, Vol. 43, May 1994, pp. 359–378.

[7] Heiskala, J., and J. Terry, *OFDM Wireless LANs: A Theoretical and Practical Guide*, Indianapolis, IN: Sams Publishers, 2002.

[8] Mody, A. N., and G. L. Stuber, "Synchronization for MIMO-OFDM," *IEEE Global Communications Conference*, San Antonio, Texas, November 2001.

[9]   Paulraj, A., R. Nabar, and D. Gore, *Introduction to Space-Time Wireless Communications,* Cambridge, UK: Cambridge University Press, 2003.

[10] Vucetic, B., and J. Yuan, *Space-Time Coding,* Chichester, UK: John Wiley & Sons, 2003.

[11] Jankiraman, M., and R. Prasad, "Hybrid CDMA/OFDM/SFH: A Novel Solution for Wideband Multimedia Communications," *ACTS Summit,* Sorrento, Italy, 1999.

[12] Prasad, R., *CDMA for Wireless Personal Communications,* Norwood, MA: Artech House, 1996.

[13] Prasad, R., and S. Hara, "Overview of Multicarrier CDMA," *Proc. of the 4th IEEE International Symposium on Spread Spectrum Techniques and Applications (ISSSTA '96),* September 1996, pp. 107–114.

[14] Linnartz, J.-P., and N. Yee, "Multi-Carrier Code Division Multiple Access (MC-CDMA): A New Spreading Technique for Communication Over Multipath Channels," Final report 1993–1994 for MICRO Project 93–101.

[15] Van Nee, R., and R. Prasad, *OFDM for Wireless Multimedia Communications,* Norwood, MA: Artech House, 2000.

[16] Fazel, K., and L. Papke, "On the Performance of Convolutionally-Coded CDMA/OFDM for Mobile Communication System," *Proc. of IEEE PIMRC '93,* September 1993, pp. 468–472.

CHAPTER 10

The Way Ahead

"The journey of a thousand miles begins with one step."

—Lao Tzu

10.1 Introduction

In this book we have covered what are essentially the principles of MIMO wireless. The study was aimed at providing the reader with the basic tools to enable further study of this extremely hot topic in communications. Hence, we have not reached the end, but rather the end of the beginning. In this chapter, we will briefly review certain essential topics meriting further study. The aim is to impart to the reader a flavor of the subject sufficient to whet one's appetite. The reader is strongly encouraged to further examine the references given at the end of this chapter on topics of interest for a more detailed knowledge and understanding. There are so many aspects to cover in MIMO that even listing them is an onerous task. However, this book will not be complete if we do not cover two extremely interesting topics—MIMO-MU and linear dispersive coding.

10.2 MIMO Multiuser

We have up until now covered MIMO-single user (MIMO-SU), wherein we have multiple antennas at the transmitter interacting with multiple antennas of *one user* at the receiver. We now examine a case comprising multiple users at the receiver, each with one or more antennas. For simplicity we shall assume that each user has only one antenna and interacts with a multiple antenna array at the base station. Obviously, in such a scenario the problem devolves to a case of MIMO broadcast channel (MIMO-BC) in the downlink and MIMO multiple access (MIMO-MAC) in the uplink (i.e., from the user to the base station).

We now examine the capacity of such systems. We shall first concentrate on the uplink, (i.e., MIMO-MAC). The following are the assumptions [1]:

- The sources are located in the far field (i.e., they are so far away that when the wavefront arrives at the base station, it is planar). The channel is a narrowband channel.
- The noise is i.i.d. Gaussian, spatially, and temporally white.

- The signals from the users are independent and have a Gaussian distribution and are temporally white.
- The channel is quasi-static (i.e., it varies from frame to frame).

We consider a system with M antennas at the base station and P users each equipped with one antenna. The vector $\mathbf{h}_i$ defines the frequency flat channel between the ith user ($i = 1, 2, \ldots, P$) and the base station and is of size $M \times 1$. We assume that s_i is a complex data symbol transmitted by the ith user with average energy $\epsilon\{|s_i|^2\} = E_{s,i}(i = 1, 2, \ldots, P)$. The average energy will not be the same for each user since each user will transmit with different power levels to compensate for the path loss in the respective channels.

The received signal vector $\mathbf{y}$ of size $M \times 1$ is given by [1, 2]

$$\mathbf{y} = \sum_{i=1}^{P} \mathbf{h}_i s_i + \mathbf{n} \qquad (10.1)$$

$$= \mathbf{Hs} + \mathbf{n}$$

where $\mathbf{s} = [s_1, s_2, \ldots, s_P]^T$ is a $P \times 1$ vector, $\mathbf{H} = [\mathbf{h}_1, \mathbf{h}_2, \ldots, \mathbf{h}_P]$ is an $M \times P$ matrix, and $\mathbf{n}$ is an $M \times 1$ ZMCSCG spatially white noise vector with covariance matrix $N_0 \mathbf{I}_m$. The elements of $\mathbf{H}$ are not normalized because each element has a different value owing to path loss differences between users. The number of antennas M must be equal to or greater than the number of users P so that the users can be spatially separated. This stems from basic antenna theory wherein we can form spatially separate beams equal in number to the number of antenna elements (antennas). Hence, the number of antennas needs to be at least equal in number to the number of users. Since the users are uncorrelated, the covariance matrix of the vector $\mathbf{s}$, $\mathbf{R}_{ss} = E\{\mathbf{ss}^H\}$ is given by

$$\mathbf{R}_{ss} = \text{diag}\{E_{s,1}, E_{s,2}, \ldots, E_{s,P}\} \qquad (10.2)$$

We define the capacity rate as that rate which can be reliably maintained by each user. The channel is deterministic (for simplicity) and we assume that it is perfectly known to the receiver (base station). The transmitters, unlike in a MIMO-SU case, do not have coordinated encoding since the different users are geographically dispersed.

10.2.1 Capacity in the Uplink

There are two basic decoding strategies under consideration at the base station—*joint decoding* and *independent decoding*. By joint decoding, we mean that the signals are decoded collectively, whereas independent decoding implies that each user is decoded independently and in parallel with the assumption that the other user is "noise." Joint decoding requires $O(2^P)$ operations, whereas independent decoding requires $O(P)$ operations, making the former extremely complex with the rise in number of users.

10.2.1.1 Joint Decoding

The signal is detected at the base station using ML detection. Let $\mathcal{T}$ be a subset of the set $\{1, 2, \ldots, P\}$ and $\mathcal{T}^c$ represent its complement [i.e., $\mathcal{T} \cap \mathcal{T}^c = \varnothing$, $\mathcal{T} \cup \mathcal{T}^c = (1, \ldots, P)$]. We denote the covariance matrix of the signals transmitted from the terminals by indexing them with $\mathcal{T}$, thus $\mathbf{R}_{ss,\mathcal{T}}$ and the corresponding channel matrix $\mathbf{H}_{\mathcal{T}}$ of size $M \times c(\mathcal{T})$ where $c(\mathcal{T})$ is the cardinality of the set $\mathcal{T}$. If we assume Gaussian signaling for each user, the error free rate maintained by the ith user is given by [1]

$$\sum_{k \in \mathcal{T}} R_k \leq \log_2 \det\left(\mathbf{I}_M + \frac{1}{N_0}\mathbf{H}_T\,\mathbf{R}_{ss,\mathcal{T}}\mathbf{H}_{\mathcal{T}}^H\right) \text{ bps/Hz} \qquad (10.3)$$

for all $2^P - 1$ possible nonempty subsets $\mathcal{T}$ of the set $\{1, 2, \ldots, P\}$. In the case of a two-user system ($P = 2$),

$$R_1 \leq \log_2\left(1 + \frac{E_{s,1}}{N_0}\|\mathbf{h}_1\|_F^2\right) \qquad (10.4)$$

$$R_2 \leq \log_2\left(1 + \frac{E_{s,2}}{N_0}\|\mathbf{h}_2\|_F^2\right) \qquad (10.5)$$

and

$$R_1 + R_2 \leq \log_2 \det\left(\mathbf{I}_2 + \frac{E_{s,1}}{N_0}\mathbf{h}_1\mathbf{h}_1^H + \frac{E_{s,2}}{N_0}\mathbf{h}_2\mathbf{h}_2^H\right) \qquad (10.6)$$

This result is shown in Figure 10.1 [1].

In Figure 10.1, the line AB represents (10.6), which is the maximum achievable sum-rate on the uplink. Point B signifies the rate given by (10.5), wherein user 2 transmits at full power and ignores user 1. Point A pertains to user 1 transmitting

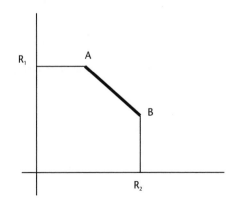

Figure 10.1 Capacity region for MIMO-MAC with joint decoding at the receiver. (*From:* [1]. © 1998, IEEE.)

at full power and at a rate defined by (10.4) and ignoring user 2. The line AB represents the time-sharing between users. If the number of users exceeds two, then the capacity region becomes polyhedral.

10.2.1.2 Independent Decoding

In this case, we attempt to recover each user's signal, treating all other signals as interfering "noise." In joint decoding each user transmits at its full rate, ignoring the other users, but the decoding is carried out in a joint manner wherein we do not treat the other users as "noise." In this case of independent decoding, we do treat the other users as "noise." In such a case the covariance matrix, $\mathbf{R}_{yy} = \epsilon\{\mathbf{yy}^H\}$, is given by

$$\mathbf{R}_{yy} = \mathbf{HR}_{ss}\mathbf{H}^H + N_0\mathbf{I}_M \tag{10.7}$$

The capacity region is defined by [1]

$$R_i \leq \log_2\left(\frac{\det(\mathbf{R}_{yy})}{\det\left(\mathbf{R}_{yy} - E_{s,i}\mathbf{h}_i\mathbf{h}_i^H\right)}\right), \, i = 1, 2, \ldots, P \tag{10.8}$$

The capacity region is defined in Figure 10.2 [1].

If the number of users exceeds two, the capacity region will be a cuboid.

We now examine the capacity variation based on the geometry of the two users and the power available to them. In Figure 10.3 we examine two cases when the two users are orthogonal to each other and when they are parallel to each other.

In Figure 10.3, when the users are orthogonal to each other, there is no mutual interference between them. In such a case, the capacity regions are maximal and the capacity region for independent decoding coincides with that for joint decoding since we can separate one signal from the other exactly. When the users are such that the array response vectors are aligned with each other, the parallel case being the extreme example, the capacities are given (10.4) to (10.6) and are defined

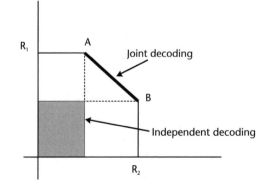

Figure 10.2 Capacity region for MIMO-MAC with independent decoding at the receiver. (*From:* [1]. © 1998, IEEE.)

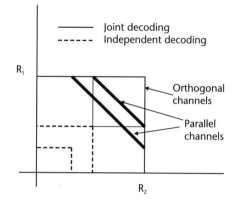

Figure 10.3 Capacity regions with different users' positions. (*From:* [1]. © 1998, IEEE.)

similar to the figure in Figure 10.1. In such a case, the users cannot be spatially separated. However one does gain a factor of M in SNR in joint decoding by using multiple antennas. This is because with M sensors, the receiver gets M replicas of the signal, which can then be added coherently, whereas noise adds up incoherently. In the limit the capacity region for independent decoding will eventually merge with that for joint decoding as the number of receivers increases [1]. Since independent decoding is more practical, such a result is important in real applications.

In a random fading channel, the capacity region is also random and a given sum-rate can be sustained only with a certain level of reliability. Figure 10.4 [2] plots the CDF of the maximum sum-rate, for a two-user MIMO-MAC system with $M = 2$ and $M = 10$. The channel is Gaussian i.i.d and the SNR is 10 dB for each user. We note that joint decoding outperforms independent decoding at all outage

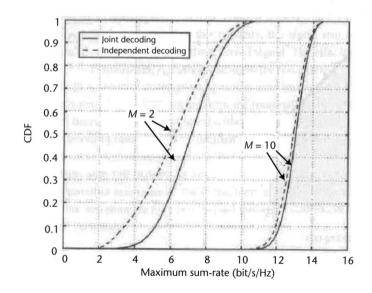

Figure 10.4 CDFs of maximum sum-rate for MIMO-MAC with joint and independent decoding at the receiver. (*From:* [2]. Reprinted with the permission of Cambridge University Press.)

levels. Also the maximum sum-rate achieved at any outage level increases with an increase in the number of base station antennas. Further, we note that the difference in maximum sum-rate achieved by the two schemes decreases with an increase in the number of base station antennas. This can be attributed to better separability of the spatial signatures with increasing M. In the limit, as $M \to \infty$, the channels become orthogonal and joint and independent decoding regions merge, as already discussed [1].

There are certain additional points to be borne in mind while evaluating MIMO-MU systems. In a MIMO-MU system if there is one stream that has a poor SNR, the system will fail. MIMO-SU has, on the other hand, the advantage of stream diversity. There is a "near-far" problem in MIMO-MU since the power control algorithms will not be enough. In such cases, the stronger channel (user) will dominate. There is no such problem in MIMO-SU. In MIMO-SU there can be cooperation between the colocated transmitting antennas for coding at the transmitter (user) and decoding at the receiver (base station). In MIMO-MU this is not possible, as the transmitters (users) are geographically separated. In MIMO-SU the capacities in forward and reverse links are similar for the same transmit power and if the channel is known at both ends. In MIMO-MU this is still a subject of research [2]. Finally, MIMO-SU suffers a small penalty if the channel is unknown to the transmitter (user). In MIMO-MU this is a much larger penalty if the user does not know the channel.

Finally, the reader may like to compare MIMO-MU technique with the hybrid technique discussed in Chapter 9. The latter was developed expressly for a large number of users. In fact, if we neglect the SFH option (for the sake of simplicity) and just integrate an OFDM system to a CDMA, as discussed in Chapter 9, and then finally integrate the CDMA-OFDM system to a MIMO system to form a CDMA-OFDM-MIMO system, the possibilities are enormous. This aspect has a lot of potential for real-life use and is relatively easy to accomplish. These aspects are worth investigating since, most important of all, we retain the existing CDMA infrastructure and do present an exciting candidate for futuristic 4G systems, as outlined in Figure 1.2.

10.3 Linear Dispersion Coding

Thus far we have examined space-time coding schemes and spatial multiplexing techniques. In space-time coding schemes the code rate r_s is less than one or equal to one (for Alamouti's scheme). At the other end of the spectrum, we have spatial multiplexing schemes that can give us a rate of M_T, where M_T is the number of transmit antennas. However, it will be appreciated that none of these codes can give us a code rate lying between these extreme values (i.e., $1 \le r_s \le M_T$. This section deals with a new concept called linear dispersion codes (LDCs), which precisely fulfill such a need. Practical modulation schemes for MIMO systems are deficient in two areas—diversity and multiplexing. Space-time coding, for example, uses specially designed code words that maximize the diversity advantage or reliability of the transmitted information. In fading channels, such codes maximize the diversity gain at the expense of a loss in capacity (except for an 2×1 Alamouti scheme) [3]. Spatial multiplexing, on the other hand, transmits independent data streams from each transmitting antenna. Multiplexing designs allow capacity to be achieved but at the expense of a loss in diversity advantage in fading channels. These aspects have already been discussed in Chapter 2. In practical systems, we require both spectral efficiency as well as high reliability. Recognizing that orthogonal space-time block codes do not achieve full-channel capacity in MIMO channels, Hassibi and Hochwald proposed a new concept—LDCs [3]. These codes use a linear matrix modulation framework in which the transmitted code word is a linear combination of certain dispersion matrixes, with the weights determined by the transmitted symbols. The key to the LDC design is that the basis matrixes are chosen such that the resulting codes maximize the ergodic capacity of the equivalent MIMO system [4]. However, the LDCs proposed in [3] only optimize the ergodic capacity; thus, corresponding good error probability performance is not strictly guaranteed [5].

To improve on this deficiency, Heath and Paulraj [4] proposed a family of LDC designs based on frame theory [6]. All types of coding discussed thus far have frame-based structures. If we choose a suitable set of parameters, frame-based LDCs have equivalent channels that achieve the full-ergodic capacity. The technique proposed in [3] is designed via a numerical optimization to maximize the mutual information between transmitter and receiver. In contrast the method proposed in [4] is a closed-form design that, *in some cases*, produces an equivalent channel that maximizes ergodic capacity. Furthermore, to ensure that a minimum diversity advantage is guaranteed (unlike in [3]), the process is further optimized by choosing code words by explicitly using the rank and determination criteria discussed in Chapter 5.

We shall briefly examine both approaches and analyze their performance in fading channels.

10.3.1 Hassibi and Hochwald Method

Suppose that there are M_T transmit antennas and M_R receive antennas and an interval of T symbols available to us during which the propagation channel is constant and known to the receiver. The transmitted signal can then be written as a $T \times M_T$ matrix **S** that governs the transmission over the M_T antennas during the interval. We assume that the data sequence has been broken into Q substreams and that $s_1, \ldots, s_Q$ are complex symbols chosen from an arbitrary, say r-PSK or r-QAM, constellation. The rate of the linear dispersion code is given by $R = (Q/T)\log_2 r$.

10.3.1.1 The Multiple-Antenna Model

In a narrowband flat fading multiple antenna communication system with M_T transmit and M_R receive antennas, the transmitted and received signals are related by

$$\mathbf{x} = \sqrt{\frac{\rho}{M_T}} \mathbf{H} \mathbf{s} + \mathbf{v} \tag{10.9}$$

where $\mathbf{x}$ is the complex received vector, $\mathbf{s}$ is the complex transmitted vector, $\mathbf{H}$ is the channel matrix, and $\mathbf{v}$ is zero-mean, unit-variance, complex-Gaussian noise $\mathcal{CN}(0, 1)$. The channel matrix $\mathbf{H}$ and transmitted vector $\mathbf{s}$ are assumed to have unit variance entries, implying that $\epsilon\{\mathrm{tr}(\mathbf{HH}^H)\} = M_T M_R$ and $\epsilon\{\mathbf{s}^H\mathbf{s}\} = M_T$. The normalization $\sqrt{\dfrac{\rho}{M_T}}$ in (10.9) ensures that ρ is the SNR at each receive antenna, independently of M_T. The reader will note that this has been the assumption throughout this book. We also assume that matrix $\mathbf{H}$ has independent $\mathcal{CN}(0, 1)$ entries.

If the channel is perfectly known to the receiver, the capacity is given by [5]

$$C(\rho, M_T, M_R) = \max_{\mathbf{R}_{ss}>0,\,\mathrm{tr}(\mathbf{R}_{ss})=M_T} \epsilon\left\{\log_2 \det\left(\mathbf{I}_{M_R} + \frac{\rho}{M_T}\mathbf{H}\mathbf{R}_{ss}\mathbf{H}^H\right)\right\}$$

(10.10)

where the maximizing covariance matrix $\mathbf{R}_{ss} = \mathbf{I}_{M_T}$ since matrix $\mathbf{H}$ has independent $\mathcal{CN}(0, 1)$ entries.

Hence,

$$C(\rho, M_T, M_R) = \epsilon\left\{\log_2 \det\left(\mathbf{I}_{M_R} + \frac{\rho}{M_T}\mathbf{HH}^H\right)\right\}$$ (10.11)

If the channel is quasi-static for at least T channel uses we obtain

$$\mathbf{X} = [x_1 \quad x_2 \quad \dots \quad x_T]^t$$
$$\mathbf{S} = [s_1 \quad s_2 \quad \dots \quad s_T]^t$$
$$\mathbf{V} = [v_1 \quad v_2 \quad \dots \quad v_T]^t$$

where $[\cdot]^T$ signifies transpose. Hence,

$$\mathbf{X}^t = \sqrt{\frac{\rho}{M_T}}\mathbf{H}\mathbf{S}^t + \mathbf{V}^t$$

We shall henceforth write this equation in its transposed form as

$$\mathbf{X} = \sqrt{\frac{\rho}{M_T}}\mathbf{SH} + \mathbf{V}$$ (10.12)

where we have omitted the transpose notation from $\mathbf{H}$ by redefining it as $M_T \times M_R$. In $\mathbf{X}$, $\mathbf{S}$, and $\mathbf{V}$, time runs vertically and space runs horizontally. Our aim is to design the signal matrix $\mathbf{S}$ such that it obeys the power constraint $\epsilon\{\mathrm{tr}(\mathbf{SS}^H)\} = TM_T$.

10.3.1.2   Coding

We call an LDC one for which

$$S = \sum_{q=1}^{Q} \left(s_q C_q + s_q^* D_q \right) \tag{10.13}$$

where $s_1, \ldots, s_Q$ are complex scalars (typically chosen from an r-PSK or r-QAM constellation) and where the C_q and D_q are *fixed* $T \times M_T$ complex matrixes. The code is completely determined by the set of *dispersion* matrixes $\{C_q, D_q\}$, whereas each individual code word is determined by our choice of the scalars $s_1, \ldots, s_Q$. The design of the code depends crucially on the choices of the parameters T, Q and the dispersion matrixes $\{C_q, D_q\}$. To choose the $\{C_q, D_q\}$ we propose to optimize a nonlinear information-theoretic criterion—the mutual information between the transmitted signals $\{s_q, s_q^*\}$ and the received signal.

We decompose s_q into its real and imaginary parts,

$$s_q = \alpha_q + j\beta_q, \, q = 1, \ldots, Q$$

and write

$$S = \sum_{q=1}^{Q} \left(\alpha_q A_q + j\beta_q B_q \right) \tag{10.14}$$

where $A_q = C_q + D_q$ and $B_q = C_q - D_q$. The dispersion matrixes $\{A_q, B_q\}$ also specify the code.

We assume that $\alpha_1, \ldots, \alpha_q$ and $\beta_1, \ldots, \beta_q$ have variance 1/2 and are uncorrelated. Thus $s_1, \ldots, s_Q$ are unit-variance and uncorrelated. Since our aim is to design the signal matrix S such that it obeys the power constraint $\epsilon\{\text{tr}(SS^H)\} = TM_T$, we introduce the following normalization on the matrixes $\{A_q, B_q\}$,

$$\sum_{q=1}^{Q} \left(\text{tr}\left(A_q^H A_q \right) + \text{tr}\left(B_q^H B_q \right) \right) = 2TM_T \tag{10.15}$$

The dispersion codes in (10.14) cover cases of orthogonal designs as well as V-BLAST. For example, in an Alamouti design, $T = M_T = Q = 2$ and

$$A_1 = \begin{bmatrix} 1 & 0 \\ 0 & 1 \end{bmatrix}, A_2 = \begin{bmatrix} 0 & 1 \\ -1 & 0 \end{bmatrix} \tag{10.16}$$

$$B_1 = \begin{bmatrix} 1 & 0 \\ 0 & -1 \end{bmatrix}, B_2 = \begin{bmatrix} 0 & 1 \\ 1 & 0 \end{bmatrix}$$

whereas V-BLAST corresponds to $Q = TM_T$ and

$$\mathbf{A}_{M_T(\tau-1)+m} = \mathbf{B}_{M_T(\tau-1)+m} = \kappa_\tau \, \eta_m^t, \qquad (10.17)$$

$$\tau = 1, \ldots, T, \, m = 1, \ldots, M_T$$

where κ_τ and η_m are T-dimensional and M_T-dimensional column vectors with one in the τth and mth, respectively, and zeros elsewhere.

Note that in V-BLAST each signal $\{\alpha_q, \beta_q\}$ is transmitted from only one antenna during one channel use. Using LDCs, however, the dispersion matrixes potentially transmit some combination of *each* symbol from *each* antenna at *every* channel use. This will give rise to transmit diversity, which is absent in a "normal" V-BLAST scheme.

10.3.1.3   Decoding

Equation (10.14) tells us that LDCs are linear in the variables $\{\alpha_q, \beta_q\}$, leading to efficient V-BLAST-like decoding schemes.

We write the block equation,

$$\mathbf{X} = \sqrt{\frac{\rho}{M_T}} \, \mathbf{SH} + \mathbf{V} = \sqrt{\frac{\rho}{M_T}} \sum_{q=1}^{Q} (\alpha_q \mathbf{A}_q + j\beta_q \mathbf{B}_q)\mathbf{H} + \mathbf{V} \qquad (10.18)$$

in a more convenient form. We decompose the matrixes in (10.18) into their real and imaginary parts to obtain

$$\mathbf{X}_R + j\mathbf{X}_I = \sqrt{\frac{\rho}{M_T}} \sum_{q=1}^{Q} [\alpha_q (\mathbf{A}_{R,q} + j\mathbf{A}_{I,q}) + j\beta_q (\mathbf{B}_{R,q} + j\mathbf{B}_{I,q})]$$

$$\times (\mathbf{H}_R + j\mathbf{H}_I) + \mathbf{V}_R + j\mathbf{V}_I$$

where $\mathbf{H}_R = \mathrm{Re}(\mathbf{H})$ and $\mathbf{H}_I = \mathrm{Im}(\mathbf{H})$. Equivalently,

$$\mathbf{X}_R = \sqrt{\frac{\rho}{M_T}} \sum_{q=1}^{Q} [(\mathbf{A}_{R,q}\mathbf{H}_R - \mathbf{A}_{I,q}\mathbf{H}_I)\alpha_q + (-\mathbf{B}_{I,q}\mathbf{H}_R - \mathbf{B}_{R,q}\mathbf{H}_I)\beta_q] + \mathbf{V}_R$$

$$\mathbf{X}_I = \sqrt{\frac{\rho}{M_T}} \sum_{q=1}^{Q} [(\mathbf{A}_{I,q}\mathbf{H}_R - \mathbf{A}_{R,q}\mathbf{H}_I)\alpha_q + (\mathbf{B}_{R,q}\mathbf{H}_R - \mathbf{B}_{I,q}\mathbf{H}_I)\beta_q] + \mathbf{V}_I$$

We denote the columns of

$$\mathbf{X}_R, \mathbf{X}_I, \mathbf{H}_R, \mathbf{H}_I, \mathbf{V}_R \text{ and } \mathbf{V}_I \text{ by } x_{R,n}, x_{I,n}, h_{R,n}, h_{I,n}, v_{R,n} \text{ and } v_{I,n}$$

and define

$$\mathcal{A}_q = \begin{bmatrix} A_{R,q} & -A_{I,q} \\ A_{I,q} & A_{R,q} \end{bmatrix}, \; \mathcal{B}_q = \begin{bmatrix} -B_{I,q} & -B_{R,q} \\ B_{R,q} & -B_{I,q} \end{bmatrix} \qquad (10.19)$$

$$\bar{h}_n = \begin{bmatrix} h_{R,n} \\ h_{I,n} \end{bmatrix}$$

where $n = 1, \ldots, M_R$. We then gather the equations in $\mathbf{X}_R$ and $\mathbf{X}_I$ to form the single real system of equations

$$
\begin{bmatrix} x_{R,1} \\ x_{I,1} \\ \vdots \\ x_{R,M_R} \\ x_{I,M_R} \end{bmatrix} = \sqrt{\frac{\rho}{M_T}} \, \mathcal{H} \begin{bmatrix} \alpha_1 \\ \beta_1 \\ \vdots \\ \alpha_Q \\ \beta_Q \end{bmatrix} + \begin{bmatrix} \nu_{R,1} \\ \nu_{I,1} \\ \vdots \\ \nu_{R,M_R} \\ \nu_{I,M_R} \end{bmatrix} \tag{10.20}
$$

where the equivalent $2M_R T \times 2Q$ real channel matrix is given by

$$
\mathcal{H} = \begin{bmatrix} \mathcal{A}_1 \overline{h}_1 & \mathcal{B}_1 \overline{h}_1 & \cdots & \mathcal{A}_Q \overline{h}_1 & \mathcal{B}_Q \overline{h}_1 \\ \vdots & \vdots & \ddots & \vdots & \vdots \\ \mathcal{A}_1 \overline{h}_{M_R} & \mathcal{B}_1 \overline{h}_{M_R} & \cdots & \mathcal{A}_Q \overline{h}_{M_R} & \mathcal{B}_Q \overline{h}_{M_R} \end{bmatrix} \tag{10.21}
$$

We now have a linear relation between the input and output vectors $\mathbf{s}$ and $\mathbf{x}$ given by

$$
\mathbf{x} = \sqrt{\frac{\rho}{M_T}} \, \mathcal{H} \mathbf{s} + \mathbf{v} \tag{10.22}
$$

where the equivalent channel $\mathcal{H}$ is known to the receiver because the original channel $\mathbf{H}$ and the dispersion matrixes $\{\mathbf{A}_q, \mathbf{B}_q\}$ are all also known to the receiver. The receiver uses (10.21) to find the equivalent channel. The system of equations between the transmitter and receiver is not underdetermined as long as

$$
Q \leq M_R T \tag{10.23}
$$

We may, therefore, use any decoding technique already in place for V-BLAST, such as successive nulling and canceling, as discussed in Chapter 6, or sphere decoding for faster ML decoding [6].

10.3.1.4 Design of the Dispersion Codes

In [3] the authors have shown that choosing random dispersion matrixes is suboptimal. It appears advantageous to minimize the average pairwise error probability obtained by choosing Gaussian $\mathbf{s}$ in (10.22). The average pairwise error has an upper bound given by [3]

$$
P_e \text{ (pairwise)} \leq \epsilon \left\{ \frac{1}{2} \det \left(\mathbf{I} + \frac{\rho}{2M_T} \mathcal{H}^{\mathcal{H}} \mathcal{H} \right)^{-1/2} \right\} \tag{10.24}
$$

We can seek to minimize the upper bound with an appropriate choice of $\{\mathbf{A}_q, \mathbf{B}_q\}$. But (10.24) is already very small. Hence, minimizing this expression will pose problems. It was shown in [3] that orthogonal space-time block codes are deficient in the maximum mutual information they support for $M_T > 2$ and $M_R > 1$. Hence, it will be better to maximize the mutual information between s and x in (10.20). To achieve this, we propose to design codes using (10.10).

10.3.1.5 The Design Method

1. Choose $Q \le M_R T$ (typically, $Q = (\min(M_T, M_R), T)$.
2. Choose $\{\mathbf{A}_q, \mathbf{B}_q\}$ that solve the optimization problem

$$C(\rho, T, M_T, M_R) = \max_{\mathbf{A}_q, \mathbf{B}_q, q = 1, \ldots, Q} \frac{1}{2T} \epsilon \left\{ \log_2 \det \left(\mathbf{I}_{2M_R T} + \frac{\rho}{M_T} \mathcal{H}\mathcal{H}^{\mathcal{H}} \right) \right\}$$

$$(10.25)$$

for an SNR ρ of interest, subject to one of the following constraints:

i. $$\sum_{q=1}^{Q} \left(\mathrm{tr}\left(\mathbf{A}_q^H \mathbf{A}_q\right) + \mathrm{tr}\left(\mathbf{B}_q^H \mathbf{B}_q\right) \right) = 2TM_T$$

ii. $$\mathrm{tr}\left(\mathbf{A}_q^H \mathbf{A}_q\right) = \mathrm{tr}\left(\mathbf{B}_q^H \mathbf{B}_q\right) = \frac{TM_T}{Q}, q = 1, \ldots, Q$$

iii. $$\mathbf{A}_q^H \mathbf{A}_q = \mathbf{B}_q^H \mathbf{B}_q = \frac{T}{Q} \mathbf{I}_{M_T}, q = 1, \ldots, Q$$

where $\mathcal{H}$ is given by (10.21) with the $\overline{h}_n$ having independent $\mathcal{N}\left(0, \frac{1}{2}\right)$ entries.

We note that (10.25) is effectively (10.10) with $\mathbf{R}_{ss} = \mathbf{I}_{2Q}$. We may take the entries of s ($\{\alpha_q, \beta_q\}$) to be uncorrelated with variance 1/2. Moreover, because the real and imaginary parts of the noise vector v in (10.20) also have variance 1/2, the SNR remains ρ. We also note that (10.25) differs from (10.10) by the outside factor (1/2T) because the effective channel is real-valued and the LDC spans T channel uses. It can be shown that the constraints are convex in the dispersion matrixes $\{\mathbf{A}_q, \mathbf{B}_q\}$. However the cost function (10.25) is neither concave nor convex in the variables $\{\mathbf{A}_q, \mathbf{B}_q\}$. Therefore, it is possible that (10.25) has a local maxima. However, the authors have found that this local maxima, if it exists at all, does not pose a serious problem. The maximization of the cost function (10.25) is achieved through constrained gradient ascent method [3].

We further note with respect to the constraints that:

- Constraint i is simply the power constraint (10.15) that ensures that $\epsilon\{\mathrm{tr}(\mathbf{SS}^H)\} = TM_T$.
- Constraint ii is more restrictive and ensures that each of the transmitted signals α_q and β_q are transmitted with the same overall power from the M_T antennas during the T channel uses.

- Constraint iii is the most stringent, since it forces the symbols α_q and β_q to be dispersed with equal energy in all spatial and temporal directions.

Constraint iii is the preferred constraint because it has the advantage of better diversity gain. Optimization of (10.25), once achieved, works well over a wide range of SNRs. As was pointed out in the beginning of this section, the design criterion used here is not directly connected with the diversity design criterion given in Chapter 5. Another interesting point is that we can premultiply the transmit vector $\mathbf{s}$ of (10.20) with a judiciously chosen orthogonal matrix. The orthogonal matrix preserves mutual information but allows us to change the dispersion code to satisfy other criteria such as space-time diversity. For example, we can use it to construct unitary $\{\mathbf{A}_q, \mathbf{B}_q\}$ from rank-one V-BLAST dispersion matrixes (10.17) so that we can subject it to constraint iii to improve performance by increasing diversity.

10.3.1.6 Performance Results of LDC versus V-BLAST

Among the many examples given by the authors in [3], we take one such pertaining to the performance of the V-BLAST algorithm. The chosen system was a $M_T = M_R = 2$, $R = 4$ system. The rate is $R = 4$, whereas the V-BLAST algorithm has $Q = 4$. The matrixes are given by (10.17). To design the LDC, the authors chose $Q = 4$ with constraint iii. To achieve $R = 4$, the modulation chosen was QPSK. The results are shown in Figure 10.5.

It can be seen from Figure 10.5, that LDC performs better than V-BLAST because it imparts a certain amount of transmit diversity, which is otherwise absent

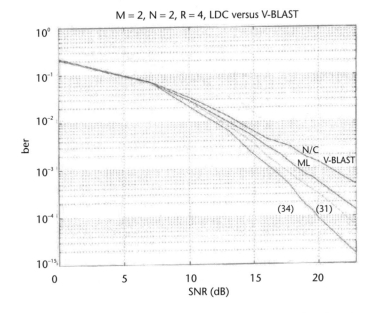

Figure 10.5 The upper two curves are BER with nulling/canceling (upper) and maximum likelihood decoding (lower). The lower two curves are the LDC for $M_T = M_R = T = 2$ and $Q = 4$ (upper, 31) and after multiplication with orthogonal matrix (lower, 34). (*From:* [3]. © 2002, IEEE.)

from V-BLAST algorithms. This occurs because the dispersion matrixes potentially transmit some combination of *each* symbol from *each* antenna at *every* channel use. The orthogonal matrix premultiplication technique performs even better because it enables us to apply constraint iii for improving diversity. Hence, in this case not only do we maximize mutual information, but we also improve on diversity. However, the degree to which the diversity is improved is less than that achieved by the method of Heath and Paulraj, which we shall discuss next. The reader meanwhile is encouraged to study [7] for more ideas on this interesting topic proposed by Hassibi and Hochwald.

10.3.2 Method of Heath and Paulraj

This method was proposed in [2] and [4].

10.3.2.1 Signal Model

Consider a $\mathbf{Q} \times 1$ vector $\mathbf{s}$ of Q complex data symbols, which is modulated by a code matrix of dimension $M_T \times Q$ and transmitted over the $M_R \times M_T$ channel $\mathbf{H}$ for each symbol period. Assume that there are T such distinct matrixes (i.e., at time $1 \le k \le T$, signal $\mathbf{X}[k]\mathbf{s}$ is transmitted, where $\mathbf{X}[k]$ is the kth code matrix). The received symbol vector at time instant k is

$$\mathbf{y}[k] = \sqrt{\frac{\rho}{M_T}}\, \mathbf{H}\mathbf{X}[k]\mathbf{s} + \mathbf{n}[k] \tag{10.26}$$

where $\mathbf{n}[k]$ is the $M_R \times 1$ ZMCSCG noise vector. We now stack the T received vectors and obtain a block signal model given by

$$\begin{bmatrix} \mathbf{y}[1] \\ \vdots \\ \mathbf{y}[T] \end{bmatrix} = \sqrt{\frac{\rho}{M_T}}\, \mathcal{H} \begin{bmatrix} \mathbf{X}[1] \\ \vdots \\ \mathbf{X}[T] \end{bmatrix} \mathbf{s} + \begin{bmatrix} \mathbf{n}[1] \\ \vdots \\ \mathbf{n}[T] \end{bmatrix} \tag{10.27}$$

or

$$\mathcal{Y} = \sqrt{\frac{\rho}{M_T}}\, \mathcal{H}\mathcal{X}\mathbf{s} + \mathcal{N} \tag{10.28}$$

where $\mathcal{Y} = [\mathbf{y}[1]^T \ldots \mathbf{y}[T]^T]^T$ is a vector of dimension $(M_R T \times 1)$, $\mathcal{H} = \mathbf{I}_T \otimes \mathbf{H}$ is a matrix of dimension $M_R T \times M_T T$, $\mathcal{X} = [\mathbf{X}[1]^T \ldots \mathbf{X}[T]^T]^T$ is a matrix of dimension $M_T T \times Q$ and $\mathcal{N} = [\mathbf{n}[1]^T \ldots \mathbf{n}[T]^T]^T$ is the stacked noise vector of dimension $(M_R T \times 1)$.

Inspection of (10.28) with (10.12) and (10.13) tells us that the derivation of the dispersion matrix $\mathcal{X}$ is similar to the approach of Hassibi and Hochwald.

10.3.2.2 Spatial Rate

The spatial rate of these codes is similar to the rate defined by Hassibi—$R = Q/T$ where Q is the number of independent symbols. If $Q = T$ we have a spatial rate of 1. If $Q = TM_T$ we have spatial rate of M_T. Hence, for $T < Q < TM_T$, we have spatial rates defined by $1 < R < M_T$.

10.3.2.3 Capacity

The ergodic capacity of this signaling scheme is given by [4],

$$C = \max_{\text{tr}(X^{\mathcal{H}}X)=M_TT} \frac{1}{T}\, \epsilon\left\{\log_2 \det\left(\mathbf{I}_{M_RT} + \frac{\rho}{M_T}\mathcal{H}XX^{\mathcal{H}}\mathcal{H}^{\mathcal{H}}\right)\right\} \qquad (10.29)$$

This expression can be derived from (10.10).

We now examine two cases of M_T, T and Q defined by the relationship in (10.23).

$$Q = M_TT$$

In this case, if we choose X such that $XX^{\mathcal{H}} = 1/M_T\mathbf{I}_{M_T}$, we obtain a capacity-optimal LDC. This capacity is defined by (10.29).

$$Q < M_TT$$

This case will be desirable due to decoding complexity, memory, or latency constraints. To accommodate this scenario, we modify the capacity-optimal design where X is a scaled unitary matrix by removing the appropriate number of columns and rescaling. We choose X such that

$$X^{\mathcal{H}}X = \frac{M_TT}{Q}\mathbf{I}_Q \qquad (10.30)$$

This makes the X matrix a tall one. Such a matrix is called a *tight frame* [8]. A tight frame allows an overcomplete representation of a signal. This increases the ratio of rows to columns, in this case, M_TT/Q. Large redundancy factors reduce the space spanned by the code words and lower the overall data rate. Obviously the resulting LDC design in terms of ergodic capacity becomes suboptimal.

The authors then search among this class of matrixes for that matrix X, which maximizes diversity based on the rate and determinant criteria, as discussed next.

Employing techniques similar to the ones discussed in Chapter 5, the authors derived the average PEP given perfect knowledge of the channel at the receiver and upper-bounded as [4],

$$P\left(\mathbf{s}^{(i)} \to \mathbf{s}^{(j)}\right) \le \epsilon\left\{e^{-\frac{\rho}{4M_T}\|\mathcal{H}X\mathbf{e}_{i,j}\|_F^2}\right\} \qquad (10.31)$$

where $\mathbf{e}_{i,j} = \mathbf{s}^{(i)} \to \mathbf{s}^{(j)}$. Simplifying,

$$P\left(s^{(i)} \rightarrow s^{(j)}\right) \leq \frac{1}{\det\left(I_{M_T M_R} + (\rho/4M_T)I_{M_R} \otimes R\right)} \tag{10.32}$$

where $R = \sum_{k=1}^{T} X_k e_{i,j} e_{i,j}^H X_k^H$. The diversity order is seen from this expression as

$$M_R \min r\left(\sum_{k=1}^{T} X_k e_{i,j} e_{i,j}^H X_k^H\right) \tag{10.33}$$

where the minimization is performed over all possible code word error vectors $e_{i,j}$. The diversity order varies between $M_R M_T$ and M_R depending on the choice of spatial signaling rate. Hence the matrix X, once it is defined for capacity as previously discussed, is then used to minimize (10.33), thereby maximizing diversity. Hence, this method designs the code with the twin metrics of diversity and capacity, unlike the approach of Hassibi, which, although it has an eye on diversity, does not explicitly design for diversity. The authors [4] have developed one such code with parameters $M_T = 4$, $M_R = 1$, $T = 4$, and $Q = 4$ and using QPSK modulation, corresponding to a spatial rate $R = 1$. An additional stipulation was that the rank of the correlation matrix R_{ss} should be at least 3. This will therefore exhibit a third order diversity since $M_R = 1$. The performance of this code was compared with the Hassibi and Hochwald code given in [3]. The linear transformation matrix for this code is shown below [4]:

$$X = \begin{bmatrix}
-0.0360 - j0.0140 & -0.1156 - j0.4268 & -0.0583 - j0.1780 & -0.0659 - j0.1637 \\
0.0048 + j0.4032 & 0.0929 + j0.0153 & -0.0975 - j0.1916 & -0.0508 + j0.1726 \\
-0.0246 + j0.0043 & 0.1657 - j0.1040 & -0.1332 + j0.2814 & -0.3343 + j0.0493 \\
-0.0394 - j0.2894 & 0.0796 + j0.0313 & -0.1593 - j0.2154 & -0.0463 + j0.2889 \\
0.1847 + j0.0048 & 0.0880 - j0.0752 & -0.0380 - j0.0685 & -0.2601 - j0.3587 \\
0.1370 - j0.1306 & -0.1296 + j0.0497 & -0.3499 + j0.2283 & 0.1243 - j0.0701 \\
-0.3409 - j0.0637 & -0.0824 - j0.3047 & -0.0908 - j0.0130 & 0.0320 - j0.1437 \\
0.1476 - j0.1949 & -0.3386 - j0.0556 & 0.2237 - j0.1041 & 0.0864 - j0.0646 \\
-0.1233 + j0.1169 & 0.1143 + j0.1568 & 0.2468 - j0.3387 & -0.0693 + j0.0554 \\
-0.1089 - j0.1982 & 0.2081 - j0.0851 & -0.0235 + j0.0812 & -0.2386 + j0.2902 \\
0.0808 - j0.2558 & 0.1048 - j0.2848 & 0.1622 - j0.1432 & 0.0890 - j0.1766 \\
-0.1831 - j0.2540 & 0.0930 + j0.2470 & -0.1391 - j0.0329 & 0.2460 - j0.0362 \\
-0.3397 - j0.2648 & -0.0574 - j0.0096 & -0.0025 + j0.1820 & -0.0800 - j0.1468 \\
0.0196 + j0.0093 & -0.3445 - j0.2295 & -0.1465 - j0.0248 & -0.1780 + j0.1563 \\
-0.2037 - j0.1250 & 0.1368 + j0.0360 & -0.1406 - j0.2795 & 0.0511 + j0.2691 \\
0.0082 + j0.0825 & -0.0188 + j0.2344 & -0.2826 - j0.1305 & -0.2044 - j0.2216
\end{bmatrix}$$

The results are shown in Figure 10.6 [4].

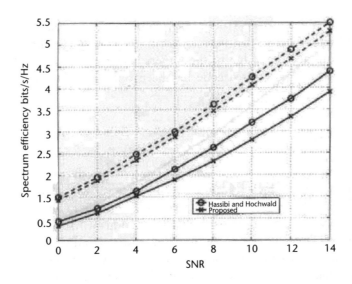

Figure 10.6   Comparison of LDC for spectral efficiency with one and two receive antennas. (*From:* [4]. © 2002, IEEE.)

Figure 10.6 shows that the capacity of the frame-based code is close but does not quite maximize ergodic capacity. This is because we have opted for $Q < M_T T$ with consequent loss in capacity. The performance difference is quite small and decreases as the number of receive antennas increases. The Hassibi and Hochwald code is capacity-optimal, whereas the proposed code is a noncapacity-optimal frame-based code.

Figure 10.7 shows us the error rate performance. The left figure shows us the code word error probability, whereas the right figure shows the bit error probability. The optimized code has a higher order diversity advantage, as is evident from the slope. At a code word error rate of 10^{-3} the proposed code has a 3 dB advantage, which grows with larger SNR. The bit error rate curve also exhibits higher diversity

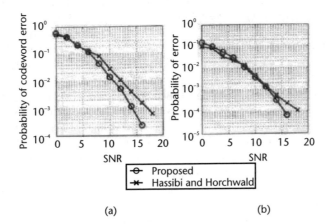

(a)   (b)

Figure 10.7   Comparison of LDC for (a) code word error probability and (b) bit error probability. (*From:* [4]. © 2002, IEEE.)

order as it has a steeper slope, especially at high SNRs. At a BER of 10^{-4} there is a 3 dB difference in the performance of the codes. This difference grows as the bit error rate decreases due to the larger diversity advantage of the proposed code. Therefore, the proposed code has been optimized for good error probability of performance.

10.4 Conclusion

This brings us to the end of our investigation into the principles of MIMO wireless. There are many more advanced topics that we would have like to cover but are unable to do so due to lack of space. One topic that comes to mind is investigating the problems connected with CCI mitigation in MIMO systems [2, 9, 10]. In a multi cell environment space-time links suffer from CCI with frequent reuse. This has a lot of bearing on field deployment of such systems. Differential space-time block codes come in useful in rapidly fading channels, as they do not require channel estimates [11, 12]. Space-time turbo trellis codes are fast emerging as codes that combine the coding gain benefits of turbo-coding with the diversity advantage of space-time coding and the bandwidth efficiency of coded modulation [13, 14]. Yet another interesting topic that bears further study pertains to parallel iterative receivers for layered space-time codes [14]. The topic of MIMO wireless, like most technical topics, requires a lot of study of technical papers. In this book every endeavor has been made to retain some of the flavor of those papers to enable the beginner to better understand technical papers on MIMO topics.

Your journey of a thousand miles has just begun. Good luck and happy journey!

References

[1]   Suard, B., G. Xu, and T. Kailath, "Uplink Channel Capacity of Space-Division Multiple-Access Schemes," *IEEE Trans. Inf. Theory,* Vol. 44, No. 4, July 1998, pp. 1468–1476.

[2]   Paulraj, A., R. Nabar, and D. Gore, *Introduction to Space-Time Wireless Communications,* Cambridge, UK: Cambridge University Press, 2003.

[3]   Hassibi, B., and B. Hochwald, "High-Rate Codes That are Linear in Space and Time," *IEEE Trans. Inf. Theory,* Vol. 48, No. 7, pp. 1804–1824.

[4]   Heath, R., and A. Paulraj, "Linear Dispersion Codes for MIMO Systems Based on Frame Theory," *IEEE Trans. Sig. Proc.,* Vol. 50, No. 10, October 2002, pp. 2429–2441.

[5]   Telatar, I. E., "Capacity of Multiantenna Gaussian Channels," *European Trans. Tel.,* Vol. 10, No. 6, November/December 1999, pp. 585–595.

[6]   Damen, O., A. Chkeif, and J. Belfiore, "Lattice Code Decoder for Space-Time Codes," *IEEE Comm. Letters,* Vol. 4, No. 5, May 2000, pp. 161–163.

[7]   Ghaderipoor, A. R., L. Beygi, and S. H. Jamali, "Analytical Survey on the Design of Linear Dispersion Space-Time Codes," *ICASSP 2002,* Vol. 3, May 2002, pp. 2433–2436.

[8]   Daubechies, I., "Ten Lectures on Wavelets," *Society for Industrial and Applied Mathematics (SIAM),* Philadelphia, PA, 1992.

[9]   Rappaport, T., *Wireless Communications: Principles & Practice,* Upper Saddle River, NJ: Prentice Hall, 1996.

[10] Giannakis, G., P. Stoica, and Y. Hua, editors, *Signal Processing Advances in Wireless and Mobile Communications, Vol. 2: Trends in Single and Multi-User Systems,* Upper Saddle River, NJ: Prentice Hall, 2000.

[11] Tarokh, V., and H. Jafarkhani, "A Differential Detection Scheme for Transmit Diversity," *IEEE J. Select. Areas Comm.,* Vol. 18, July 2000, pp. 1169–1174.

[12] Hughes, B. L., "Differential Space-Time Modulation," *IEEE Trans. Inf. Theory,* Vol. 46, No. 7, November 2000, pp. 2567–2578.

[13] Firmanto, W., et al., "Space-Time Turbo-Trellis Coded Modulation for Wireless Data Communications," *Eurasip Journal on Applied Signal Proc.,* Vol. 2002, No. 5, May 2002, pp. 459–470.

[14] Vucetic, B., and J. Yuan, *Space-Time Coding,* Chichester, UK: John Wiley & Sons Ltd., 2003.

Wideband Simulator: Description and Explanatory Notes

A.1 Introduction

This appendix deals with the details of the OFDM packet-based simulator supplied with this book. It is also called wideband simulator. The software has by now become huge and merits a separate appendix in its own right to clarify the coding aspects to the interested reader. To properly understand the MIMO-OFDM system, it is essential that the reader be thoroughly familiar with the software. It is not possible to explain this software line-by-line but the salient points are covered in this appendix. It is presumed that the reader has sufficient command of MATLAB to follow the coding.

A.2 Files Listing

The overall files listed in this software are given below:

channel.m

finish.m

form_matrix.m

get channel_ir.m

get n_antennas.m

get_bits_per_symbol.m

receiver.m

launchsim.m

rx_bpsk_demod.m

rx_deinterleave.m

rx_demodulate.m

rx_diversity.m

rx_estimate_channel.m

rx_gen deintlvr_patt.m

rx_qpsk_demod.m

rx_radon hurwitz.m

rx_convert_to_freq.m

LoadSimConsts.m

single_packet.m

transmitter.m

tx_add_cyclic_prefix.m

tx_diversity.m

tx_convert_to_time.m

tx_gen_intlvr_patt.ml

tx_gen_preamble.m

tx_interleaver.m

tx_modulate.m

tx_radon_hurwitz.m

txfour.m

tx_round_ofdm_syms.m

verify_params.m

input_options.m

launch.m

start_simulation.m

These files pertain to the SISO and space-time block coding operations for a generic OFDM-based packet transmission system. In addition to these files, the following files were added to enable V-BLAST operations:

Q. m	tx_gen_preamble_vblast.m
rx_demodulate_vblast.m	vblast.m
rx_convert_to_freq_vblast.m	vblast_method.m

The program is started from the MATLAB command line by typing "Launch." Henceforth, the term "command line" will be understood to be the MATLAB command line. This invokes the *launch.m* file. This is the file that forms the GUI. The details here are irrelevant, since teaching the GUI formation is beyond the scope of this book. The layout of the GUI is self-explanatory and the reader can position the arrow cursor to the statements on the GUI in blue color. The tool tips facility of the GUI will reveal details on what needs to be entered. The interesting point to note on the GUI is that it will plot the "raw" packet error rates (PER)/BER. The "raw" error rates pertain to results without taking into account any type of coding. This simulator pertains to an uncoded system. There is also a "Save File" option for saving the file in the current directory.

A.2.1 Zero-Padding

A zero-padding facility has been provided with this simulator. This facility allows the user to select different OFDM sizes and earmark any desired number of sub-carriers for carrying data and pad the remaining, including the center dc carrier, with zeros. The minimum number of subcarriers allowed is 16, which is the length of the cyclic prefix. This facility enables the user to evaluate the performance of OFDM-based systems with various types of channel estimation techniques and so on in the presence of zero padding. The reader should bear in mind that zero padding is necessary to avoid ICI in OFDM systems and that using all the subcarriers available in an OFDM system is not realistic. The zero-padding facility is invoked when the user clicks the check box. This removes the interleaving check box and replaces it with an edit box for entering the number of data carriers. Zero padding and interleaving are mutually exclusive. This has been done to simplify coding. The reader can modify the code if it is deemed necessary to include interleaving with zero padding.

In these simulations we assume a guard interval of 800 ns (16 samples), the bandwidth being 20 MHz. Since this simulator is for an indoor wireless LAN, where the maximum expected delay is typically 200 ns, this is adequate. This length of the cyclic prefix is unchanging irrespective of the length of the FFT. However, the error involved thereby is smaller than 1 dB if the OFDM symbol duration is kept at about five times the guard time. Hence, we lose less power to the guard band with higher FFT sizes and more with lower FFT sizes. Strictly speaking, this leads to inaccuracy in estimating the SNR value. This aspect has not been taken into account in the simulations, since the error involved is marginal and has minimal effects on the results. Therefore, even for the lowest FFT size provided (i.e., 64 points), the OFDM symbol duration (this includes the guard band) is 80 samples long.

A.2.2   Convolutional Coding

Convolutional coding has been provided with this simulator. There are three coding rates, 1/2, 2/3 and 3/4 rates. A half rate is the minimum. The user can, however, obtain plots without the effect of convolutional coding if desired.

We now examine each file in the order of execution of the program. Each packet in this simulator comprises 10 OFDM blocks (i.e., number of data carriers × the number of bits (depending on the type of modulation) × 10, where 10 is an arbitrary number.

A.3   SISO Mode

In this mode we obtain a pure SISO system. The first file we need to look at is the *launchsim.m* file. This is the main file of the simulator. It is invoked when the "start" button on the GUI is pressed. This invokes the *start_simulator.m* file, which in turn runs the *launchsim.m* file.

- *launchsim.m:* The first command in this file is "*LoadSimConsts*" at line 6. This invokes the *LoadSimConsts.m* file. This file is the main input file for the simulator.
- *LoadSimConsts.m:* The file is generally self-explanatory, but certain aspects need to be clairified. The expressions are in the "struct" form of MATLAB. The reader is advised to check this form in case of doubt. At line 7 we assign the chosen number of data carriers to a variable. At line 8 we check whether the user has opted for zero-padding. If yes then we verify whether the chosen number is more than 16. The figure 16 is the lower limit because of the cyclic prefix, which comprises 16 samples. We then verify that the entered number of carriers is an even number. The idea is to evenly divide the zero pads on either side of the data carriers. Finally, if we have selected the zero-pad option, we select the long training symbols to match the chosen number of data carriers or else the long training symbols will be equal to the chosen FFT length. The sampling frequency for this system is derived from IEEE 802.11a as 20 MHz. Hence, the bandwidth is 20 MHz.
- *launchsim.m:* We now return to *launchsim.m*. The remaining lines are purely initialization. We start with a packet count of one. The next function that is invoked is the *single_packet.m*. While invoking this function for each value of SNR, we not only forward the value of the current SNR, but also the "SimulationParameters" file. This file basically contains the selected options exercised by the user on the GUI. In this connection we need to note that there are two files, *input_options.m* and *verify_params.m*. As their name implies, the former reads the entries from the GUI, whereas the latter checks the correctness of these entries. The mechanics of this are beyond the scope of this book. The reader is advised to check the MATLAB guide

books in this regard. The outputs from the *single_packet.m* file comprise "raw" as well as "inf" outputs, the former being without convolutional coding and the latter with the coding in place. The remaining entries in the *launchsim.m* file pertain to plotting routines. The file name is entered from the file section of the GUI. The file is saved in the current directory. It is recommended that the user first run a few short simulation runs, to ensure that the files are being saved correctly. The program ends with printing the elapsed time for the full simulation on the command line.

- *single_packet.m:* At line 4, we derive the channel impulse response. This is based on the user's choice as entered on the GUI. The K-factor is entered as a number and not in dBs. This is required in case the Rician option is exercised. The formula used is the same as the formula given in Chapter 2. K = 0 is the same as pure Rayleigh fading. [The number of antennas and type of system under simulation (SIMO, MISO, MIMO) is known by the program in the *get_channel_ir.m* file, by using the *get_n_antennas.m* file.] At line 7 we invoke the *transmitter.m* file. The input parameters are the "SimulationParameters." The user can at anytime insert a breakpoint in the program and type "SimulationParameters" on the command line. This will list out the user options. This is a useful feature to examine in case we get strange results. The output parameters from the *transmitter.m* file are the input bits (*tx_inf_bits*), (*tx_raw_bits*) as well as the transmitted signal (*txsignal*). We shall return to this file after examining the rest of the program.

- *transmitter.m:* This file builds up the transmitted signal. In line 18, random bits are selected depending on the data carrier length option (i.e., with or without the zero-padding option and FFT size). Lines 26 to 30 pertain to convolutional coding and puncturing. Line 36 rounds the number of bits into a multiple of OFDM symbols to maintain integer numbers. Line 39 pertains to interleaving. The theory behind this is explained in Chapter 8 and in [1]. Thereafter the signal is modulated. Since this is a SISO mode, no diversity is exercised. At line 52, the final frame is sent for the IFFT operation in *tx_convert_to_time.m* file.

- *tx_convert_to_time.m:* In this file at line 12 we investigate the zero-padding option. If this option is exercised, the specified data carriers are padded with zeros on either ends of the OFDM symbol and the center dc carrier is also set to zero. Thereafter, at line 28, we divide the entire frame into multiples of the FFT length. At line 31 we interchange the subcarrier sets. At line 37 we reshape the entire data frame into groups of the chosen FFT length. This implies that the basic frame defined by the packet length is much longer than the frame set one IFFT can handle. Hence, transmission of one packet results in a number of frames (length of each being defined by the chosen FFT size) being transmitted. At line 39 we rearrange the subcarriers and, finally, at line 43 we once more make it one long frame, so that during simulation the entire frame is transmitted (actually multiple frames of "FFT lengths" each).

- *transmitter.m:* At line 55, we invoke the function *tx_add_cyclic_prefix.m*. In this function we basically unravel the constituent frames from the entire long frame (time_syms) and add cyclic prefixes to each of them. After this operation we once again make it into one long stream. At line 58, we invoke the *tx_gen_preamble.m* function.

- *tx_gen_preamble.m:* This is an important step. We now add the preamble that carries with it the long training symbol for channel estimation, as explained in Chapter 8. At lines 6 and 7 we assign variables to the number of transmit and receive antennas, respectively, for ease of programming. Remember that in OFDM systems, after IFFT, the information being transmitted is in time domain. Hence, it becomes necessary for us to also convert the training symbols to time domain before appending them to the long frame. At line 12 we carry out this exercise with the long training symbols. As usual we first invoke the *SimulationConstants* structure for the long training symbol and we then convert it into time domain. At line 13 we check if *SimulationParameters.TxDiv = 1*. This is one only if the user has checked the Tx. checkbox in the MIMO section of the GUI. In our case, we did not since we are talking about a SISO operation. Hence, we proceed to line 16 to form the "long_trs_signal" and append a 32-length cyclic prefix to two long training symbols. This is borrowed from the IEEE 802.11a standard. Finally at line 39 we form the preamble. Note that this preamble already exists in the time domain. We now return to the *transmitter.m* file.

- *transmitter.m:* At line 68 we form the final frame for transmission. Note the math for power correction. This completes the *transmitter.m* file.

- *channel.m:* In this file, we deal with the interaction of the transmitted signal with the channel. The type of channel selected is up to the user and is based on GUI entries. The transmitted signal is convolved with the channel impulse response ("cir") between lines 10 and 17. We next jump to line 144, where we calculate the noise variance and add it to the signal at line 147. We now proceed to the *receiver.m* file.

- *receiver.m* At line 13 we (assuming perfect synchronization) detach the cyclic prefix from the received signal. Since the cyclic prefixes are different for single and multiple transmit antennas, we use an *if* loop. Since we are now in the time domain, we need to convert to frequency domain. We invoke the *rx_convert_to_freq.m* file at line 21.

- *rx_convert_to_freq.m:* From lines 10 to 19 we extract the received long training symbols from the overall time signal. The coding shows this exercise for the two transmit antenna and four transmit antenna options also. Do not forget that the long training symbols for the SISO case are $2 \times fft\_length$ samples long, as shown in line 12. The extracted long training symbols are converted to frequency domain at line 30 and reordered at line 31 and 32. Do not forget that we had initially reordered the transmitted signal in the *transmitter.m* file. This operation, therefore, annuls that move. At line 38 we examine the zero-pad option. If there is a zero-pad option, we remove the zero-padding to make one contiguous signal. Now these subcarriers contain training symbols for each of them (i.e., we now have the channel information for the path each of the subcarriers has taken in its passage through the channel). These are now required for channel estimation, which will be done later on in a separate file. Similarly we extract the data symbols from the time signal. At line 55 we split the extracted data stream into groups of *fft_length* + *16*. Remember that *fft_length* + *16* samples imply *fft_length* plus the 16-sample cyclic prefix. We then reshape the extracted data symbols (samples) at lines 58 and 59. Finally, we remove the guard interval from each of them at line 61. At line 71 we convert the extracted data stream to frequency domain. At line 74 we reorder to undo the change we made at the transmitter and similar to the operation we conducted earlier in this file for the training symbols. At line 64 we examine the zero-pad option and, if "true," we extract the contiguous samples by removing the zero-padding. We now return to *receiver.m* file.

- *receiver.m:* At line 24 we invoke the *rx_estimate_channel.m* file for channel estimation and pass it the frequency domain training symbols information we recovered from the original time signal at line 21.

- *rx_estimate_channel.m:* The operation up to line 21 pertains to least squares estimation, as explained in Chapter 6 and 7, but for a single transmit antenna. The coding from lines 22 to 35 pertain to multiple transmit antennas. In each case, we examine the FFT estimation option. Lines 14 to 17 and 28 to 31 are the option for using the FFT method of channel estimation. This was explained in Chapters 6 and 7. Lines 39 to 42 pertain to perfect channel knowledge. At line 44, we examine the zero-pad option and extract only those channel estimates that have a direct relationship to their respective data carriers.

- *receiver.m:* At line 27 we carry out diversity processing. In *rx_diversity.m* file, for a SISO channel, we correct for the channel shifts based on our channel estimates (*channel_est*). Remember that these channel estimates can be estimates or perfect channel knowledge depending on our user option on the GUI. The *rx_diversity.m* coding is trivial and easy to follow. The data information that we recover after this operation has now been corrected by *rx_diversitc.m* file for channel shifts. It is, therefore, ready for demodulation. The rest of the *receiver.m* file now concentrates on deinterleaving and demodulation to obtain raw bits. The rest of the coding involves deconvolution and is self-explanatory. We now return to the *single_packet.m* file.

- *single_packet.m:* We have just completed line 13. The rest of the code determines the "raw / inf" errors. These values are passed on to the main file *launchsim.m*.

- *launchsim.m:* We have finished executing lines 58-59. We now load the returned values to the respective counters and flush the event queue (*drawnow*). We now repeat the entire cycle for the next packet.

A.4   SIMO Mode

This will occur only if the transmitter diversity check boxes on the MIMO block of the GUI [OFDM/MIMO(MRC/STC)] is left unchecked (this implies no transmit diversity or just a single transmit antenna). In this mode we trace the programming of a 1×2 system. To achieve this mode, check the box next to receiver diversity. This is the only box that need be checked. We will only discuss the differences from the SISO program. The SIMO mode uses MRC at the receiver. The first difference occurs in the *receiver.m* file.

- *receiver.m*: At line 27, the *rx_diversity.m* file is invoked.
- *rx_diversity.m*: In this file at line 40, MRC is implemented.

The rest of the signal processing is the same as for SISO.

A.5   MISO/MIMO Mode

This will occur only if the transmitter diversity box is checked (for a transmit diversity of two) or $\times 4$ (for a transmit diversity of four) and the receiver diversity boxes are also checked (for MIMO) or left unchecked (for MISO). In our case we will examine an 2×2 Alamouti scheme. We check the transmit diversity box and the receiver diversity check boxes. Once again we will only discuss the differences from SISO.

From Chapter 4 we know that the Alamouti transmission matrix is given by

$$S = \begin{bmatrix} s_1 & s_2 \\ -s_2^* & s_1^* \end{bmatrix} \tag{A.1}$$

We now follow the *transmitter.m* file.

- *transmitter.m*: At line 47 we invoke the *tx_diversity.m* file.
- *tx_diversity.m*: At line 5 we invoke the *tx_radon_hurwitz.m* file. Radon-Hurwitz is the basic theorem from which the orthogonal STBC codes are derived [1]. Alamouti is a special 2×2 case of Radon-Hurwitz theorem.
- *tx_radon_hurwitz.m*: At line 12 we create the s_1 symbol stream. This is $S(1, 1)$ of (A.1). At line two we create the s_2 symbol stream. This is $S(1, 2)$ of (A.1). Both of these are alternate symbols from the input *mod_syms* stream. At line 16 we create the $S(2, 1)$ stream and at line 17, the $S(2, 2)$ stream. Finally we create *ofdm_sym_out* matrix, which has two rows. The first row contains $S(1, 1)$ and $S(1, 2)$ alternating and the second row contains $S(1, 2)$ and $S(2, 2)$ alternating. We now return to *transmitter.m* file.
- *transmitter.m*: The output of line 47 is a two-row matrix. At line 58 we invoke the *tx_gen_preamble.m* file.
- *tx_gen_preamble.m*: The initial coding in this file is crucial. At lines 21 and 23 we form two rows of training symbols, since this is a 2×2 system. If you examine the coding, we note that these training sequences are orthogonal in time (i.e., each sequence is transmitted, in turn, by adding one symbol length zeros before the stream from the second antenna). The rest follows as for the SISO case but the options are for a 2×2 case. We now return to the *receiver.m* file.

- *receiver.m:* Line 27 invokes the *rx_diversity.m* file.
- *rx_diversity.m:* At line 15, the first row of the received matrix invokes the *rx_radon_hurwitz.m* file. At line 17 the second row does the same.
- *rx_radon_hurwitz.m:* At lines 15 and 18 each row of the received data stream is processed as per theory given in Chapter 4. This is the decoded data that is now returned to the *rx_diversity.m* file.
- *rx_diversity.m:* At line 42 the original data stream is recovered in frames of "data carrier" rows each. Finally these are normalized and returned to the *receiver.m* file.

There are no other changes in the coding.

A.6   V-BLAST Mode

This mode is invoked when the V-BLAST check box is checked in the MIMO block. All combinations are not incorporated. Only 1×2, 2×2, 2×3, and 2×4 combinations have been incorporated. However, once the reader understands the coding, any combination can be incorporated. The reader must be aware from Chapter 6 that V-BLAST in a 2×2 combination has a receive diversity order of only 1 (from $M_R - M_T + 1$) for ZF receivers. It is very inefficient but has been introduced here from the point of view of study. In these simulations perfect channel knowledge is assumed. For simplicity, we have excluded interleaving. The reader is free to include any of these facilities to see the final effect once the coding becomes familiar. In view of these extensive changes it becomes more convenient to add a fresh set of coding (admittedly very inefficient, but easier to understand) as an else-option in an if-then-else loop. This is what is done in files *transmitter.m* and *receiver.m*. Let us choose a combination of 1×2 for purposes of explanation.

- *transmitter.m:* At line 70 we exercise the V-BLAST option. We then assign *SimulationParameters.M* to variable M at line 73. This saves on needless typing as this variable will be used everywhere in the program. At line 76 we round the "tx_bits" to "rdy_to_mod_bits." The rest of the coding is easy to follow. We basically split the "*mod_syms*" into two streams and process it thereafter as normal. We next examine the *channel.m* file.
- *channel.m:* At line 18 we exercise the V-BLAST option. At lines 22 and 24 we invoke the *get_channel_ir.m* file, which will get us two separate channels. Remember that this is a spatial multiplexing algorithm and, hence, we need to impart to the simulation two distinct and separate channels. The returns are multitap, but we select only the first tap, since this is a single-tap narrowband channel. We then convolve the channel with the *tx_signal*. At lines 29 and 30 we save the channel information to upload it in the receiver (perfect channel knowledge) prior to decoding. The rest of the processing is as usual. We now examine the *receiver.m* file.
- *receiver.m:* We invoke the *form_matrix_.m* file at line 51.
- *form_matrix_.m:* At lines 8 to 13 the channel matrix is formed from the files saved earlier.

- *receiver.m:* The rest of the files are straightforward until line 55, when we invoke the *vblast_method.m* file.
- *vblast_method.m:* At line 14 we invoke the V-BLAST algorithm.
- *vblast.m:* Lines 14 to 26 implement the ZF algorithm, as described in Chapter 6. At line 20 we invoke the *Q.m* file for slicing operation, as explained in Chapter 6.

The rest of the coding is similar to SISO.

Finally, it is pointed out that if we choose the zero-pad option and use 52 data carriers, we approximate the PHY layer of an IEEE 802.11a system with perfect timing and frequency synchronization and no carrier frequency error or any phase noise in the carrier oscillators.

Reference

[1]   Heiskala, J., and J. Terry, *OFDM Wireless LANs: A Theoretical and Practical Guide,* Indianapolis, IN: Sams Publishing, 2002.

Narrowband Simulator

B.1 Introduction

This appendix deals with the details of the narrowband simulator supplied along with this book. This is not one simulator but a series of programs based on GUIs as well as plain coding covering the operation of the MIMO algorithms in a frequency flat channel. The programs are distributed on chapter basis in directories as shown below:

Chapter 2: Capacity calculations.
Chapter 4: Space-time block coding.
Chapter 5: Space-time trellis coding.
Chapter 6: V-BLAST algorithm.

B.2 Description

Each directory has a "readme.txt" file. The reader is advised to study this file prior to using the software. In Chapters 2 and 6, the main files are indicated in the respective "readme.txt" files. If the reader types "help *file_name*" on the command line, invoking the main file, comprehensive help information is displayed. The programs in Chapters 4 and 5 are GUI-based. Once again, just like in the wideband case, titles marked in blue carry complete information on how to enter inputs into the GUI. This information is displayed when the cursor is placed on the title. The file for the simulation runs can be saved by checking the "Save File" checkbox. In such an event, the file is saved in the current directory. It is noted that measuring frame error rates using STTC is a tedious procedure and requires at least 10,000 packets to obtain a satisfactory curve. This requires a fast machine.

List of Acronyms

3G	third generation
4G	fourth generation
A/D	analog-to-digital
ADSL	asymmetric digital subscriber line
AMPS	Advanced Mobile Phone Service
AOA	angle of arrival
AOD	angle of departure
AWGN	additive white Gaussian noise
BER	bit error rate
BPSK	binary phase shift keying
CBPSK	complementary BPSK
CCI	cochannel interference
CDF	cumulative distribution function
CDMA	code division multiple access
COFDM	coded orthogonal frequency division multiplexing
CP	cyclic prefix
CRC	cyclic redundancy check
CSMA/CA	carrier sense multiple access/collision avoidance
CW	continuous wave
D/A	digital-to-analog
D-BLAST	diagonal Bell Labs layered space-time
dc	direct current
DE	diagonal encoding
DFT	digital Fourier transform
DQPSK	differential QPSK
DS	direct sequence
DSL	digital subscriber line
DSP	digital signal processing
FDMA	frequency division multiple access
FEC	forward error correction
FFT	fast Fourier transform
FH	frequency hopping

FIR	finite impulse response
GSM	global system for mobile
HE	horizontal encoding
ICI	intercarrier interference
IEEE	Institute of Electrical and Electronic Engineers
IFFT	inverse fast Fourier transform
i.i.d.	independent identically distributed
IIR	infinite impulse response
IMTS	improved mobile telephone service
ISI	intersymbol interference
ISO	International Organization for Standardization
LHS	left-hand side
LOS	line-of-sight
LSE	least squares estimate
LST	layered space-time
MAI	multiple access interference
MIMO	multiple-input multiple-output
MIMO-BC	MIMO broadcast channel
MIMO-MAC	MIMO multiple-access channel
MIMO-MU	multiple-input multiple-output multiuser
MIMO-SU	multiple-input multiple-output single user
MIPS	million instructions per second
MISO	multiple-input single-output
ML	maximum likelihood
MLSE	maximum likelihood sequence estimation
MMSE	minimum mean square error
MRC	maximum ratio combining
MSI	multistream interference
OFDM	orthogonal frequency division multiplexing
OSI	Open System Interconnect
OSTBC	orthogonal space-time block code/codes/coding
OSUC	ordered successive cancellation
PAPR	peak-to-average power ratio
PDF	probability density function
PEP	pairwise error probability
PER	packet error rate
PSK	phase shift keying
QAM	quadrature amplitude modulation
QoS	quality of service
QPSK	quadrature phase shift keying
RF	radio frequency

RHS	right-hand side
RMS	root mean square
SC	single carrier
SDMA	space division multiple access
SER	symbol error rate
SFH	slow frequency hopping
SI	self interference
SIMO	single-input multiple-output
SINR	signal to interference and noise ratio
SISO	single-input single-output
SM	spatial multiplexing
ST	space-time
STBC	space-time block code/codes/coding
STC	space-time coding
STTC	space-time trellis code/codes/coding
SUC	successive cancellation
SVD	singular value decomposition
TCM	Trellis coded modulation
TDD	time division duplexing
TDMA	time division multiple access
UMTS	universal mobile telecommunications system
VCO	voltage controlled oscillator
VE	vertical encoding
WSS	wide sense stationary
WSSUS	wide sense stationary uncorrelated scattering
XPC	cross-polarization coupling
XPD	cross-polarization discrimination
ZF	zero forcing
ZMCSCG	zero mean circularly symmetric complex Gaussian

List of Symbols

$\approx$	approximately equal to
$*$	convolution operator
$\otimes$	Kronecker product
$\odot$	Hadamard product
$\mathbf{0}_{m,n}$	$m \times n$ all zeros matrix
$\lvert a \rvert$	magnitude of the scalar a
$\mathbf{A}^\dagger$	Moore-Penrose inverse (pseudoinverse) of $\mathbf{A}$
$[\mathbf{A}]_{i,j}$	ijth element of matrix $\mathbf{A}$
$\lVert \mathbf{A} \rVert_F^2$	squared Frobenius norm of $\mathbf{A}$
$\mathbf{A}^H$	conjugate transpose of $\mathbf{A}$
$\mathbf{A}^T$	transpose of $\mathbf{A}$
$c(X)$	cardinality of the set X
$\delta(x)$	Dirac delta (unit impulse) function
$\delta[x]$	Kronecker delta function, defined as
	$\delta[x] \begin{cases} 1 & \text{if } x = 0 \\ 0 & \text{if } x \neq 0, x \in \mathcal{Z} \end{cases}$
$\det(\mathbf{A})$	determinant of $\mathbf{A}$
$\operatorname{diag}\{a_1, a_2, \ldots, a_n\}$	$n \times n$ diagonal matrix with $[\operatorname{diag}\{a_1, a_2, \ldots, a_n\}]_{i,i} = a_i$
ϵ	expectation operator
$f(x)$	PDF of the random variable X
$F(x)$	CDF of the random variable X
$\mathbf{I}_m$	$m \times m$ identity matrix
$\min(a_1, a_2, \ldots, a_n)$	minimum of $a_1, a_2, \ldots, a_n$
$Q(x)$	Q-function, defined as $Q(x) = \left(1/\sqrt{2\pi}\right) \int_x^\infty e^{-t^2/2}\, dt$
$r(\mathbf{A})$	rank of the matrix $\mathbf{A}$
$\mathcal{R}$	real field
$\operatorname{Re}\{\mathbf{A}\}\ \operatorname{Im}\{\mathbf{A}\}$	real and imaginary part of $\mathbf{A}$, respectively
$\operatorname{tr}(\mathbf{A})$	trace of $\mathbf{A}$
$\operatorname{vec}(\mathbf{A})$	stacks $\mathbf{A}$ into a vector columnwise

$(x)_+$ defined as $(x)_+ = \begin{cases} x & \text{if } x \geq 0,\ x \in \mathcal{R} \\ 0 & \text{if } x < 0,\ x \in \mathcal{R} \end{cases}$

$\mathcal{Z}$ integer field

About the Author

Mohinder Jankiraman received his B. Tech. degree in electronics and telecommunications from the Naval Electrical School in Jamnagar, India, in 1971. Subsequently, he served as an electrical officer in the Indian Navy for many years. In 1982 he was seconded to research work in military electronics. He took part in a number of military research projects and won a number of awards for technology development in India. His research has spanned several disciplines, emphasizing signal processing, development of naval mines, torpedoes, sonars, radar, and communication systems. He retired from the Indian Navy in 1995 with the rank of a Commodore and joined the International Research Centre for Telecommunication-Transmission and Radar (IRCTR) in the Delft University of Technology, Netherlands, in 1997.

Dr. Jankiraman completed his Master of Technology in Design (MTD) degree in 1999, graduating cum laude, and went on to complete his Ph.D. from Aalborg University, Denmark, in September 2000. He then worked with Summitek Instruments, Denver, Colorado, for about a year as a senior RF engineer on the design and development of passive intermodulation measurement (PIM) analyzers, before joining the University of Aalborg, Denmark, as assistant research professor in June 2002. During this phase, Dr. Jankiraman worked extensively in the area of OFDM-based communication systems and cell phone location systems for the European Commission. He returned to the United States in June 2003, and is presently a technical consultant in wireless communications, based in Dallas, Texas. He is a senior member of IEEE.

Index

The Artech House Universal Personal Communications Series

Ramjee Prasad, Series Editor

Wireless Intelligent Networking, Gerry Christensen, Paul G. Florack, and Robert Duncan

Wireless LAN Standards and Applications, Asunción Santamaría and Francisco J. López-Hernández, editors

Wireless Technician's Handbook, Second Edition, Andrew Miceli

For further information on these and other Artech House titles, including previously considered out-of-print books now available through our In-Print-Forever® (IPF®) program, contact:

Artech House
685 Canton Street
Norwood, MA 02062
Phone: 781-769-9750
Fax: 781-769-6334
e-mail: artech@artechhouse.com

Artech House
46 Gillingham Street
London SW1V 1AH UK
Phone: +44 (0)20 7596-8750
Fax: +44 (0)20 7630-0166
e-mail: artech-uk@artechhouse.com

Find us on the World Wide Web at:
www.artechhouse.com